Advances in Polymer Processing 2020

Christian Hopmann · Rainer Dahlmann
Editors

Advances in Polymer Processing 2020

Proceedings of the International Symposium on Plastics Technology

Springer Vieweg

Editors
Christian Hopmann
Institute of Plastics Processing (IKV)
RWTH Aachen University
Aachen, Germany

Rainer Dahlmann
Institute of Plastics Processing (IKV)
RWTH Aachen University
Aachen, Germany

ISBN 978-3-662-60808-1 ISBN 978-3-662-60809-8 (eBook)
https://doi.org/10.1007/978-3-662-60809-8

Planung/Lektorat: Thomas Lehnert
This Springer Vieweg imprint is published by the registered company Springer-Verlag GmbH, DE part of Springer Nature.
The registered company address is: Heidelberger Platz 3, 14197 Berlin, Germany

Editorial

This edited volume contains the selected papers presented at the International Symposium on Plastics Technology held in Aachen/Germany in March 2020. Lateral thinkers from all over the world have been invited to present their view on the future of plastics technology in an era of major societal and technological changes. On the one hand the growing expectations regarding sustainability and recyclability while keeping efficiency and functionality challenge researchers and industry. On the other hand, digitization and additive manufacturing are powerful drivers for even disruptive changes in processing and production, which offer huge opportunities in plastics technology and stimulate growing interest in both, plastics technology research and industry. The papers presented on the symposium offer a multi-perspective view on these opportunities. They provide scientific insights and knowledge, but as well intent to create awareness and to stimulate critical thinking. In consequence, the papers, which have primarily been written for research experts in industry and science, might as well be valuable for students.

Aachen, Germany
November 2019

Christian Hopmann
Rainer Dahlmann

Contents

Scientific Committee

Olga Sousa Carneiro is Associate Professor with habilitation at the Department of Polymer Engineering (DEP) of University of Minho, Portugal. From 2007 until present she took several duties: member of the committee that proposed the product design graduation program; national coordinator of the Engineering Design and Advanced Manufacturing focal area of the MIT Portugal Program and Director of the corresponding Ph.D. Program; Head of DEP; member of the administration board of two interface units (PIEP and CEiiA). She teaches polymer topics in several graduation, M.Sc. and Ph.D. programs. Her research activity is carried out at IPC – Institute for Polymer and Composites, being focused on polymer processing, polymer rheology and development of new products. She is involved in some tens of scientific projects and applied research projects carried out in collaboration with national and international companies. She edited two books and is author or co-author of 14 book chapters and 70 papers published in international journals.

Rainer Dahlmann is Associate Professor at the RWTH Aachen University and member of the scientific board of the IKV – Institute for Plastics Processing in Industry and Craft at RWTH Aachen, Germany. After studying Physics, he received a doctoral degree at the faculty for mechanical engineering at RWTH Aachen University in the field of plasma and surface technology. He is Head of the Centre for Analysis and Testing of Plastics (KAP) at IKV, which he built up in 2003 and which he directs since then. Furthermore, he is member of the institute management. Dahlmann's fields of activity and research are plasma technology, additive manufacturing, materials technology, testing and analysis, circular economy as well as failure analysis for plastics products.

Dahlmann is chairman of the expert committee "Failure Analysis for Plastics Products" of the Association of German Engineers (VDI) and member of several other expert committees. He participates in research projects and is a board member of the Collaborative Research Centre TR87. Professor Dahlmann gives the lectures "Material Science of Plastics", "Functionalisation of Plastics Surfaces" and "Additive Manufacturing in Plastics Processing" at RWTH Aachen University.

Saeed Farahani is currently working at Clemson University Centre for Automotive Research (CUICAR) as a research fellow. He received his Ph.D. in Automotive Engineering from Clemson University, USA, and his M.Sc. and B.Sc. degrees in Mechanical Engineering from Sharif University in Tehran, Iran. His research is at the intersection of multi-material design, integrated manufacturing processes and data-driven production systems in plastic and composite industries. His recent publications span topics in modelling of hybrid processes and enhancing of quality control and optimization procedures using Industry 4.0 related technologies.

Walter Friesenbichler is Full-Professor for Injection Moulding of Polymers at Montanuniversität Leoben in Austria since July 2010, where he has carried out research and taught since graduating from Montanuniversität in 1984. Professor Friesenbichler's research has centred on three areas: applied rheology, injection moulding and polymer nanocomposites. Since 1985 he has carried out research on injection moulding simulation – standard and special injection moulding technologies such as rubber injection moulding, injection moulding compounding, exjection, rapid heat cycle moulding and applied rheology using extruders and injection moulding machines surface structures of injection moulded parts and systematic development of injection moulded parts. The experimental effort in applied rheology has been aimed at polymer melts, wall slipping PVC compounds, rubber compounds, metal and ceramic feedstocks, PNCs and wood plastic composites. Recent research is focused on optimization of rubber injection moulding, rheology of thermoplastics and rubber compounds with regard to pressure dependence of viscosity, extensional rheology and viscoelastic melt behaviour injection moulding simulation. Prof. Friesenbichler is member of the Austrian Physical Society, the Polymer Processing Society and the Association of German Engineers (VDI). He received the "Golden Batch of Honour" of the Association of Austrian Plastics Processors in 2011 and the "Herman F. Mark Medal" in 2016.

Professor Christian Hopmann holds the Chair for Plastics Processing and is director of the industry based IKV – Institute for Plastics Processing in Industry and Crafts at RWTH Aachen University, Germany. He is co-founder of the AZL – Aachen Centre for Lightweight Production and vice-dean of the Faculty for Mechanical Engineering. After studying mechanical engineering, he received his doctoral degree from RWTH Aachen University and participated in the program for executive development at the International Institute for Management Development (IMD), Lausanne, Switzerland. Following a senior vice-director position at IKV, he started his industrial career in 2005 at the plastics processing company RKW SE, mainly as head of production and latterly as managing director of RKW Sweden A.B. in Helsingborg, Sweden.

Hopmann holds numerous mandates as reviewer, author, editor and speaker at international conferences. He received the innovation award of Germany's federal state North Rhine-Westphalia, is visiting professor of the Beijing University of Chemical Technology, Beijing, China, and fellow of the Society of Plastics Engineers, Bethel, USA.

Guido Tosello is Associate Professor at the Department of Mechanical Engineering of the Technical University of Denmark. He is senior lecturer, research manager and consultant for the manufacturing industry. Tosello's principal research interests are the analysis, monitoring, optimization and simulation of precision moulding processes at micro and nano scales. Tosello is the recipient of the "Technical University of Denmark Best Ph.D. Research Work 2008 Prize" for his Ph.D. thesis "Precision Moulding of Polymer Micro Components"; of the 2012 Alan Glanvill Award from the Institute of Materials, Minerals and Mining (IOM3), London, United Kingdom, for research of particular merit in the field of polymeric materials; of the Young Research Award 2014 from the Polymer Processing Society, Akron, USA, for research excellence in polymer processing within six years from Ph.D. graduation; and of the Outstanding Reviewer Award 2016 of the Institute of Physics, London, United Kingdom, for his contribution to the Journal of Microengineering and Micromechanics. He is the editor of the book 'Micro Injection Molding' published in 2018 by Carl Hanser Verlag.

Volker Schöppner Since 2007 Volker Schöppner is Professor at the Polymer Processing Faculty of Mechanical Engineering, University of Paderborn (UPB), Germany. After finishing his mechanical engineering degree in 1988, he worked for a German extrusion machine manufacturer and for a German automotive supplier. He started his academic career with his Ph.D. thesis on the topic of "Simulation of the Plasticizing Unit of Single-Screw Extruders" and habilitated with a thesis on "Process Engineering of Extrusion Lines". Volker Schöppner is a member of the Polymer Processing Society (PPS) and German Delegate at the International Institute of Welding

(IIW, Topic: Welding of plastics), Paris, France. Furthermore, he is a member of the German Scientific Alliance of Polymer Technology (WAK), Kaiserslautern, Germany, and dean of the Faculty of Mechanical Engineering at UPB.

Lih-Sheng (Tom) Turng received his B.Sc. degree in mechanical engineering from the National Taiwan University, Taipei City, Taiwan, and his M.Sc. and Ph.D. degrees from Cornell University, Ithaca, USA, (with the Cornell Injection Moulding Program, CIMP). He worked at C-MOLD developing injection moulding simulation software for 10 years before joining the University of Wisconsin–Madison (UW-Madison), Madison, USA, in 2000. His research encompasses injection moulding, nanocomposites, multi-functional materials, bio-based polymers and tissue engineering. Turng holds the Kuo K. and Cindy F. Wang Professorship and the Vilas Distinguished Achievement Professorship. He is the Co-Director of the Polymer Engineering Centre at UW-Madison, a fellow member of the American Society of Mechanical Engineers (ASME), New York City, USA, and the SPE, the recipient of the 2018 Wisconsin Alumni Research Foundation (WARF) Innovation Award, 2015 Plastics Educator of the Year Award from the SPE Milwaukee Section, 2011 Engineer of the Year award from the SPE Injection Moulding Division, and an Honoured Service Member of the SPE. Professor Turng has published over 240 refereed journal papers since 2000, six books, three edited books in multiple languages, ten book chapters, four special issues journals, a dozen magazine and encyclopaedia articles, 18 patents and patent applications, 180 peer-reviewed conference papers and over 250 conference presentations and posters that led to 17 best papers/posters awards.

Photoluminescent Tracer Effects on Thermoplastic Polymer Recycling

Jörg Woidasky[1(✉)], Jannick Schmidt[1], Maximilian Auer[1], Immo Sander[2], Alexander Schau[2], Jochen Moesslein[3], Pascal Wendler[3], Daniel Kirchenbauer[3], Dirk Wacker[3], Guojun Gao[3], Andrey Turshatov[4], Bryce S. Richards[4], Stefan Wiethoff[5], and Claus Lang-Koetz[1]

[1] Institute for Industrial Ecology, Pforzheim University, Tiefenbronner Straße 65, 75175 Pforzheim, Germany
joerg.woidasky@hs-pforzheim.de
[2] Werner & Mertz GmbH, Rheinallee 96, 55120 Mainz, Germany
[3] Polysecure GmbH, Sankt Georgener Str. 19, 79111 Freiburg, Germany
[4] Institute of Microstructure Technology, Karlsruhe Institute of Technology, Hermann-von-Helmholtz-Platz 1, 76344 Eggenstein-Leopoldshafen, Germany
[5] Der Grüne Punkt – Duales System Deutschland GmbH, Frankfurter Straße 720–726, 51145 Köln, Germany

Abstract. German packaging law demands an increase in the plastic packaging recycling quota from 36% to 63% by 2022. Application of inorganic crystalline fluorescent substances for tracer-based sorting (TBS) provides an innovative approach to meet this recycling goal. TBS is already applied in industry to separate ground polyvinylchloride (PVC) window profiles with and without fiber reinforcement with a sorting capacity of 0.5–2.5 t/h. In Germany, a TBS recycling system for post consumer packaging is being developed, starting with bottles as a model product. More than 80% of end-of-life plastic bottles from households show colors and transparency in favor of TBS. Selected trivalent lanthanide elements – erbium (Er^{3+}), holmium (Ho^{3+}), and thulium (Tm^{3+}) – serve as effective tracer materials. These markers result in bright emission lines in green, red and near-infrared (NIR, at 800 nm) when excited with 980 nm laser light. This process of upconverting light – such that the energy of the emitted photons is greater than that of incident photons – is a unique anti-Stokes process. Tracer substances as a part of the printing ink used on labels or directly on the packaging were identified to be most efficient for industrial application. Mixtures of these tracer substances were applied on samples with a 100 $\mu g/cm^2$ concentration. In total 15 samples were tested, and experiments demonstrated that 11 marker combinations out of the 15 can be identified, based on the ratio of the individual signals. Using tracers as additives in the ppm range renders products identifiable by optical identification regardless of the polymers' physical or chemical properties, thus sorting for product properties becomes possible. Examples include distinguishing food from non-food packaging, or identification of hazardous goods, even if identical packaging materials are being used.

C. Hopmann and R. Dahlmann (Eds.): *Advances in Polymer Processing 2020*, pp. 1–13, 2020. https://doi.org/10.1007/978-3-662-60809-8_1

Keywords: Plastics sorting · Fluorescent marker materials · Recycling · Upconversion

1 Polymers in the European Circular Economy

The aim of the circular economy is the minimization both of raw materials and energy input as well as less waste generation in industrial systems by repeatedly using products or materials [1]. The circular economy is defined as a system keeping the value of products, materials and resources in the economy for as long as possible, with minimal waste generation. This is regarded as an essential contribution to sustainability [2], especially to the United Nations Sustainable Development Goals (UN SDG) with its goal 12 "Sustainable consumption and production" [3].

While polymers are an important material in the economy today, they are a significant source of waste as well. Global production of plastics amounted to 348 million metric tons (t) in 2017 [4], and it is expected to reach about 640 million t by 2035 [5]. In Europe the total plastic converter demand was at 51.2 million tons of plastics in 2017 (Table 1), while 27.1 million tons of plastic waste were collected (2016 figure; [4]). This waste is mostly sent to landfill (27.3% or 7.4 million t), or incinerated (41.6% or 11.3 million t), meaning that a mere 31.1% of all plastic waste in Europe was collected for recycling (8.4 million t) [4].

Table 1. European plastic use in 2017 [4].

European data for 2017	Plastic converter demand
Total mass [t]	51.2 million*
Percentage share of industries	
Packaging	39.7%
Building and construction	19.8%
Automotive	10.1%
Electronics	6.2%
Agriculture	3.4%
Household/Leisure/Sports	4.1%
Others	16.7%

*EU 28 plus Norway and Switzerland

Packaging is the most important industrial sector for plastics consumption in Europe. Out of the five most important polymers, four – high-density polyethylene (HDPE), low-density polyethylene (LDPE), polypropylene (PP), and polyethylene terephthalate (PET) – have their largest market volume in the packaging sector, while the largest polyvinylchloride (PVC) market is building and construction [6]. Consequently, the most important source of plastic waste in Europe is the packaging sector with its mainly short-lived products made from polyolefins (LDPE, HDPE and PP) and PET.

Although in Europe approximately 3000 companies are dealing with recycling, recycled plastics in Europe only cover 6% of the total plastic demand. About 100 companies are processing recycled plastics [5]. The annual capacity of the European recycling industry in the early 2010 years was estimated to be 5 million t of mechanical recycling, and 50,000 t for chemical recycling [7]. This corresponds to the 2016 figures with 8.4 million tons of EU plastics recycled, with a share of domestic recycling (inside EU) of 63% or 5.3 million t [4].

Unlike in other recycling businesses, legislation plays a pivotal role for business development for plastics recycling. For example, European Union (EU) legislation requires avoidance of waste in the first place, and then subsequently emphasises the re-use of polymer products [2], but, in the long run, all plastic products will be converted into waste that has to be recycled, and eventually disposed of. Based on the 2015 European Union (EU) Action plan for a circular economy [2], plastics were, amongst others, identified to be key priority material streams for a circular economy. In 2018, the EU Circular economy package was adopted by the European Commission, including a Europe-wide strategy for plastics in the circular economy. Key elements of this plastics circular economy include improvement of design to support recycling, separate collection to ensure quality inputs for recycling, modern and sufficient sorting and recycling plants and markets available for absorbing the recycled material [5].

In Germany, higher recycling quotas for plastic packaging have already been implemented by the German packaging law [8], which demands the increase in the plastic packaging recycling quota from 36% to 63% by 2022. To meet this challenging recycling goal, new technologies will be needed to produce more secondary polymers with a sufficiently high quality for materials recycling. One of the promising options to support recycling is the use of inorganic crystalline fluorescent tracer substances, which provide identification and sorting options independently from the packaging materials' properties.

2 Photoluminescent Tracers

Photoluminescence (PL) is triggered by the interaction of electromagnetic irradiation (light) responding electronic excitation and relaxation processes. One important physical effect observed here is the down-shifting (DS) PL effect caused by internal conversion from a higher excited energy level to a lower emissive energy level (or shorter wavelength irradiation versus longer wavelength emission producing Stokes type emission). Down-shifting is one of the PL mechanisms depicted in Fig. 1(a).

However, it is actually the mechanism (b) in Fig. 1 that is predominantly used with the materials described here, whereby radiative excitation occurs twice via the intermediate level E1 in PL tracer substances, thus allowing the absorption of two photons with the same wavelength to reach the excited state E2. Relaxation is reached by radiative emission of only one high energy package to the ground state G. Due to that nature of converting lower to higher energy it is named up-conversion (UC). Up-conversion can thus be understood as photoexcitation at a longer wavelength in the near-infrared region (NIR) followed by an emission of a shorter wavelength in the visible region (VIS). As shown in Fig. 1(b), two NIR photons are needed to produce one VIS photon.

The electronic states or levels to which excitation occurs should be stable so no immediate relaxation can ensue. Electrons involved in these transitions should therefore not be interacting with the host crystal. Lanthanide ions represent ideal candidates with these qualities. This is because the 4f-electrons that are built up in this period of the table of elements are "inner electrons" that are efficiently shielded from their outer 5d-counterparts with which chemical bonding occurs so that electron-phonon coupling to f-f transitions is reduced. The f-levels of the trivalent lanthanide ions are very large in number stemming from a fine and hyper-fine structure resulting from mutual electrostatic and spin-orbit interaction.

Moreover, each transition between states of the single ion occurs with its specific probability that reflects in the value of the cross section. It may be entirely different and much more effective in another ion. Fortunately, there are a number of energy transfer processes that enable using an ion with a high cross section as a very effective absorber and transfer this energy to a very effective emitter. This is the reason why ytterbium (Yb^{3+}) is used as the dopant to absorb irradiative light effectively and then transfer this energy once or multiple times to either erbium (Er^{3+}) or holmium (Ho^{3+}) or thulium (Tm^{3+}) as co-dopant and effective emitter ions. The dopant is also called the activator and the co-dopant the sensitizer.

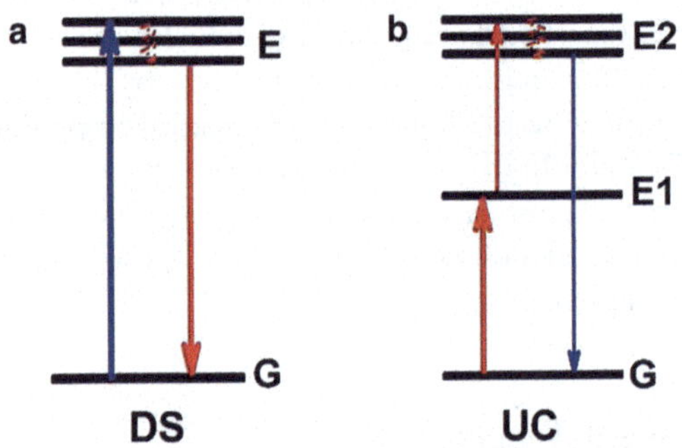

Fig. 1. Basic fluorescence mechanisms in phosphors: a) Down-shifting or Stokes emission where the photon excitation energy is larger than the emission energy due to non-radiative relaxation processes in the host crystal; b) Up-conversion absorbs two photons of lower energies via level E1 as a stepping-stone and emits one photon at higher energy.

All dopant and co-dopant pairs were thoroughly investigated to find the optimal doping concentrations for a maximal emission signal. These inorganic compounds provide the basis for efficient tracer materials together with structural and particle morphology control by means of X-ray diffraction and electron microscopy.

3 Tracer Based Plastic Sorting Experiences and Challenges

Application of fluorescent markers which emit merely one distinct emission, or the application of substances emitting multiple distinct emission lines ("taggers") for polymer sorting processes has been discussed since the 1990s [9]. Research on polymer packaging and polymer particle sorting showed the applicability of tracers which can be excited by X-rays [10] both for macro-sorting (sorting of entire products such as bottles) and micro-sorting (sorting of smaller particles such as ground polymers), providing identification options for high speed identification and automated sorting even for black polymers. Tracer substances applied were based on rare earth oxides and used in concentrations below 1000 ppm. Fluorescent tracers which are excitable at 310–370 nm and emit in the visible spectrum (VIS) are commercially available, and they have been used for macro-sorting (bottle sorting) with conveyor speeds of up to 3.5 m/s with a sorting purity of up to 95% successfully [11]. Here, 0.5–5 ppm of binary markers have been applied in the polymers.

Moreover, in a lab scale sorting plant micro-sorting with tracer substances in polymer flakes (POM, PBT, ASA) was carried out up to 300 kg/h and with even better sorting results (sensitivity of 99.4% and precision of 99.5%) [9]. In this case, both excitation and fluorescence signals were in the visible spectrum.

The key challenge of fluorescence tracer application for particle or product identification is to achieve a sufficient signal-to-noise ratio. Tracer concentration in the material should be minimal for economic and environmental reasons, so this will decrease the signal. As an alternative, UC PL has been identified to provide an excellent signal-to-noise ratio, using excitation in the invisible infrared spectrum (IR, at 980 nm), and generating emissions mainly in the visible spectrum. If applied in a dark environment without visible light, only the tracer substance will be visible, and an excellent signal-to-noise ratio can be achieved. This technology has neither been tested for nor applied to end-of-life polymer sorting yet, so it provides the technical basis for this contribution.

PL tracer substances are characterized by specific emission lines and lifetime (PL decay time) properties. Typical substances used to this end are organic molecules or lanthanide elements complexes using e.g. europium and terbium [9, 10]. Polymer tracers can only be detected on the product surface [10], and in polymers typically tracer substance concentrations are between 0.5 and 20 ppm. Minimum tracer concentrations between 25 and 100 ppm for UV tracers and from 100 to 1,000 ppm for X-ray excitable tracers have been reported for identification of black plastics [12]. Experiments with ABS showed no change in mechanical properties of the material as long as the fluorescence tracer concentration did not exceed 250 ppm [9], and no photo-degradation after UV irradiation exposure was observed in PP with 1000 ppm tracer concentration in the μm range [10].

4 Application: PVC Profiles from the Building Industry

In 2018, more than 33,000 t of used PVC building products such as window profiles, roller shutters and doors in Germany have been recycled into secondary products. Moreover, almost 65,000 t of PVC production waste have found their way into secondary applications [13].

The circular economy of PVC window profiles is one of the use cases for tracer-based sorting, as some window profiles are made from a combination of different

PVC types which are partly reinforced with glass fibers. After the products use phase both PVC types have to be separated, for proper recycling, as glass fiber reinforced polymers have different processing and abrasion properties than unfilled resins. The co-extruded product shows a material fit, i.e. dismantling it is not an option for separation of reinforced from glass-fiber free portions of the profiles. The PVC profiles are ground to a particle size of 3–15 mm. No distinction can be made between reinforced and glass-fiber free PVC in current near infrared (NIR) sorting systems, as this identification method relies on the polymer matrix properties only. To solve this challenge, the glass fiber reinforced PVC was provided with fluorescent tracers (Fig. 2). A tracer concentration of 100 ppm proved sufficient to exceed the threshold of the noise of the camera used for identification, so this is the lower limit of tracer concentration in this case.

Fig. 2. PVC Profile with fluorescent tracer.

The sorting process in this application is binary, i.e. it is only examined if a tracer substance is present. The PVC sorting machine which was specifically designed and built for PVC particle sorting uses fanned-out laser beams to excite the tracer substances in the PVC particles. It detects the fluorescent particles through a camera system and separates them from the unmarked particles by means of compressed air pulses. The throughput of the machine lies between 0.5 and 2.5 t/h, with a conveyor speed for the flakes of 3 m/s. The purity of the sorted PVC material lies beyond 99%.

5 Application: Polymer Packaging

Post-consumer plastic bottles have been selected as model products for the tracer-based sorting application in waste management. To provide detailed information on end-of-life plastic bottle packaging, a total of 35 randomly selected households from Baden-Wuerttemberg/Germany was asked to provide their end-of-life plastic bottles which they discarded during one week in June 2018. The total mass of material collected was 9.86 kg, with 8.4 kg of bottles, along with miscellaneous bottle closures weighing 1.338 g. In total 236 bottles were collected and characterized amongst others for their contents, their materials (both the bottle and the caps), or their color. Visual inspection using a reference chart with RAL colors was used for color identification. Material identification (Table 2) was based on packaging marking and on attenuated total reflection Fourier transform infrared spectroscopy (ATR-FTIR) measurements using a Bruker Alpha FTIR with Bruker Optics ATR-Polymer Library.

Table 2. End-of-life polymer bottle properties from households.

Product	Total packaging Mass share [%]	Material	Bottles only Mass share [%]	Caps&Closures only Mass share [%]
Food applications	42.0	PE-HD	27.7	16.7
Non-food:		PE-LD	2.2	2.8
Cleaning agent	9.7	PET	57.9	–
Washing agent	8.9	PP	10.0	78.9
Sanitary products	28.5	Others	2.2	1.5
Others	10.9	Not identifiable	–	0.1

As a result, clear material or light colors that are advantageous for tracer identification (crystal clear and RAL Deutsches Institut für Gütesicherung und Kennzeichnung colors 1000–3000 and 9000), amount to 87.2% of all bottles.

Besides adding tracer substances to the packaging material, other options for tracer application exist such as applying a tracer as a coating or on labels. Expert judgement was made on these options, and a tracer substance application in the label print was identified to be the most advantageous approach.

Currently, experiments on re-recycling of tracers containing PE and PET are being carried out, to quantify the effects both on tracer and resin performance.

6 Tracer Substances Regime

If only one tracer substance in the polymer matrix is used, this will result in binary information only, but a combination of miscellaneous tracer substances will enable multiple codes. These codes may be used for identification of several products or substances in parallel, such as e.g. combinations of colors with food- or non-food applications.

To this end, experiments with tracer substances on a support material (like on packaging labels) with a total concentration of 100 μg/cm² (for material efficiency reasons) were carried out to determine how many fractions of plastic waste can be separated if multiple tracer substances are applied. As this value translates into a tracer concentration of about 6000 ppm in the material, a sufficient PL signal could be expected here. The total concentration of 6000 ppm was distributed to different individual tracer substances: depending on the combination of tracers, individual emission line combinations were obtained. The sum of all tracers on the label never exceeded 6000 ppm. A maximum tracer concentration of 500 μg/cm² was found to be the upper limit for this application for economic reasons.

The emission lines of the tracers were between 400 nm (blue) and 800 nm (IR). As the blue emission line exhibited a low intensity, only three different tracer substances with their specific emission lines (green, red and IR) proved applicable for combinatory experiments (Fig. 3). The tracers used are marker 1 (green PL), marker 2 (red PL) and marker 3 (IR PL). Figure 3 shows that the tracers often emit several colors. Although other UC tracers may be synthesized from other atoms or host lattices, the peaks in green, red or IR will remain in the same places of the spectrum.

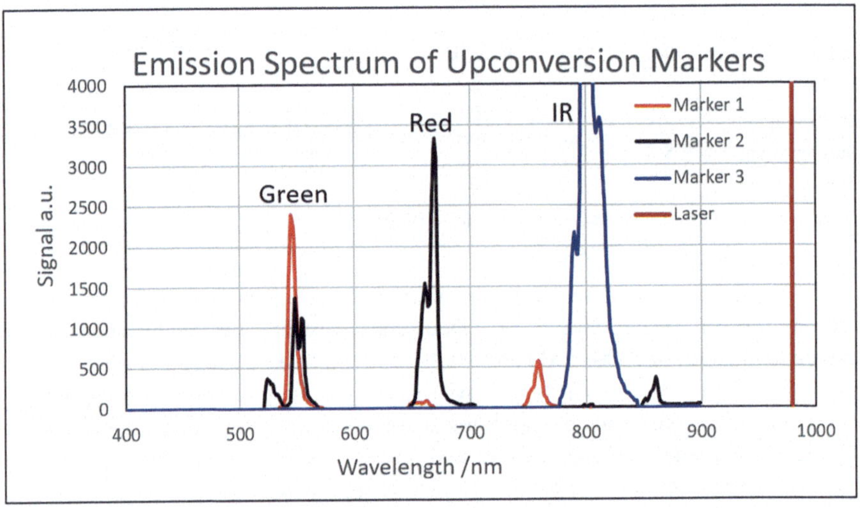

Fig. 3. Emission spectra of the green, red, and IR tracers.

In the next step a combination of these tracers in percentage increments was undertaken, for example combining 25% of marker 1 with 25% of marker 2 and 50% of marker 3. Depending on these increments, a theoretical number of combinations can be obtained as shown in Table 3. It was decided to use the 25% increment with the three tracers identified above (bold formatting in Table 3) for the first experiments, resulting in a maximum of 15 tracer combinations. With three tracers and a 12.5% increment, 45 tracer combinations would have been possible.

The tracers were mixed with commercially available white wall paint. This paint was applied on metal sheets with a concentration of 100 μg/cm² in a 125 μm layer.

1 cm^2 pieces were cut as samples from these coated metal sheets. These samples were rotated by a motor in a measurement chamber to simulate movement during the identification step, such as being moved on a belt with a speed of up to 3 m/s, or free falling. Excitation was achieved by a line laser, with a power density on the tracers at approximately 35 mW/mm^2. The signal was measured with three photodiodes equipped with filters to measure the individual signals in green, red and IR only. The laser irradiation was filtered in all three cases, but not measured. The distance between tracer and photodiode was adjusted to 30 cm.

Table 3. Tracer combination options.

%- increments	12,5%	20%	25%	33,33%	50%
↓Number of tracer substances	Potential combinations = Number of potential sorting fractions				
2	9	6	5	4	3
3	45	21	**15**	10	6
4	165	56	35	20	10
5	495	126	70	35	15
6	1287	252	126	56	21

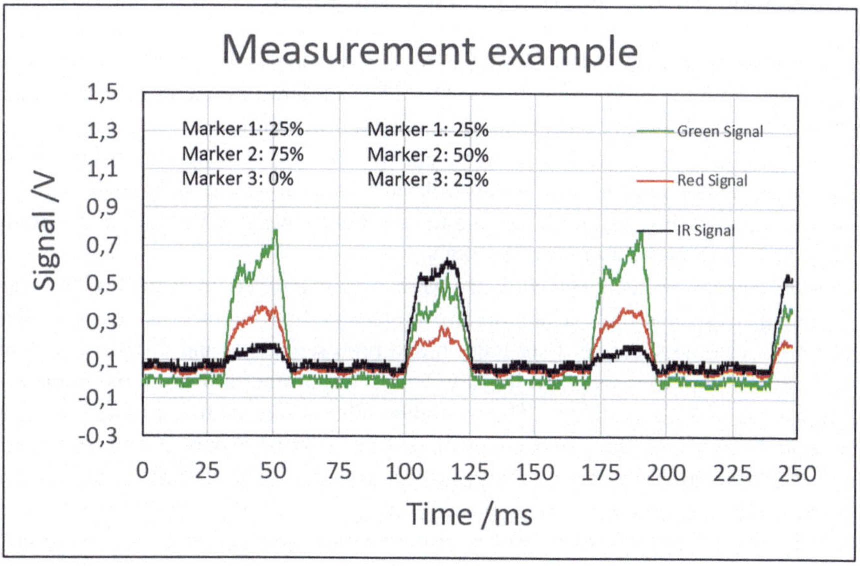

Fig. 4. Measurement results for two selected tracer combination samples.

Figure 4 shows the signal of two selected tracer samples, each combining two or three individual tracer substances as identified in the figure. Due to the experimental setup the two samples were measured in parallel and rotated through the identification (sensor) area, producing signals of one tracer combination from 30 to 60 ms (duplicated at 160–195 ms) of the reading period, and from 100 to 125 ms (restarting at 235 ms) of the second tracer sample. The signal difference between minimum and maximum of the individual signals was used for evaluation. In the given case, for "marker 1: 25% + marker 2: 75%" the results were green = 0.82, red = 0.368, IR = 0.14. As the absolute signal readings depend on the geometrical conditions of the measurement such as the distance (length) between tracer and the photodiodes the ratios of the individual readings are used for the differentiation of the tracers i.e. green/red, green/IR and red/IR. It would be green/red = 0.82/0.368 = 2.23 in the experiment shown.

Experiments showed so far that eleven out of 15 tracer combinations can be identified, while the remaining four combinations are too similar to be distinguished. In the experiment discussed, "marker 1: 75% + marker 2: 25%" is indistinguishable from "marker 1: 50% + marker 2: 50%". The reason is that marker 2 generates both a red and a green signal, and these signals are subject to some natural variation. Moreover, matrix effects and effects of sample preparation might play a role here. Future experiments are planned to demonstrate that the distinction of up to 45 tracer combinations is possible. Potential improvements are expected from new tracer substances that have less mixed signals, and improvements of the measurement technique.

7 Implementation of Tracer Based Sorting in Industry

The importance of stakeholder involvement in R&D activities and innovation management has been recognized as crucial [14, 15]. Numerous stakeholders from tracer producers to brand owners have both to be involved for the implementation of tracer-based sorting in industry and need to be convinced. The consideration of the future value chain of the innovation is highly relevant, too. However, stakeholder involvement goes beyond this, and it allows one to comprehensively record and map possible chances and barriers of any new technology.

In the current project, stakeholders were first examined in a methodologically structured analysis and the relevant organizations and their interaction were visualized from a life cycle perspective. Especially brand representatives and distributors, waste management companies and recyclers are seen as important partners in the implementation process of the technology. They amongst others were involved in an interactive workshop. In this way, the perspectives of relevant interest groups have been embedded in the innovation process of the project at an early stage. Based on this interaction and its results, optimization proposals were generated in parallel to the technical research and development work, to adjust the technical development focus and additional aspects of work (cf. [16]).

As of today, the most important barriers are seen in the three areas "process/facilities", "markers" and "benefit distribution via various stakeholders". From the stakeholders' point of view, the most important drivers can be found in the three thematic areas "quality/recycled goods quality", "process/facilities" and "regulation/legal framework". Regarding regulation, especially increased recycling rates stipulated by regulations are seen as a key driver. It is interesting to note that "process/facilities", i.e. the integration of the process into the current sorting process, on the one hand is seen as the strongest barrier, but on the other was also identified to be a strong driver, too [17].

The implementation of tracer-based sorting requires the collaboration of numerous stakeholders and will lead to a shift in value creation in packaging recycling. Consequently, the development of concrete business models is currently an important focus of the project. The individual business interests of all stakeholders have to be sufficiently accommodated. Business model development is thus an important component in providing the technologically promising innovation with actual opportunities for implementation. It renders business model development to be a core element in the implementation of innovation.

8 Summary

In the research project "MaReK" ("Markerbasiertes Sortier- und Recyclingsystem für Kunststoffverpackungen"), industry and universities are cooperating under a grant of the German Federal Ministry for Education and Research (BMBF) to develop a tracer-based sorting and recycling system for packaging, in order to support high-quality material recycling. This research is motivated by the low portion of plastic (packaging) materials recycled in Europe and Germany. The main findings and current conclusions of the research include:

- Application of tracers for waste management purposes: The research consortium is the first to apply up-conversion tracer materials for post-consumer packaging recycling and to publish experimental results. Using tracers as additives in the ppm range renders products identifiable by optical identification processes regardless of the physical or chemical properties of the packaging material, whereas current packaging sorting relies on packaging material chemistry and geometrical properties only.
- Development of anti-Stokes photoluminescence (PL) tracer substances: Materials such as doped gadolinium oxysulfide micro-phosphors provide distinct emission lines of different colors in the visible spectrum. Three substances emitting in green and red and infrared are available for application as of today, and more substances are currently under development.
- Excellent signal-to-noise ratio: IR is used to excite the anti-Stokes PL, resulting in distinct visible PL signals that are both well applicable for product sorting and also do not suffer from auto-fluorescence. If applied in a dark environment, only the tracer emission signals are visible.
- Modifications of tracer-oriented packaging design: Highest tracer material efficiency can be achieved when tracer substances are used as additives in printing ink on labels or on the packaging surface directly. Tracer concentrations of 100 $\mu g/cm^2$ on the packaging surface are more than sufficient for identification. As of today, already about 80% of all recyclable packaging bottles are compatible with tracer application requirements.
- Sorting code development: Tracer combinations and their performance for identification of moving objects was measured and proved sufficient to distinguish eleven tracer combinations. Both the number of tracer combinations and the sensor technique applied will be subject to further research. Tracer concentration far below the 1000 ppm range will become possible for technical applications.

Consequently, tracer materials provide new opportunities for polymer product identification and marking. This technological approach goes beyond current NIR sorting capabilities, as it allows distinguishing product properties, e.g. distinguishing food from non-food packaging, even if identical polymers are being used for packaging. Moreover, identification of numerous products in parallel, i.e. in only one sorting step becomes possible, so from a technical point of view the current, multi-step sorting processes may be replaced. A single-step pilot identification facility using tracer-based sorting for experimental purposes has been in operation in Freiburg/Germany since September 2019.

Acknowledgements. Substantial part of the research reported in the above contribution was made possible by a grant of the German Federal Ministry for Education and Research (BMBF) as a part of the framework program "Research for Sustainable Development" (FONA3) on the topic "Plastics in the environment" with grants no. 033R195A-E under supervision of the project executing organization Jülich (PTJ). The authors are thankful to Mrs. Gabriella Loveday (Pforzheim University) for proofreading this contribution.
The sole responsibility for this text is with the authors.

References

1. Geueke, B., Groh, K., Muncke, J.: J. Clean. Prod. **193**, 491–505 (2018)
2. European Commission: Closing the loop – An EU action plan for the circular economy (COM(2015)614 final) (2015). Accessed 7 Aug 2018
3. United Nations: Transforming our world: the 2030 agenda for sustainable development (A/RES/70/1) (United Nations)
4. Plastics Europe: Plastics – The facts 2018: an analysis of European plastics production, demand and waste data. Plastics Europe, Brussels (2018)
5. European Commission: A European strategy for plastics in a circular economy (COM(2018) 28 final) (2018). Accessed 6 July 2018
6. Ragaert, K., Delva, L., van Geem, K.: Waste Manage. (New York, N.Y.) **69**, 24–58 (2017)
7. Ignatyev, I.A., Thielemans, W., Vander Beke, B.: ChemSusChem **7**, 1579–1593 (2014)
8. Gesetz zur Fortentwicklung der haushaltsnahen Getrennterfassung von wertstoffhaltigen Abfällen: VerpackG
9. Brunner, S., Fomin, P., Kargel, C.: Waste Manage. (New York, N.Y.) **38**, 49–60 (2015)
10. Bezati, F., Froelich, D., Massardier, V., Maris, E.: Resour. Conserv. Recycl. **55**, 1214–1221 (2011)
11. Ahmad, S.R.: Environ. Technol. **25**, 1143–1149 (2004)
12. Dvorak, R., Kosior, E., Moody, L.: Development of NIR detectable black plastic packaging. Final report (2011)
13. Rewindo: Kunststofffenster-Recycling in Zahlen 2018, Bonn (2018)
14. Charter, M., Clark, T.: Sustainable Innovation: Key Conclusions from Sustainable Innovation Conferences 2003–2006. Organised by The Centre for Sustainable Design, Farnham (2007)

15. Cancino, C.A., La Paz, A.I., Ramaprasad, A., Syn, T.: Technological innovation for sustainable growth. An ontological perspective. J. Clean. Prod. **79**, 31–41 (2018)
16. Gasde, J., Preiss, P., Lang-Koetz, C.: Integrated innovation and sustainability analysis in collaborative R&D projects. In: Proceedings of the ISPIM Innovation Conference – Celebrating Innovation: 500 Years, Since da Vinci
17. Gasde, J., Klinke, C., Woidasky, J., Lang-Koetz, C.: Integrierte Innovations- und Nachhaltigkeitsanalyse im Bereich Sortierung und Verwertung von LVP-Abfällen Tagungsband des Wissenschaftskongress „Abfall- und Ressourcenwirtschaft", pp. 99–103. Deutsche Gesellschaft für Abfallwirtschaft e.v.; Innsbrück University Press, Innsbrück (2019)

An Efficient Strategy Based on Hyperspectral Imaging for Brominated Plastic Waste Sorting in a Circular Economy Perspective

Giuseppe Bonifazi[1], Ludovica Fiore[1], Pierre Hennebert[2], and Silvia Serranti[1(✉)]

[1] Department of Chemical Engineering, Materials & Environment, Sapienza – University of Rome, Via Eudossiana 18, 00184 Rome, Italy
silvia.serranti@uniroma1.it
[2] INERIS, French National Institute for Industrial Environment and Risks, CS 10440, 13592 Aix-en-Provence Cedex 03, France

Abstract. Plastic is one of the materials mostly used in many industrial sectors, as those of packaging, construction, agriculture, etc. Furthermore, polymers are also extensively used in electrical and electronic equipment (EEE) manufacturing.

Plastic waste originating from EEE (WEEE) is a challenge for recycling due to presence of various hazardous elements (i.e. additives) representing a polluting source for the environment and a risk factor for the human health. Among these, there are those containing brominated flame retardants (BFRs), largely utilized to increase fire resistance, avoiding or delaying the flames, thus allowing to the plastics-based-manufactured-products to respect safety requirements. However, plastics with high BFRs levels cannot be recycled and therefore must be removed from the recycling stream. In a circular economy perspective, it is thus necessary to develop a system able to efficiently separate plastic waste into homogeneous fractions based on the BFRs content. To improve the circularity of plastics, it is thus essential to recover and recycle bromine-free plastics to avoid the loss of available secondary raw materials.

The aim of this study was to develop an efficient, reliable and sustainable approach based on hyperspectral imaging (HSI) operating in SWIR range (1000–2500 nm) coupled with chemometrics, specifically addressed to identify plastics containing BFRs. In order to investigate the possibility of reliable sorting, plastic scraps from cathode ray tube with different contents of BFRs were investigated. The concentration of bromine per scrap was preliminary measured by X-ray fluorescence in order to validate the results obtained by HSI. The recyclable low-bromine fraction was successfully identified using the developed strategy based on HSI.

Keywords: Brominated flame retardants · Plastic waste · Hyperspectral imaging

C. Hopmann and R. Dahlmann (Eds.): *Advances in Polymer Processing 2020*, pp. 14–27, 2020. https://doi.org/10.1007/978-3-662-60809-8_2

1 Introduction

The world plastic production almost reached 350 million tonnes in 2017 [1]. The large productions of plastic materials lead to a constant increase in the waste that must be managed. The European Union is promoting the transition to a circular economy approach. This represents a more sustainable alternative to the traditional linear economy based on a take-make-consume-throwaway. Through the circular economy, waste prevention and reuse of products is promoted. Alternatively, waste can be recycled or used (i.e. final waste not economically feasible to be recovered with the actual available technologies) to generate energy [2].

In the perspective of the circular economy, the recycling of plastics can reduce the environmental impact, the depletion of resources and therefore the use of energy and materials per unit of output. Recycling is the final result of the intermediate stages of collection, sorting and processing of polymers. The sorting phase is fundamental to identify the type of polymer and the possible presence of additives [3]. Furthermore, the presence of substances deliberately added in plastic products (e.g. plasticizers, stabilizers, pigments and flame retardants) affect the recovery performance that can be obtained by recycling processes [4]. In particular, brominated flame retardants (BFRs) are added to many plastic products and electrical/electronic equipment to make them less flammable [5]. There are about 75 different commercial BFRs [6], characterized specific properties and uses. Five classes can be identified: i) Polybrominated diphenyl ethers (PBDEs: $C_{12}H_{(10-x)}Br_xO$ ($x = 1, 2, ..., 10 = m + n$)) used in plastics, textiles, electronic castings, circuitry; ii) Hexabromocyclododecanes (HBCDDs: $C_{12}H_{18}Br_6$) used in thermal insulation in the building industry; iii) Tetrabromobisphenol A (TBBPA: $C_{15}H_{16-x}Br_xO_2$ where $x = 1–4$) and other phenols used in printed circuit boards, thermoplastics (mainly in TVs); iv) Polybrominated biphenyls (PBBs: $C_{12}H_{(10-n-m)}Br_{(n+m)}$ where $n + m = 1–10$) formerly used in appliances, textiles, plastic foams and v) other brominated flame retardants [7].

Due to their toxicity, the European Union has limited and/or banned some BFRs. They have potential to cause endocrine disruption (i.e. TBBPA), can dramatically alter liver enzymes and thyroid homeostasis (i.e. HBCDDs) and are associated with developmental, reproductive and neuro-toxicity (i.e. PBDEs). These substances must be replaced with safe or safer materials to ensure a significant reduction in health and environmental risks [8].

Even if plastic waste from electrical and electronic equipment (WEEE) are collected separately and submitted to recycling processes, its content of BFRs would be likely to pose risks to health or environment, especially when their preliminary handling (i.e. collection, storage, hand sorting, etc.) and the further processing (i.e. automatic sorting/separation strategies definition) are not correctly implemented and/or carried out [9]. Restricting the use of these dangerous substances in new products is the present way but the legacy of these substances in previously produced items, in a circular economy perspective, imposes to manage these wastes by removing the fraction containing dangerous substances from the waste flow addressed to recycling. Plastic materials with high BFR levels should be identified and separated from other recycling materials, for this purpose, effective methods for the identification of BFR are necessary. In order to quantify bromine content in WEEE, several studies, implementing different detection strategies, have been carried out. The BFRs present in the WEEE polymeric fractions were analyzed by Raman spectroscopy [10] and by chromatography [11]. Television sets and personal computer monitors were studied using a portable X-ray fluorescence device (XRF) [12].

In facilities bath density separation (typically density >1.1 kg/l), and X-ray transmission density separation (typically Br concentration >15,000 mg/kg) are used [13]. Nevertheless, depending on WEEE category, about 30% of the dense scraps contain less than 1000 mg Br/kg, due to other (non-hazardous) additives [14]. As the brominated plastics are landfilled or burnt, this fraction is lost for recycling. Additionally, the sorting of light insulation material (expanded polystyrene and polyurethane foam) cannot be done by density, despite they can contain HBCDD [15, 16]. HSI approach can represent a solution for identification and quality control of different plastic waste [17–19]. It could be used for primary sorting, or in a second sorting step, to recover the fraction of dense plastics that are not brominated.

The aim of this study was to develop and evaluate the efficiency and reliability of HSI for identifying WEEE plastic with BFRs, at laboratory scale. For this purpose, a two-classes model (high Br > 2000 mg/kg and low Br < 2000 mg/kg) was performed based on European Committee for Electrotechnical Standardization (CENELEC) standards on WEEE treatment. The CENELEC standard sets a requirement of maximum 2000 mg/kg of bromine in the plastics that are sent for recycling [20].

2 Materials and Methods

2.1 Materials

Dense plastic scraps from cathode ray tubes devoted to industrial incineration were supplied by a company handling hazardous waste. 350 individual scraps were analyzed for total bromine by portable X-ray fluorimeter (XRF). From this set, 55 scraps chosen randomly with sizes between 2 and 10 cm and characterized by different color, thickness, weight were used for HSI investigation. To build the classification model, the scraps were divided in two datasets: 30 scraps for training and 25 scraps for validation. An example of some of the plastic waste particles selected inside the two sample sets is reported in Fig. 1.

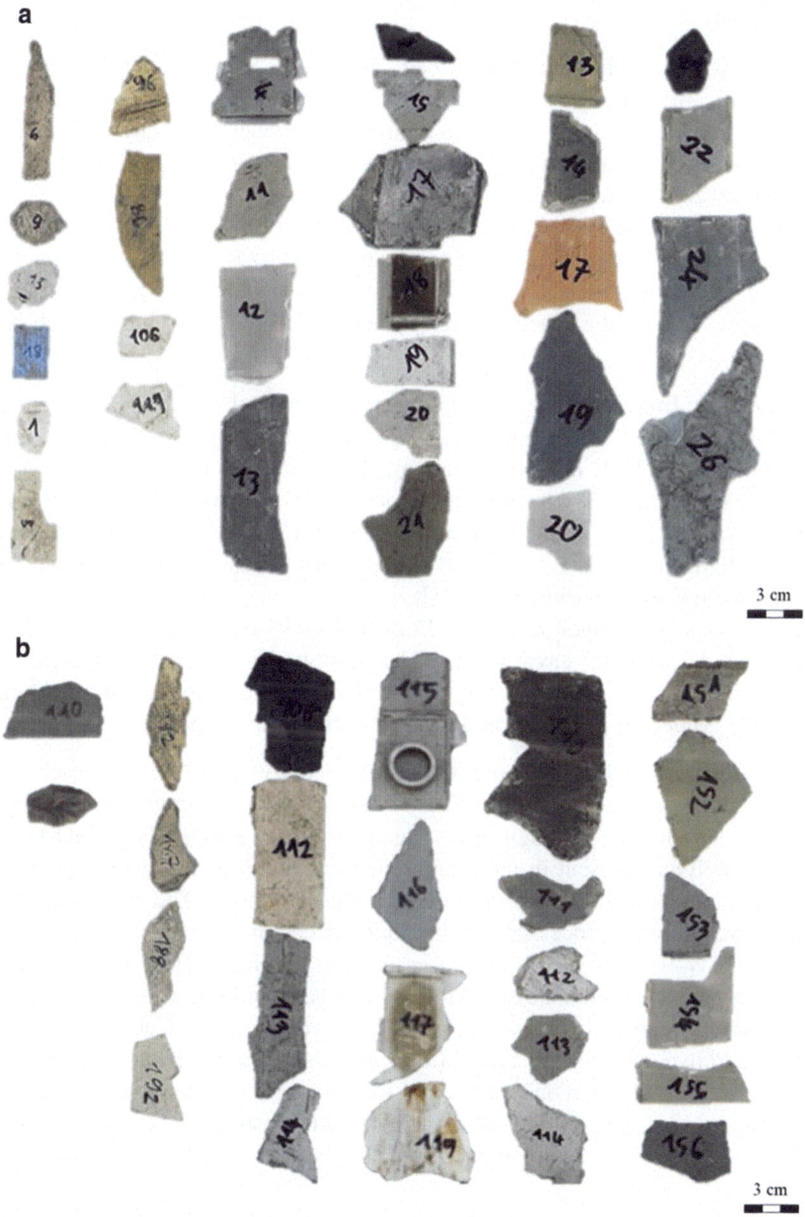

Fig. 1. Example of the reference plastic waste particles of shredded shells of cathode ray tubes (television apparatus), adopted to define the training (a) and the validation dataset (b).

2.2 Methods

Portable X-Ray Fluorimeter. A hand-held XRF Niton XL2 with benchtop stand and software dedicated for plastics (with automatic thickness correction) was used for Br measurements. Measurement time was 1 min. Independent measurement of plastic (PE and PS) reference material of 4 mm thickness (JSM P711-1 and JSAC 0651 to 0655) showed consistency of the measurement.

The analysis of the Br content was useful to divide the samples into training and validation datasets and to compare the results obtained by HSI based on Partial Least-Squares Discriminant Analysis (PLS-DA) classification. More in details, these analyses are necessary for the spectra training phase which requires samples with known composition.

Hyperspectral Imaging. The samples were acquired by hyperspectral imaging using SISU Chema XL™ Chemical Imaging Workstation (Specim, Finland) equipped with ImSpector™ N25E imaging spectrograph, working in the SWIR range (1000–2500 nm). A 31 mm lens with a field of view of 50 mm was adopted, the spectral resolution was 6.3 nm. The hyperspectral image of a sample is collected in the form of a three-dimensional data set (called a hypercube of data) which has two spatial dimensions and a spectral dimension [21].

Spectral data were analyzed using PLS-Toolbox™ (Eigenvector Research, Inc.) running into the Matlab® environment (The Mathworks, Inc.). Hyperspectral imaging acquisitions were performed at RawMaLab of the Department of Chemical Engineering, Materials and Environment (Sapienza – University of Rome, Italy).

Spectral Signatures and Principal Component Analysis (PCA). On each plastic sample, Regions of Interest (ROI) were tracked in order to acquire the raw spectra embedding particles spectral signatures. This preliminary analysis is useful to identify the type of polymer present in the training and validation dataset and, especially for this study, to attempt the recognition of brominated plastics. The spectra of the different plastics were preprocessed using scatter correction methods and spectral derivatives. Raw SWIR spectral data preprocessing was applied [22] in order to highlight the differences between the classes. More in details, in order to decrease the effect of light scattering, Standard Normal Variate (SNV) was used. 1st Derivative can eliminate both additive and multiplicative effects in the spectra. Derivation technique uses smoothing in order to not reduce the signal-to-noise ratio and, therefore, to emphasize spectral ranges [23]. Finally, the data is centered with respect to their average through Mean Centering (MC). This method eliminates the data offset not important for the analysis.

Principal Component Analysis (PCA) was then applied on the preprocessed hypercubes. PCA is a data matrix processing whose purpose is to represent the variation occurring in a set of variables using a smaller number of "principal" components [24]. The principal components are linear combinations of the original variables and are calculated to express the maximum variance contained in the data set. The multivariate nature of the data is represented by a new space with a small number of dimensions [25, 26].

Through the PCA it is possible to detect similarities between the samples; in particular samples characterized by similar spectral features are grouped in the score plot. Therefore, through the PCA, it is possible to verify and to obtain the distribution in subgroups of the data set, without knowing the nature of the samples.

Partial Least-Squares Discriminant Analysis (PLS-DA). Partial Least-Squares Discriminant Analysis (PLS-DA) is utilized to develop a classification model: a classification rule is built by selecting samples (training set) whose classification is known [27]. The classification rule is then validated using an independent data set. The discriminating method classifies the samples in one of the available classes following the Bayes rule, i.e. assigning the sample to the class corresponding to the highest probability [28].

To measure the efficiency of the model, it is statistically validated using the sensitivity, specificity and class error for calibration (Cal) and for cross-validation (CV). More in details, sensitivity, specificity and class error are expressed as [29]:

$$Sensitivity = \frac{TP}{TP + FN} \tag{1}$$

$$\text{Specificity} = \frac{\text{TN}}{\text{TN} + \text{FP}} \tag{2}$$

$$Class\,error = 1 - \left(\frac{Sensitivity + Specificity}{2}\right) \tag{3}$$

Where the true positive (TP) and the true negative (TN) indicate the pixels that were appropriately assigned as belonging (TP), or not belonging (TN), to a precise class. FP and FN represent false positives and false negatives, respectively, indicating the pixels that were erroneously assigned as belonging (FP), or not belonging (FN), to a precise class. Sensitivity (the fraction of particles of positive exact classification) and specificity (the fraction of particles of negative exact classification) take values between 0 (poor classification) and 1 (perfect classification). The class error is the global fraction of particles with inexact classification.

Venetian Blinds was used as a cross-validation method for estimating the performance of a predictive model and to select the number of Latent Variables (LVs). Cross-validation was carried out by grouping the samples. Each cross-validation group is taken out from the training set, one at a time. Each time, the model is calibrated on the remaining training samples and then used to predict samples of the cross-validation group [28]. In the Venetian Blinds methods, each test set is determined by selecting every s^{th} object in the data set (s is the number of data splits, which must be less than half of the total number of objects in the data set), starting at objects numbered 1 onwards [30].

3 Results

3.1 X-Ray Fluorescence

The main characteristics of the analyzed plastic samples, in terms of Br content, are shown in Tables 1 and 2, with reference to the training and validation sets, respectively.

Table 1. Br content measured by XRF for plastic waste samples used as a training set.

ID	Color	Thickness (mm)	Weight (g)	Br (mg/kg)	ID	Color	Thickness (mm)	Weight (g)	Br (mg/kg)
Br 6	Grey	2	2.84	151836	I 15	Grey	5	2.59	5
Br 9	Grey	3	3.45	114398	I 17	Grey	2	14.36	8
Br 13	Grey	1	1.06	65041	I 18	Grey	2	3.56	6
Br 18	Blue	4	4.19	111693	I 19	Grey	1	2.88	8
Br 1	White	2	1.95	100733	I 20	Grey	2	2.45	5
Br 8	Grey	2	1.03	88251	I 21	Grey	2	5.05	5
Br 96	White	2	3.12	108079	N Br 13	White	3	3.83	7
Br 98	Grey	2	4.61	135483	N Br 14	Grey	3	3.8	14
Br 106	White	3	1.29	78048	N Br 17	Brown	2	7.11	11
Br 119	White	2	1.44	83674	N Br 19	Grey	3	8.84	9
I 4	Grey	2	6.15	6	N Br 20	Grey	2	2.68	14
I 11	Grey	2	4.5	4	N Br 21	Grey	1	1.22	12
I 12	Grey	1	5.4	6	N Br 22	Grey	2	4.86	11
I 13	Grey	2	7.84	4	N Br 24	Grey	3	8.47	9
I 14	Grey	3	1.29	4	N Br 26	Grey	3	11.29	149

The results clearly outlined the great difference of composition, with regards to Br presence in the samples. More in details, in the training dataset there are 10 plastic waste samples with high bromine content (from 65041 to 151836 mg/kg) and 20 samples with low content (from 4 to 149 mg/kg). Instead, in the validation dataset there are 6 samples with high bromine content (from 118633 to 83885 mg/kg) and 19 with low content (from 3 to 309 mg/kg).

Table 2. Br content measured by XRF for plastic waste samples used as a validation set.

ID	Color	Thickness (mm)	Weight (g)	Br (mg/kg)	ID	Color	Thickness (mm)	Weight (g)	Br (mg/kg)
Br 110	Grey	3	2.57	110272	I 119	Grey	1	4.48.	122
Br 115	Grey	3	1.16	83885	N Br 110	Grey	2	9.43	8
Br 182	Grey	2	3.01	118633	N Br 111	Grey	3	2.99	7
Br 187	Grey	2	2.08	112381	N Br 112	Grey	3	2.47	6
Br 188	Grey	2	2.16	110187	N Br 113	Grey	3	2.71	15
Br 192	Grey	2	2.41	102853	N Br 114	Grey	2	4.22	6
I 108	Grey	2	5.60	9	I 151	Grey	2	2.76	4
I 112	Grey	2	8.26	9	I 152	Grey	2	4.36	4
I 113	Grey	1	5.07	12	I 153	Grey	3	3.51	3
I 114	Grey	1	2.62	4	I 154	Grey	3	3.78	4
I 115	Grey	2	10.28	6	I 155	Grey	2	3.03	3
I 116	Grey	2	4.22	4	I 156	Grey	3	3.86	309
I 117	Grey	4	6.43	77					

3.2 Hyperspectral Imaging

Spectral Signatures and Principal Component Analysis (PCA). Starting from the results obtained by X-Ray Fluorescence, the samples were divided into two classes based on the bromine content (high Br > 2000 mg/kg and low Br < 2000 mg/kg), this concentration limit being recommended in [20].

The average raw reflectance spectra of the two classes of plastics characterized by high and low bromine content are reported in Fig. 2. The spectral features of the investigated plastic samples indicate that they are all constituted by polystyrene (PS); the absorption of C–H_2 and C–H in the third region of harmonics (1043, 1151, 1214, 1308 nm), of C–H_2 and δC–H_2 in the second harmonic region (1352, 1415, 1648 nm) and of C–H group in the first combination region (1817, 1868, 1918, 2012, 2074 nm) were identified [31].

The corresponding pre-processed spectra of two classes after the application on Standard Normal Variate (SNV), 1st Derivative and Mean Center (MC) algorithms, are shown in Fig. 3.

After pre-processing, PCA was applied on the calibration dataset. The PCA results (PC1-PC3 score plot) are reported in Fig. 4. PC1 and PC 3 explain 33% and 17% of the total variance, respectively. The spectral data of the polymer samples are dividend in two groups based on their bromine content. More in detail, the PC3 negative values identifies the samples with high content while the PC3 positive values those with low bromine content.

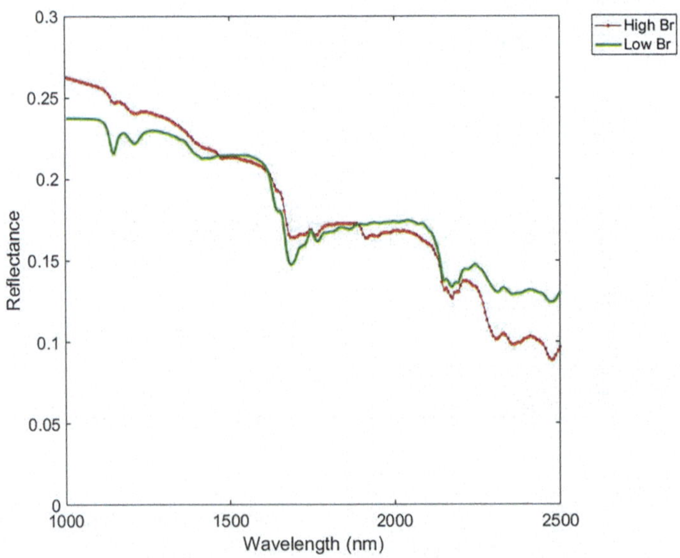

Fig. 2. Average raw reflectance spectra of the plastic waste samples acquired by HSI in the wavelength range 1000–2500 nm, representative of the two classes characterized by high (Br > 2000 mg/kg) and low (Br < 2000 mg/kg) brominated flame retardants content.

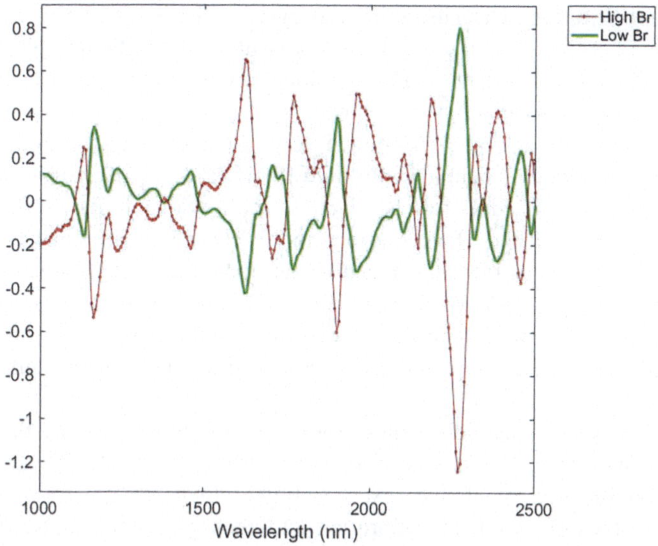

Fig. 3. Average pre-processed reflectance spectra of plastic waste samples acquired by HSI in the wavelength range 1000–2500 nm, representative of the two classes characterized by high (Br > 2000 mg/kg) and low (Br < 2000 mg/kg) brominated flame retardants content. Applied preprocessing are: Standard Normal Variate (SNV), 1st Derivative and Mean Center (MC) algorithms.

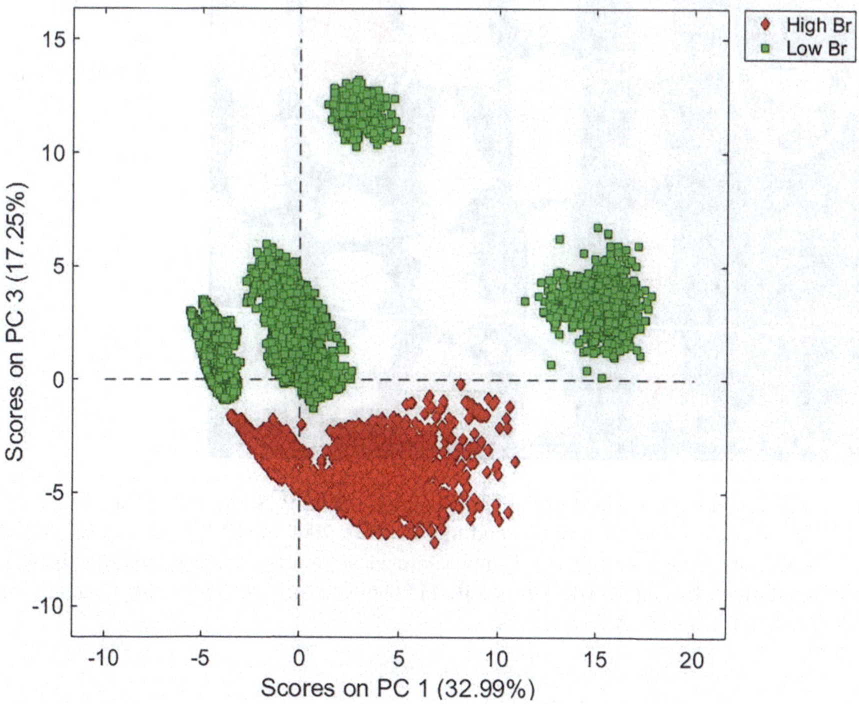

Fig. 4. PC1-PC3 score plot of the training waste plastic samples characterized by the presence of high (Br > 2000 mg/kg) or low (Br < 2000 mg/kg) brominated flame retardants content.

Partial Least-Squares Discriminant Analysis (PLS-DA). After setting class by PCA, the PLS-DA classification model was utilized to classify plastic samples belonging to the validation dataset. The classification results are reported in Fig. 5.

The two-classes model (class 1: plastic with high Br > 2000 mg/kg and class 2: plastic with low Br < 2000 mg/kg) gave very good results: 22 samples out of 25 were correctly classified. 2 samples with high Br (Br 110: 110272 mg/kg and Br 115: 83885 mg/kg) are classified as low Br, and 1 sample with low Br (N Br 110: 8 mg/kg) is classified as high Br. The three incorrectly classified samples have a flat and noisy spectrum, with very low reflectance levels (maximum 0.13). This class of samples, characterized by a dark color, could constitute a third class to be introduced in a further study. For high Br, 4 scraps are correctly identified and 2 scrap is incorrectly identified; for low Br, 18 scraps are correctly identified and 1 scraps are incorrectly identified.

The sensitivity and specificity values range from 0.98 to 0.99 and the class error values, lower than 0.01 (Table 3), confirm the quality of the model. So, the frequency of correct classification of all the pixels is higher than 98%, with an equivalence between the values obtained for calibration and cross-validation, indicating that the models are not under or over-fitted.

Fig. 5. Prediction map obtained after the application of the PLS-DA two-classes model. The classes are: high Br (Br > 2000 mg/kg) and low Br (Br < 2000 mg/kg) plastics, reported in the figure in red and green, respectively. Samples circled in yellow are incorrectly identified: the samples with high Br content (Br 110 and Br 115) are classified as low Br while sample with low Br (N Br 110) is classified as high Br.

Table 3. Sensitivity, specificity and class error values of PLS-DA model developed for recognition of the different plastic samples with high and low bromine content.

		High Br	Low Br
Sensitivity	Calibration	0.984	0.999
	Cross validation	0.984	0.999
Specificity	Calibration	0.999	0.984
	Cross validation	0.999	0.984
Class error	Calibration	0.008	0.008
	Cross validation	0.008	0.008

4 Conclusion

The EU strategy for plastics containing BFRs, in the circular economy, requires that they must be sorted and separately managed. This study investigated the possibility to apply a HSI sensing architecture to identify plastics with high and low bromine content. Starting from the acquired spectra of the different samples, a full chemometric recognition approach, based on PLS-DA classification and modelling, was developed. A two-classes model was carried out, to identify plastics with high (Br > 2000 mg/kg) and low (Br < 2000 mg/kg) Br content. Classification results were very good, with 95% of non-brominated particles identified without error.

The proposed approach can be used as a primary or complementary detection engine inside a waste plastic recycling plant to sort plastics with BRFs, according to concentration limits required by law. The study also shows the need to introduce a third class constituted by samples characterized by darker colors. Additionally, this method could be used for expanded polystyrene (EPS) or polyurethane foam (PUF) from construction and demolition waste that cannot be sorted for bromine by density.

References

1. Plastics Europe: Plastics – the facts 2018. An analysis of European plastics production, demand and waste data. https://www.plasticseurope.org (2018). Accessed 26 June 2019
2. EC: Circular economy. Implementation of the circular economy action plan. https://ec.europa.eu/environment/circular-economy/index_en.htm. Accessed 26 June 2019
3. Raj, F.: Recycling of Polymers. Methods, Characterization and Applications. Wiley-VCH (2017)
4. Faraca, G., Astrup, T.: Plastic waste from recycling centres: characterization and evaluation of plastic recyclability. Waste Manage. **95**, 388–398 (2019)
5. Alaee, M., Arias, P., Sjödin, A., Bergman, A.: An overview of commercially used brominated flame retardants, their applications, their use patterns in different countries/regions and possible modes of release. Environ. Int. **29**, 683–689 (2003)
6. Tohka, A., Zevenhove, R.: Brominated flame retardants – A nuisance in thermal waste processing? In: TMS Extraction and Processing Division Meeting on Recycling and Waste Treatment in Mineral and Metal Processing: Technical and Economic Aspects, Sweden (2002)

7. Covaci, A., Harrad, S., Abdallah, M.A.-E., Ali, N., Law, R.J., Herzke, D., de Wit, C.A.: Novel brominated flame retardants: a review of their analysis, environmental fate and behavior. Environ. Int. **37**, 532–556 (2011)
8. EC: Directive 2002/95/EC of the European Parliament and of the Council of 27 January 2003. The restriction of the use of certain hazardous substances in electrical and electronic equipment. Official Journal of the European Union. https://eur-lex.europa.eu/LexUriServ/LexUriServ.do?uri=OJ:L:2003:037:0019:0023:EN:PDF (2002). Accessed 8 July 2019
9. EC: Directive 2011/65/EC of the European Parliament and of the Council of 8 June 2011. The restriction of the use of certain hazardous substances in electrical and electronic equipment (recast). Official Journal of the European Union. https://eur-lex.europa.eu/legal-content/EN/TXT/?uri=celex%3A32011L0065 (2011). Accessed 8 July 2019
10. Kikuchi, S., Kawauchi, K., Ooki, S., Kurosawa, M., Honjho, H., Yagishita, T.: Non-destructive rapid analysis of brominated flame retardants in electrical and electronic equipment using Raman spectroscopy. Anal. Sci. **20**(8), 1111–1112 (2004)
11. Schlummer, M., Brandl, F., Maeurer, A., van Eldik, R.: Analysis of flame retardant additives in polymer fractions of waste of electric and electronic equipment (WEEE) by means of HPLC-UV/MS and GPC-HPLC-UV. J. Chromatogr. A **1064**, 39–51 (2005)
12. Aldrian, A., Ledersteger, A., Pomberger, R.: Monitoring of WEEE plastics in regards to brominated flame retardants using handheld XRF. Waste Manage. **36**, 297–304 (2015)
13. Hennebert, P., Filella, M.: WEEE plastic sorting for bromine essential to enforce EU regulation. Waste Manage. **71**, 390–399 (2018)
14. Hennebert, P.: Sorting of waste for circular economy: sampling when (very) few particles have (very) high concentrations of contaminant or valuable element (with bi- or multi-modal distribution). In: Proceedings of the 17th International Waste Management and Landfill Symposium (Sardinia 2019), Cagliari, Italy (2019a)
15. Pivnenko, K., Granby, K., Eriksson, E., Astrup, T.F.: Recycling of plastic waste: screening for brominated flame retardants (BFRs). Waste Manage. **69**, 101–109 (2017)
16. Hennebert, P.: Literature evidence that plastics of construction and textiles should be sorted for brominated flame retardants before recycling. In: Proceedings of the 17th International Waste Management and Landfill Symposium (Sardinia 2019), Cagliari, Italy (2019b)
17. Bonifazi, G., Capobianco, G., Serranti, S.: A hierarchical classification approach for recognition of low-density (LDPE) and high-density polyethylene (HDPE) in mixed plastic waste based on short-wave infrared (SWIR) hyperspectral imaging. Spectrochim. Acta **A198**, 115–122 (2018)
18. Serranti, S., Gargiulo, A., Bonifazi, G.: Characterization of post-consumer polyolefin wastes by hyperspectral imaging for quality control in recycling processes. Waste Manage. **31**, 2217–2227 (2011)
19. Serranti, S., Palmieri, R., Bonifazi, G., Cózar, A.: Characterization of microplastic litter from oceans by an innovative approach based on hyperspectral imaging. Waste Manage. **76**, 117–125 (2018)
20. CENELEC CLC/TS 50625-3-1: Requirements for the collection, logistics and treatment of WEEE – Part 3-1: Specification relating to depollution – General. CENELEC, Brussels, Belgium (2015)
21. Harsanyi, J.C.: Hyperspectral image classification and dimensionality reduction: an orthogonal subspace projection approach. IEEE Trans. Geosci. Remote Sens. **32**(4) (1994)
22. Xu, L., Zhou, Y.P., Tang, L.J., Wu, H.L., Jiang, J.H., Shen, G.L., Yu, R.-Q.: Ensemble preprocessing of near-infrared (NIR) spectra for multivariate calibration. Anal. Chim. Acta **616**, 138–143 (2008)
23. Rinnan, Å., van den Berg, F., Engelsen, S.B.: Review of the most common preprocessing techniques for near-infrared spectra. TrAC Trends Anal. Chem. **28**, 1201–1222 (2009)

24. Shang, H.L.: A survey of functional principal component analysis. AStA Adv. Stat. Anal. **98**, 121–142 (2014)
25. De Juan, A., Maeder, M., Hancewicz, T., Duponchel, L., Tauler, R.: Chemometric tools for image analysis. In: Infrared and Raman Spectroscopic Imaging, pp. 65–109 (2009)
26. Jolliffe, I.Y.: Principal Component Analysis. In: Springer Series in Statistics, 2nd edn. (2002)
27. Barker, M., Rayens, W.: Partial least squares for discrimination. J. Chemom. **17**, 166–173 (2003)
28. Ballabio, D., Consonni, V.: Classification tools in chemistry. Part 1: linear models. PLS-DA. Anal. Methods **5**, 3790 (2013)
29. Amigo, J.M., Babamoradi, H., Elcoroaristizabal, S.: Hyperspectral image analysis. A tutorial. Anal. Chim. Acta **896**, 34–51 (2015)
30. Eigenvector: Using cross-validation. http://wiki.eigenvector.com/index.php?title=Using_Cross-Validation (2016). Accessed 10 May 2019
31. Workman Jr., J., Weyer, L.: Practical Guide and Spectral Atlas for Interpretive Near-Infrared Spectroscopy. CRC Press, Boca Raton (2012)

Melt-Processing of Biopolymer Composites with Nanocellulose Additives

Pieter Samyn(✉) and Hesam Taheri

Institute for Materials Research (IMO-IMOMEC), Applied and Analytical Chemistry, Hasselt University, 3590 Diepenbeek, Belgium
pieter.samyn@uhasselt.be

Abstract. Cellulose fibers are advantageous reinforcing materials for bio-based composites because of their availability and good mechanical properties. Several chemical processes have been developed to turn the native fibers into micro- (MFC) or nanofibrillated (NFC) cellulose fibers. These nanocomponents are believed to even have stronger reinforcing capacity because of surface interaction effects at the nanoscale and formation of a fine web structure. During further processing of fibrillated cellulose in combination with a biopolymer matrix, however, there is evidence of agglomeration due to the hydrophilic nature of the fibers and incompatibility with the polymer matrix. In our approach, surface modification of the micro- and nanofibrillated cellulose is done by the in-situ deposition of hydrophobic nanoparticles onto the fiber surface, allowing to tune the required hydrophobicity of the cellulose additives and to make them compatible with extrusion applications, as demonstrated for PLA biopolymers.

Keywords: Biopolymers · Nanocellulose · Melt-blending

1 Introduction

Cellulose fibers are advantageous reinforcing materials for the formulation of bio-based composites because of their availability and good mechanical properties. The hierarchical structure of macro-scale cellulose fibers organized in the plant cell walls contains multiple bundles of cellulose microfibrils that have alternative amorphous and crystalline domains. Several chemical processes have been developed to turn the native fibers into micro- (MFC) or nanofibrillated (NFC) cellulose fibers [1]. These nanofibers are believed to even have stronger reinforcing capacity than native cellulose fibers because of surface interaction effects at the nanoscale and formation of a fine web structure. During processing of fibrillated cellulose in combination with a hydrophobic biopolymer matrix, however, there is evidence of agglomeration due to the highly hydrophilic nature of the fibers leading to incompatibility with the polymer matrix. Several methods have been developed for the surface hydrophobization of cellulose fibers [2] and nanofibers [3].

C. Hopmann and R. Dahlmann (Eds.): *Advances in Polymer Processing 2020*, pp. 28–36, 2020. https://doi.org/10.1007/978-3-662-60809-8_3

In view of replacing petroleum-based products with bioplastics, especially poly(lactic acid) (PLA) and polyhydroxyalkanoates (PHB) are of interest. However, those biopolymers do not have suitable mechanical properties that allow them to compete directly with the main-stream plastics such as PE, PP, PS, and PVC. Plasticizers already exist to tailor the flexibility of biopolymers and to meet a variety of application requirements, but this mostly comes at the cost of strength and stiffness. Recently, different approaches such as copolymerization, blending, and addition of filler materials have been tried to achieve the flexibility and toughness of PLA, but the main focus has been centralized on blending with low molecular weight and low glass transition temperature polymers such as polyethylene glycol (PEG), polypropylene glycol (PPG), polyhydroxybutyrate to improve flexibility [4]. However, plasticizers have many limitations such as a huge decrement of the glass transition temperature Tg that may affect the processing and molding [5]. More recently, there is a remarkable interest using MFC/NFC into polymer matrices to produce nanocomposites due to significant property enhancements by incorporation of only a few weight percent of the NFC/MFC. With use of nanofillers like MFC/NFC, cellulose nanowhiskers (CNWs) or clay, the remarkable property enhancements make their nanocomposites superior candidates for materials application in the food packaging, electronics, and automotive industries [6]. The nanocomposite properties depend on the quality of nanoparticles dispersion in a polymer matrix [7]. However, the relationship between mechanical properties of nanocomposites and nanoparticle dispersion still remains unresolved and is often difficult to control [8].

As a main challenge, the MFC/NFC nanofillers fail to disperse and remain aggregated into the polymer matrix. In case of aggregation of nanofillers, the modulus and strength of the nanocomposite could be comparable to those of the pure polymer but toughness and elongation could even get worse [9]. Therefore, the dispersibility of MFC/NFC on a nanoscopic or molecular level has been noticed as the critical issue in nanocomposite science and technology. In recent work, the hydrophobic-modified NFC was obtained by grafting hydrophobic monomers on NFC to improve the compatibility between NFC and the biopolymer during blending and improve its dispersiblity. However, most approaches are based on the use of solvent casting techniques in order to produce nanocomposites, while it remains more challenging to incorporate the nanofillers during continuous melt-processing steps. In order to achieve these goals, understanding of thermodynamic behavior of PLA and its composites requires quantitative thermal and molecular mobility analysis [10]. Moreover, the rheological behavior and thermal properties can be significantly altered by stereo-sequence distribution, architecture and molecular weight [11].

In contrast with traditional chemical surface modification, the hydrophobization of micro- and nanofibrillated cellulose in this study is done by deposition of hydrophobic nanoparticles, allowing to tune the required hydrophobicity of the cellulose additives and to make them compatible with PLA for, e.g., extrusion applications. In order to get more fundamental insight in the melt-blending characteristics of the surface-modified MFC/NFC and the PLA, rheological measurements were done.

2 Materials and Methods

Different morphologies of micro- and nanofibrillared cellulose were produced from hardwood pulp in a Microfluidizer M-110EH (Microfluidics) using different geometries of the interaction chamber (200 and 87 μm) under various pressures (350–1700 bars) and number of passages (5–50). As such, 2% fiber suspensions with gradually decreasing fiber diameter of 500 nm to 50 nm were obtained, including both microfibrillated cellulose (MFC) and nanofibrillated cellulose (NFC) grades. A commercial grade of PLA (3001D) was delivered by NatureWorks LLC (USA) with MFI = 22 at 210 °C and density of 1.24 g/cm^3. The polymer pellets were first grained in a Hellweg table granulator M 50/80 and subsequently dried for 24 h at constant temperature of 80 °C before further processing or characterization.

The fibrillated cellulose fibers have been hydrophobized by an in-situ reaction for synthesis of organic nanoparticles of poly(styrene-co-maleimide) or SMI, in analogy with a previously reported protocol [12]. The fibers were loaded as separate batches in an autoclave reactor together with poly(styrene-co-maleic anhydride) precursor copolymer at weight ratio (maleic anhydride)/fiber = 1/1 and ammonium hydroxide at weight ratio NH_3/(maleic anhydride) = 1.01. During imidization, the nanoparticles were formed and immediately deposited onto the fiber surface by chemical interaction. The native MFC/NFC fibers and modified MFC fibers (mMFC) were freeze-dried for further processing.

The PLA nanocomposites with three types of nanocellulose fibers were made by melt-processing. The blending was done by using a lab-scale parallel co-rotating twin-screw compounder (Process 11, Thermo Scientific, Germany), by premixing the required material fractions as dry powders and subsequent feeding to the hopper. The temperature profile in different compartments of the compounder was optimized during preliminary processing of the native PLA, ranging between 195 °C (feeding zone) and 215 °C (near die). The obtained blends were subsequently compression molded into films on a hydraulically heated press (LaboPress P 200 S, Vogt Maschinenbau GmbH).

Characterization was done through rheological measurements on the aqueous suspensions of single MFC/NFC fibers before and after surface-modification, and on the molten polymer blends with different fiber concentrations between 0 and 20 wt.-%. The thermal characteristics were evaluated through differential scanning calorimetry (Q200 TA Instruments). The mechanical properties were determined from a stress-strain experiment on a standard mechanical test bench (Zwick, Germany). Morphologies were studied on a tabletop TM3000 scanning electron microscope (Hitachi, Germany).

3 Results and Discussion

3.1 Morphology

The morphology of cross-sections taken from cryogenically fractured films with different fiber percentages is shown in Fig. 1, for pure PLA and PLA nanocellulose biocomposites. The native material is obviously prone to rather brittle type of fracture

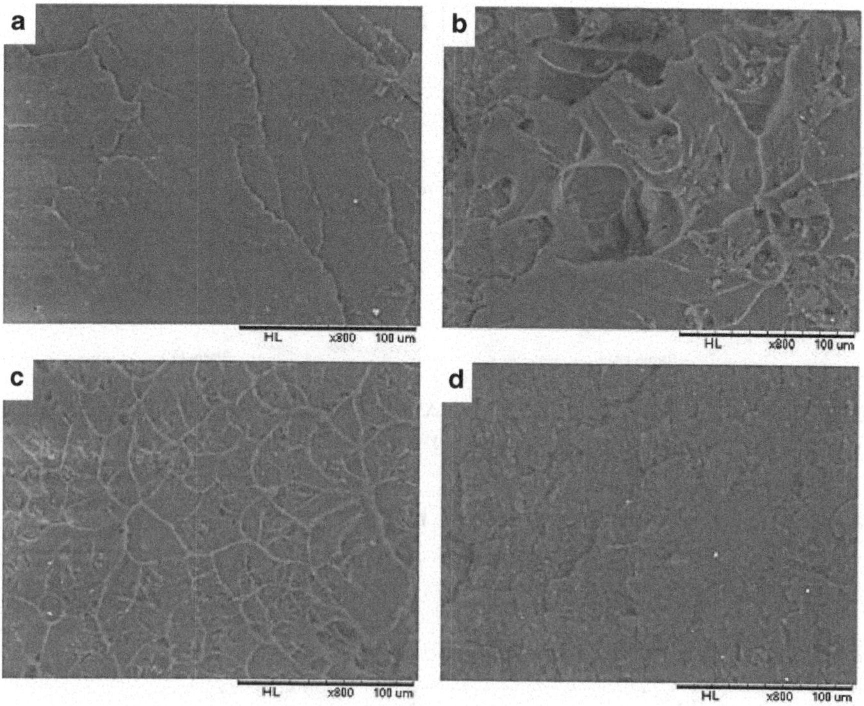

Fig. 1. SEM micrographs on fractured surfaces illustrating morphology of (a) PLA, (b) PLA + 1% MFC, (c) PLA + 5% mMFC, (d) PLA + 2% NFC

with clearly aligned fracture surfaces and flat surface morphology. It is found that the unmodified MFC fibers can be homogeneously dispersed into the matrix up to 1 wt.-%, the unmodified NFC fibers up to 2 wt.-% and the modified MFC fibers up to 5 wt.-%. The unmodified MFC fibers are prone to poor adhesion and fiber pull-out at higher concentrations. The occurrence of smooth fracture surfaces and formation of a single fiber network of the modified mMFC ensures good dispersive and distributive mixing and provides best mechanical performance. The role of the single NFC fibers is less visible, but the performance depends more on a good dispersion and high density of the fiber network, resulting in a more ductile aspect of the fracture surface.

3.2 Melt-Rheology

The rheological properties of neat PLA and PLA nanocomposites loaded with unmodified and modified MFC/NFC were measured to study the visco-elastic behavior and viscosity of molten nanocomposites at 180 °C. As an example, the rheological data for PLA and NFC/PLA blends is shown in Figs. 2 and 3 to illustrate the evolution of loss moduli and storage moduli over a given strain and frequency range.

The strain sweep test (Fig. 2) was done at constant frequency 1 Hz to find out the response of nanocomposites at different strain and region of linear behavior. Based on

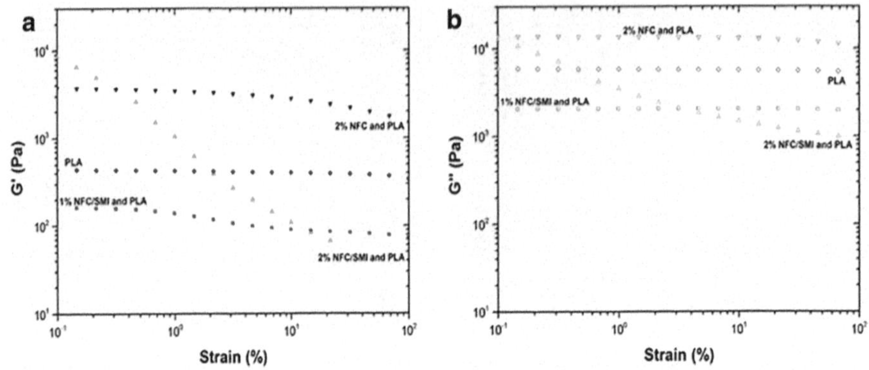

Fig. 2. Strain sweep test of neat PLA and PLA nanocomposites loaded with unmodified and modified MFC/NFC filler, a) storage modulus, b) loss modulus.

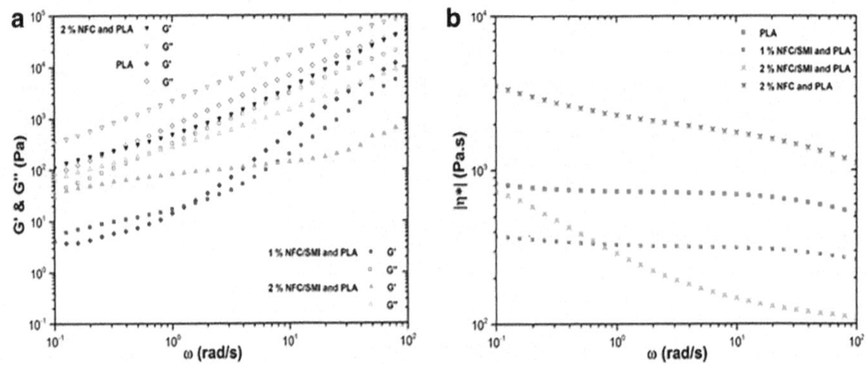

Fig. 3. Frequency sweep test of neat PLA and PLA nanocomposites loaded with unmodified and modified MFC/NFC filler, a) storage modulus, b) loss modulus.

this test, the rheological behavior of nanocomposites loaded with low concentration of unmodified or modified MFC/NFC shows a linear behavior with $G'' > G'$ for the whole range of strain. This linear behavior with $G'' > G'$ can also be observed for neat PLA, but the overall G' and G'' trends for 0.25 wt.-% unmodified and modified MFC/NFC are lower than for neat PLA. The homogeneous mixtures of PLA/MFC can be obtained depending on the strain and frequency regimes during oscillatory rheometry. With the increment of unmodified filler component (1 wt.-%) in PLA, the overall trends of G' and G'' show the larger values in comparison with neat PLA but the linear behavior remains unchanged with $G'' > G'$. Unlikely, the modified filler (1 wt.-%) shows a strain softening behavior with increment of strain. This behavior can occur due to the influence of surface-modified mMFC that behaves as a plasticizer. However, with increment of filler concentration (5 wt.-%) the strain softening behavior was slightly pronounced. According to the strain sweep test, the linear regime for G' and G'' at intermediate strain 1% was observed. Optimum concentrations of modified fibers to obtain good dispersive and distributive mixtures were at around 1–2%. The hydrophobic surface modification of the fibers clearly have a positive effect on lowering the G′ and G″ relatively to the pure PLA and the 2% unmodified fiber mixtures.

The frequency sweep tests (Fig. 3) were done for neat PLA and different nanocomposites to observe the response of the system. For neat PLA, both G′ and G″ increased systematically as a function of the applied frequency. However the increment of G′ is faster than G″ confirming the gradual tendency towards a solid-like behavior for molten PLA. This increment can be related to the high molecular weight of polymer as a general characteristic. The addition of unmodified and modified MFC/NFC fillers has a direct effect on the increment of G′ and G″ as a function of frequency. In this regard, the low concentration of both unmodified and modified filler shows very close plots for G′ and G″. Interestingly, the PLA nanocomposite with 1 wt.-% unmodified MFC/NFC filler shows a greater G′ (hundred time) at low and intermediate frequency compared to neat PLA while the PLA nanocomposite with 5 wt.-% modified mMFC filler does not show this behavior. The viscous behavior is more pronounced with increment of modified MFC/NFC filler concentration confirming the effects of deposited SMI nanoparticles at higher temperature (180 °C).

3.3 Thermal Properties

A selection of results from DSC analysis for different PLA nanocomposites with 0.25, 1 wt.-% unmodified MFC/NFC and 0.25, 0.5, 1, 5 wt.-% modified MFC/NFC is shown in Fig. 4.

For both unmodified and modified MFC/NFC PLA blends, an alteration of glass transition T_g, crystallization temperature T_c and melting peaks T_m was observed. The increment of MFC/NFC filler from 0.25 to 1 wt.-% affected the melting peak of nanocomposite: the narrow symmetrical endotherm of the nanocomposite with 0.25 wt.-% MFC/NFC filler changed into a broad melting with two shoulders or maxima for the nanocomposite with 1 wt.-% of MFC/NFC filler. This behavior was also observed

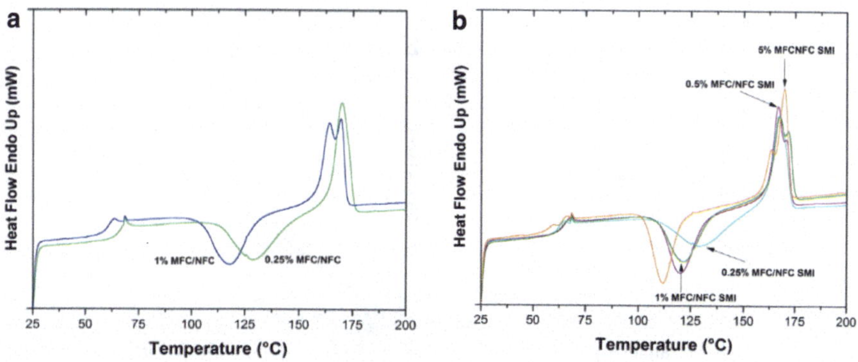

Fig. 4. DSC analysis of PLA nanocomposites with fibrillated cellulose, (a) unmodified fibers at 0.25 wt.-% and 1 wt.-%, (b) modified fibers at 0.25–5 wt.-%.

for nanocomposites with 0.5, 1 and 5 wt.-% modified MFC. A double melting endotherm for polymer blends has also been reported due to a melting and recrystallization phenomena [13]. The melting behavior of unmodified and modified MFC/NFC nanocomposites is reminiscent of a transcrystalline layer that coexists in the vicinity of cellulose nanowhiskers [14], which likely explains the occurrence of two maxima in the melting endotherm. However, the broadening of the melting endotherms is more pronounced with increment of the MFC/NFC filler. The presence of MFC/NFC filler increases the degree of crystallinity of the neat PLA matrix. The degree of crystallinity can also be decreased due aggregation of MFC/NFC filler with different concentrations. It can be an interpretation that MFC/NFC nanofillers act as nucleating agents for PLA crystallization.

3.4 Mechanical Properties

The results of mechanical tensile testing on the melt-processed PLA nanocomposite films are evaluated and given in Fig. 5. The modulus of elasticity determined by the slope of the curve seems not to be mainly influenced by the nanocellulose additives, while the ultimate strain and strength at break depends on the type of filler. The addition of 1 wt.-% unmodified MFC increases the elongation and strength in respect to the unfilled material, while the higher percentages of unmodified MFC cause deterioration of the mechanical properties. For MFC, there is almost no increase in strength and a decrease in elasticity at higher loading percentages, indicating weak interactions between the fillers and polymer matrix. The addition of modified mMFC fibers seems to provide better mechanical properties, with a significant increase in strength and elongation at 1 wt.-% modified mMFC, while higher percentages of nanocellulose fillers further deteriorate the properties. However, the elasticity of the native PLA seems to be retained or slightly improved after blending with mMFC. From the mechanical testing data, it has been cleared that the presence of SMI nanoparticles are one of the key factors to reach dispersive mixing, while the unmodified MFC only reached distributive mixing during melt-processing in the extrusion line.

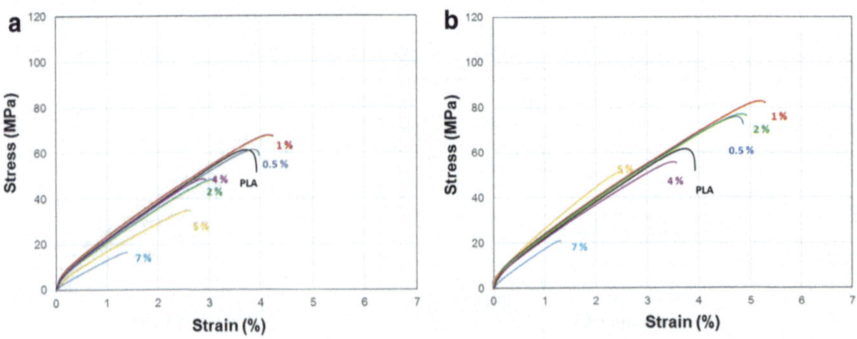

Fig. 5. Mechanical stress-strain curve of PLA nanocomposites with fibrillated cellulose, (a) unmodified MFC fibers at 0.5–7 wt.-% (b) modified mMFC fibers at 0.25–7 wt.-%.

4 Conclusion

The melt-blending of fibrillated cellulose materials within a PLA matrix has been demonstrated during compounding on a continuous mini-extruder and subsequent compression molding. The benefits of blending hydrophobized fibrillated cellulose result in a maximum percentage of 5 wt.-% fibers that can be homogeneously mixed with a good dispersion and distribution within the polymer matrix. For the unmodified fibers, lower percentages could be blended with less interface compatibility resulting in fiber pull-out. The effects of nanoparticle deposits on the fibers are expressed in beneficial melt-rheological parameters with a significant drop in viscosity of the nanocomposite melt, indicating that the surface-modified fibers act as a plasticizer. The fillers affect the crystallization properties and melting endotherm of the PLA, suggesting their role as nucleating agent resulting in higher crystallinity. The latter is reflected also in improvement in mechanical properties, mainly in the case of surface-modified fibers.

References

1. Siro, I., Plackett, D.: Microfibrillated cellulose and new nanocomposite materials: a review. Cellulose **17**, 459–494 (2010)
2. Samyn, P.: Wetting and hydrophobic modification of cellulose surfaces for paper applications. J. Mater. Sci. **48**, 6455–6498 (2013)
3. Jonoobi, M., Harun, J., Mathew, A.P., Hussein, M.Z., Oksman, K.: Preparation of cellulose nanofibers with hydrophobic surface characteristics. Cellulose **17**, 299–307 (2009)
4. Paul, M.A., Alecandre, M., Degee, P., Henrist, C., Rulmont, A., Dubois, O.: New nanocomposite materials based on plasticized poly(L-lactide) and organo-modified montmorillonites: thermal and morphological study. Polymer **44**, 443–450 (2003)
5. Ahmed, J., Varshney, S.K., Auras, R., Hwang, S.W.: Thermal and rheological properties of l-polylactide/polyethylene glycol/silicate nanocomposites films. J. Food Sci. **75**, 97–108 (2010)
6. Treece, M.A., Oberhauser, J.P.: Soft glassy dynamics in polypropylene-clay nanocomposites. Macromolecules **40**, 571–582 (2007)
7. Mackay, M.E., Tuteja, A., Duxbury, P.M., Hawker, C.J., Van Horn, B., Guan, Z., Chen, G., Krishnan, R.S.: General strategies for nanoparticle dispersion. Science **311**, 1740–1743 (2006)
8. Akcora, P., Kumar, S.K., Moll, J., Lewis, S., Schadler, L.S., Li, Y., Benicewicz, B.C., Sandy, A., Narayanan, S., Ilavsky, J., Thiyagrajan, P., Colby, R.H., Douglas, J.: Gel-like* mechanical enforcement in polymer nanocomposite melts. Macromolecules **43**, 1003–1010 (2010)
9. Ginzburg, V.V., Weinhold, J.D., Jog, P.K., Srivastava, R.: Thermodynamics of polymer- clay nanocomposites revisited: compressible self-consistent field theory modeling of melt-intercalated organoclays. Macromolecules **42**, 9089–9095 (2009)
10. Pyda, M.: Quantitative thermal analysis of carbohydrate–water systems. In: Lorinczy, D. (ed.) The Nature of Biological Systems as Revealed by Thermal Methods. Kluwer Academic Publisher, Amsterdam
11. Ahmed, J., Jhang, X.J., Song, J., Varhney, S.K.: Thermal characteristics of polylactides. J. Therm. Anal. Calorim. **95**, 957–964 (2009)

12. Rastogi, V.K., Stanssens, D., Samyn, P.: Reaction efficiency and retention of poly(styrene-co-maleimide) nanoparticles deposited on fibrillated cellulose surfaces. Carbohydr. Polym. **141**, 244–252 (2016)
13. Su, Z., Li, Q., Liu, Y., Hu, G.H., Wu, C.: Multiple melting behaviour of poly(lactic acid) filled with modified carbon black. J. Polym. Sci. **B47**, 1971–1980 (2009)
14. Gray, D.G.: Transcrystallization of polypropylene at cellulose nanocrystal surfaces. Cellulose **15**, 297–301 (2008)

Continuous Chemical Recycling of Polystyrene with a Twin – Screw Extruder

Philipp Schäfer[1(✉)], Christian Hopmann[1], Martin Facklam[1], Laura Hollerbach[2], Tristan Kolb[3], Andreas Schedl[3], Hans-Werner Schmidt[4], Franziska Nosić[5], and Bianca Wilhelmus[5]

[1] Institute for Plastics Processing, RWTH Aachen University, Aachen, Germany
Philipp.Schaefer@ikv.rwth-aachen.de
[2] Department of Processing and Recycling, RWTH Aachen University, Aachen, Germany
[3] Neue Materialien Bayreuth GmbH, Bayreuth, Germany
[4] Macromolecular Chemistry and Bavarian Polymer Institute, University of Bayreuth, Bayreuth, Germany
[5] INEOS Styrolution Group GmbH, Frankfurt am Main, Germany

Abstract. The increasing scarcity of resources and growing environmental awareness require higher recycling rates for plastic waste. Common techniques to do that are mechanical recycling, thermal recycling and chemical recycling, which is also called feedstock recycling. From all three techniques, chemical recycling is the only one which can produce new materials that correspond to the quality of conventional virgin material. However, the technique is limited to suitable polymers, e.g. polystyrene, which can be depolymerised at elevated temperatures. For an efficient industrial scale-up, a continuous process is desirable. In our work, we present such a continuous process for the recycling of polystyrene from post-industrial waste. A co-rotating, tightly intermeshing twin-screw extruder in high-temperature design is used together with a vacuum separation system with three degassing openings. By determining a stable process point a continuous depolymerisation of polystyrene is technically realised.

The atmospheric oxygen and moisture are removed via the first degassing opening of the extruder. The degradation products of the depolymerisation process are then degassed through the second and third opening. The degradation products are passed through a water-cooled condenser where they are liquefied. The styrene yield is maximised by tuning the process parameters barrel temperature, screw speed and configuration, mass throughput and degassing design. Analysis of the products reveals a considerable influence on increasing recovery rates with increasing barrel temperature, decreasing throughput and longer residence time. A longer residence time is realised by a lower throughput and an optimised screw configuration. We anticipate our process as a very promising technique to efficiently and economically scale-up the chemical recycling of poly-styrene waste.

Keywords: Polystyrene · Chemical recycling · Twin-screw extruder

C. Hopmann and R. Dahlmann (Eds.): *Advances in Polymer Processing 2020*, pp. 37–49, 2020. https://doi.org/10.1007/978-3-662-60809-8_4

1 Introduction

In the plastics processing industry, the transition from a linear to a circular economy is becoming increasingly important. This is due to climate change, environmental pollution, population growth and resource dependency. According to a study of PlasticsEurope Deutschland e. V., Frankfurt am Main, Germany, the amount of plastic packaging waste collected increased by 12% from 14.9 million tonnes in 2006 to 16.7 million tonnes in 2016 (EU28+NO/CH) [1]. At the same time, the recycling rate went up by 74% and the thermal recycling of plastic packaging waste by 71%. Landfill was reduced by 53% during the same period. However, a continuous increase in global plastics production can be observed. Mature recycling concepts are needed to minimise the ecological consequences of inappropriate disposal. There are three basic types of recovery of plastics: thermal recovery, mechanical recycling and chemical recycling [2].

Thermal recovery is currently the predominant one in Europe. It does not require mono-fraction material feedstock [1]. The plastic waste is incinerated and the released energy is used for power generation. A big drawback of this technique is the irreversible loss of the material in the material cycle.

In the case of mechanical recycling, sorted plastic waste is washed, melted and regranulated. It must be taken into account that a deterioration of the plastic quality occurs if the material is not sorted [3]. Additionally, in every form of material recycling, the polymers are thermally and mechanically damaged by re-melting. This leads to a deterioration of the mechanical properties [3].

The chemical recycling of plastic waste – often also called feedstock recycling – describes the splitting of macromolecules into individual parts or monomers. After purification, the monomers obtained can be used for the synthesis of new polymers. Using this technique, polymers with the same quality as the virgin material can be obtained [4].

Most polymers do not form a reactive monomer during thermal degradation. The thermal degradation of polyethylene, polypropylene and polyesters results in non-specific products such as waxes, light oils and gases. Nevertheless, polystyrene (PS) and polymethyl methacrylate (PMMA) are polymers which can be split back into their monomers beyond a critical temperature and are thus suitable for depolymerisation [5].

The mechanisms of depolymerisation of PS and PMMA differ fundamentally. For PMMA, the macromolecule is split at the end of the chain [6]. A complete thermal degradation under nitrogen atmosphere leads to a monomer concentration (methyl methacrylate) of >98% in the condensate [7]. In the case of PS, the chain shortening can take place both at the end and statistically distributed within the chain [8]. The cracking of the chain along the macromolecule leads to two different radical chain ends, which show different decay mechanisms (Fig. 1, left side). The end of the chain with the primary radical can split-off styrene monomer by regeneration of a comparable end of the radical chain (depolymerisation, Fig. 1, path 1). The end of the chain with the secondary radical can form a tertiary radical via backbiting and a 1,5-transfer. This tertiary radical can split off styrene, dimers and trimers by recovering a comparable end of the chain (Fig. 1, path 2) [9].

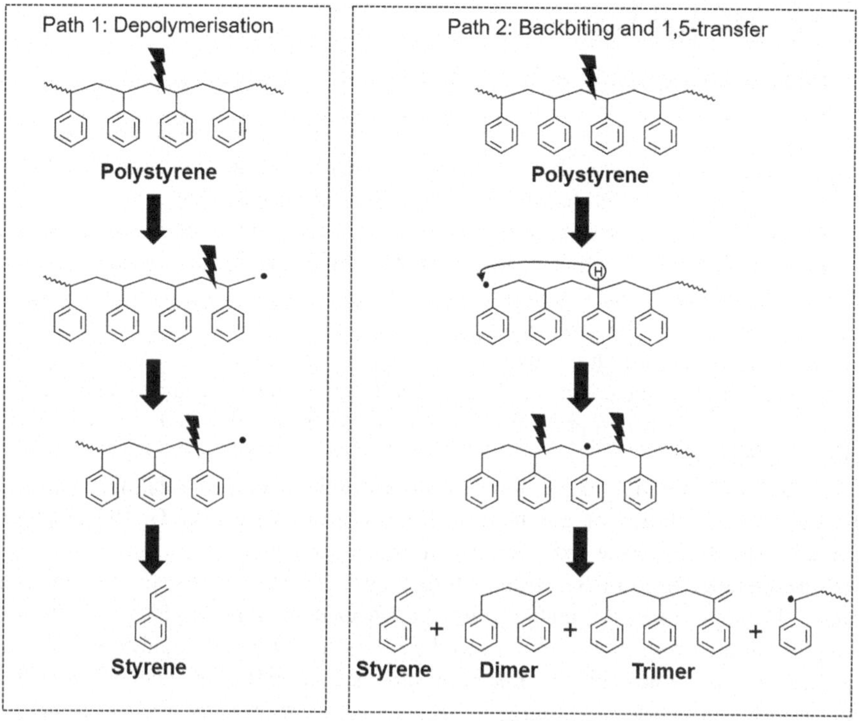

Fig. 1. Mechanism of thermal degradation of PS [7]

Chemical recycling of plastics in screw reactors was already the subject of research in the 1990s [10]. The focus was on the use of PMMA. The first discontinuous tests for the depolymerisation of PMMA were carried out using a high-temperature kneader and corresponding degassing/condenser plant. These tests served to adjust the required barrel temperatures, residence time and cooling temperatures of the subsequent modules [11]. A recovery of 85% of the material in the form of condensate was achieved at a barrel temperature of 400 °C and a residence time of up to 45 min. The continuous realisation was then carried out on a twin-screw extruder with a screw diameter of 30 mm and a length-to-diameter ratio of 37 [12]. In the continuous trials, a condensate yield of up to 60% was achieved at a constant throughput of 3 kg/h, barrel temperatures of up to 380 °C and rotation speeds of up to 200 rpm.

The following investigations focus on the chemical recycling of PS by means of thermal degradation in the twin-screw extruder. Specifically, the suitability of the continuous extrusion process for the depolymerisation of PS is verified. Important parameters for that are the residence time and the energy input. In addition, the influence of the process conditions on the thermal degradation and the composition of the condensate obtained is investigated.

2 Materials and Methods

2.1 Methodical Approach

The model substance that is used to perform the investigations is general purpose polystyrene, type PS 156F from INEOS Styrolution Group GmbH, Frankfurt am Main. The material is not pre-dried before being processed in the extruder.

Preliminary analyses focus on the degradation behaviour of PS at different temperatures using thermogravimetric analysis (TGA) (Fig. 2). It becomes clear that temperatures above 400 °C are required to gain a significant mass loss and thus a thermal decomposition. It must be taken into account that in a twin-screw extruder the maximum residence time at high throughputs is less than one minute. However, very good mixing and a high mechanical load are possible due to the co-rotating intermeshing twin screws, so that a comparable reduction of chain length can be expected with shorter residence time.

The continuous depolymerisation of PS is carried out in a co-rotating, intermeshing twin-screw extruder of the type ZSK26Mc from Coperion GmbH, Stuttgart, Germany. The twin-screw extruder has a screw diameter of 26 mm and a high length-to-diameter ratio of 58 in order to achieve a long residence time of the plastic melt. The equipment is capable to handle barrel temperatures of 450 °C. In order to enable a high shear energy input into the melt, the screw configuration features a dispersive mixing zone with five kneading blocks each and a left-handed element in addition to the obligatory melting zone (Fig. 3). Degassing using a vacuum pump in barrel four reduces moisture and oxygen, which could influence thermal degradation. Volatile components such as monomers, dimers and trimers are continuously removed in two further degassing zones in the barrels eight and thirteen and condensed in a separator. Here, the vacuum pressure is limited to 600 mbar in order to prevent the low-viscous melt from boiling up and allowing plastic melt to enter the vacuum system. The implemented melt pump enables a reliable sealing of the extruder

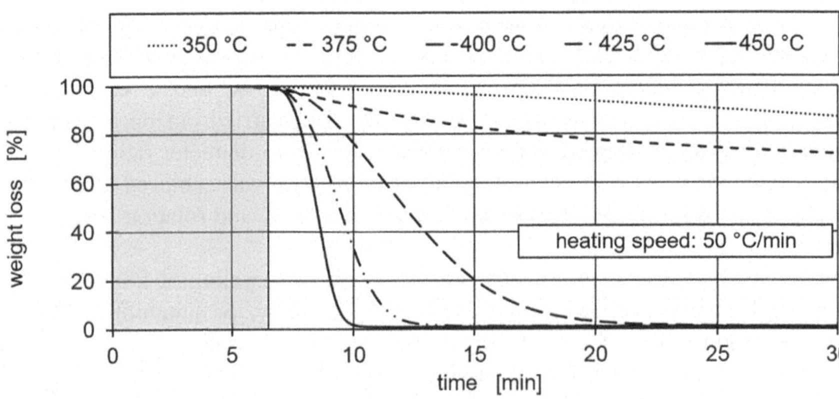

Fig. 2. Thermogravimetric analysis of the isotherm degradation behaviour at different set temperatures

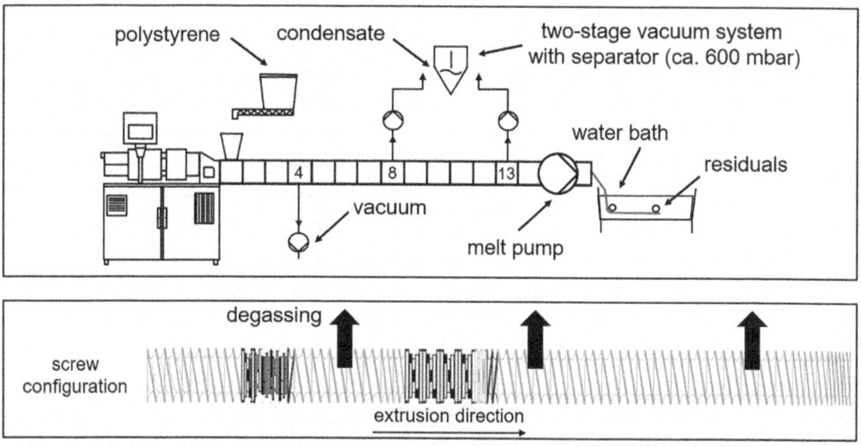

Fig. 3. Schematic experimental setup and screw configuration for thermal degradation of PS in twin-screw extruders

and simultaneously reduces the backlog length of the melt before exit. The speed, throughput and barrel temperature profile are varied in three stages (Table 1).

The process of thermal degradation of polystyrene in twin-screw extruders is limited by different boundary conditions. In addition to the residence time, which is restricted by the length-to-diameter ratio of the screws and also by the mass throughput, the processing temperature also influences the molecular weight reduction and thus the formation of styrene. As the degradation increases with higher temperatures, the highest possible barrel temperatures should be used. However, styrene has a self-ignition temperature of 490 °C [13]. In order to prevent self-ignition, a maximum barrel temperature of 430 °C is used for the first investigations of the degradation process. Due to the high temperature, the polystyrene has a very low viscosity in the extruder. As a consequence, the boiling melt is able to enter the vacuum system, which could cause clogging of the pipelines. For this reason, a vacuum of only 600 mbar is used in the first tests. In the selected process window, the melt does not boil up and a reliable operating point is reached. In general, low absolute pressures achieve an increased degassing performance [14]. The degassing process is also supported by continuous surface renewal. The aim of the first experiments is to carry out thermal degradation of polystyrene at a stable and reliable operating point. The optimisation of the process will follow in further investigations.

Table 1. Varied process parameters for continuous depolymerisation of PS in a twin-screw extruder

	–	0	+
Throughput [kg/h]	10	15	20
Screw speed [rpm]	900	1000	1100
Barrel temperature [°C]	max. 400	max. 415	max. 430

2.2 Characterisation of Residues and Condensate

Continuous depolymerisation in the twin-screw extruder produces residues that pass completely through the extruder. The volatile components (condensate) are removed and collected via multi-stage degassing system. The degree of degradation is determined via melt flow rate (MFR) measurement of the residues. As the degree of degradation increases, the molecular weight decreases and thus the MFR increases. The MFR value is measured at a temperature of 170 °C and 2.16 kg. In addition, the condensate yield is the focus of the analysis. It is assumed that the amount of condensate discharged by the multi-stage vacuum degassing corresponds to the difference between the mass flow of polystyrene supplied and the mass flow of the residues. The amount of condensate in the separator is not weighed, since considerable amounts of condensate remain in the piping system and can therefore only be taken into account to a limited extent. At selected test points, the collected condensate is analysed with an Agilent 7890A gas chromatograph from Agilent, Santa Clara, United States. A flame ionisation detector was used as detector.

3 Results and discussion

The following section examines whether the results obtained during discontinuous depolymerisation can be transferred to continuous depolymerisation of PS in the twin-screw extruder. In particular, the limited residence time of the plastic melt in the twin-screw extruder and lower mass temperatures could lead to a lower thermal degradation compared to discontinuous depolymerisation. Also the degradation in the twin-screw extruder is not only caused by thermal stress but also by mechanical energy input. The functional principle of the twin-screw extruder allows a constant renewal of the surface so that the degassing of volatile components is supported. This study focuses on the influence of the process conditions on the depolymerisation of PS in the twin-screw extruder.

Figure 4 describes the influence of the set barrel temperature profile and the screw speed on the specific mechanical energy input, the mass temperature and the minimum residence time of the plastic melt in the twin-screw extruder. The minimum residence time is determined by a colour change. The specific mechanical energy input is determined according to the equation in Fig. 4. M_D stands for the applied torque at the screw, n stands for the screw speed and $\dot{m}$ is the material throughput. Due to the high barrel temperatures of up to 430 °C, the melt viscosity in the extruder is low, so that the specific mechanical energy input cannot be significantly increased. At the same time, the degassing of the volatile components under vacuum leads to a strong heat extraction. As a consequence, the measured mass temperature is below the barrel temperature. Here, it is also evident that an increase in speed does not lead to an increase in mass temperature. Instead, a slight decrease in mass temperature can be observed with an increase in screw speed from 1000 to 1100 rpm. Since an increased surface renewal takes place at high screw speeds, the degassing performance and at the same time the evaporative heat extraction are increased. As expected, a variation of the barrel temperature profile does not result in a change of the minimum residence time of

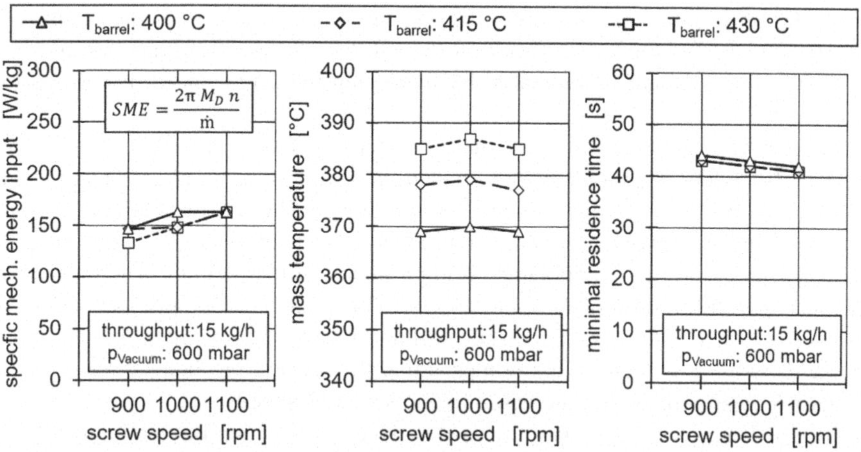

Fig. 4. Influence of barrel temperature profile and screw speed on minimum residence time of melt, mass temperature and specific mechanical energy input

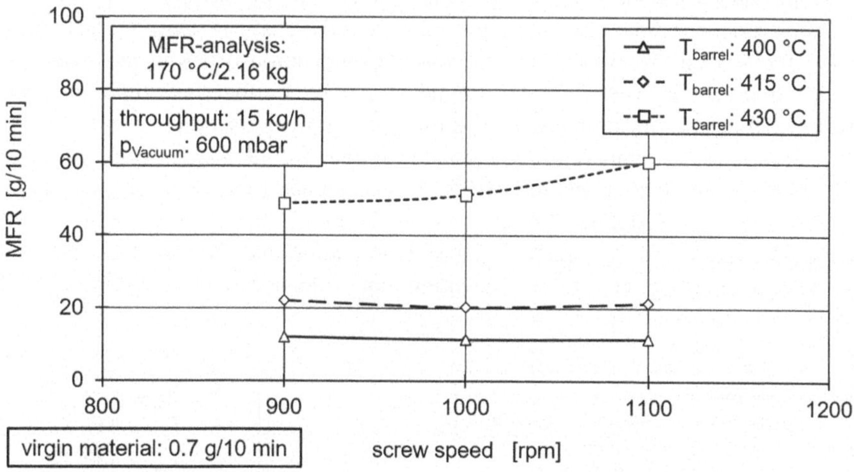

Fig. 5. Influence of the barrel temperature profile and screw speed on the MFR

the plastic melt in the twin-screw extruder. At high barrel temperatures, a high degree of decomposition is to be expected due to the high thermal load. At the same time, high mass temperatures support the degassing process, since high diffusion rates and low viscosities are present.

MFR measurements are used to estimate the degree of decomposition depending on the process conditions in the twin screw extruder. Figure 5 shows the corresponding results of the MFR measurements as a function of the barrel temperatures and screw speeds used. It is evident that the MFR increases significantly with increasing thermal stress due to an increase in the barrel temperature. At a barrel temperature of 430 °C, MFR values of up to 60 g/10 min can be achieved. At the same time, no uniform influence of the screw speed on the MFR is discernible. Thus, the mechanical energy input cannot be increased by increasing the screw speed due to the low viscosity.

Figure 6 describes the influence of the barrel temperature profile and the screw speed on the condensate yield. Higher condensate yields are achieved with an increasing barrel temperature profile, as the material is subjected to greater thermal stress. In addition, at higher barrel temperatures there is a lower melt viscosity, which supports the degassing of volatile components through higher diffusion rates. The condensate yield increases from 6.8% (400 °C) to 19.2% (430 °C) at a screw speed of 1000 rpm. However, a higher screw speed does not lead to a further improvement in the condensate yield, as the low melt viscosity means that the specific mechanical energy input and mass temperature are not increased by higher screw speeds.

The condensate collected was analysed by Neue Materialien Bayreuth GmbH. The results are shown in Fig. 7. Irrespective of the screw speed used, the styrene monomer content is approximately 80%. The remaining condensate consists approximately of one-third each of dimers, trimers and other components. As described above, it cannot be ensured that the condensate collected in the separator is a representative sample of all decomposition products. High-molecular compounds in particular can remain in the piping system and are not fully detected during the analysis. Despite these uncertainties, the analysis shows that the thermal decomposition of PS in the twin-screw extruder enables the production of styrene monomer. Further investigations would show to what extent this monomer can be used for repolymerisation.

In addition to the temperature, the residence time will also significantly influence the thermal decomposition degree of PS. The residence time of the plastic melt is determined by the throughput. Figure 8 shows the influence of the throughput on the specific mechanical energy input, the mass temperature and the minimum residence time. With increasing throughput, the minimum residence time is reduced. At the

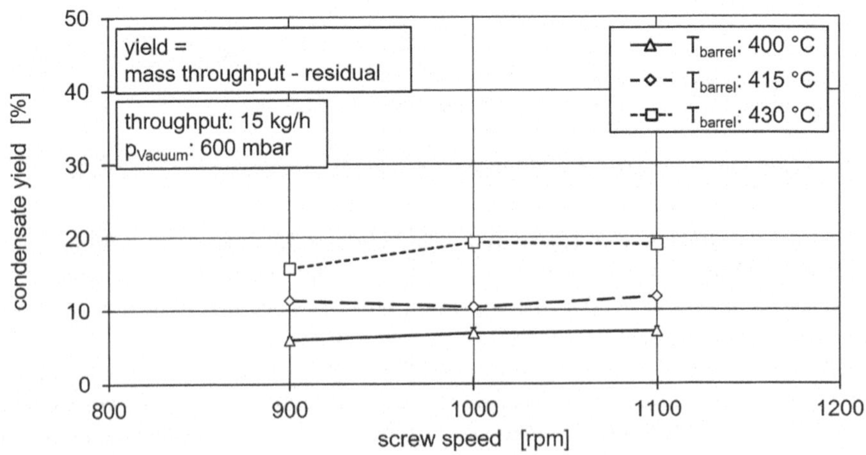

Fig. 6. Influence of barrel temperature profile and screw speed on condensate yield

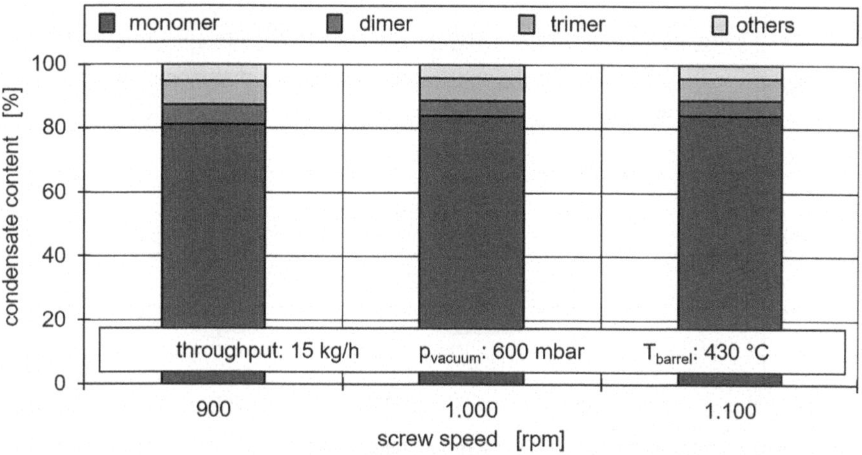

Fig. 7. Content of the condensate as a function of screw speed

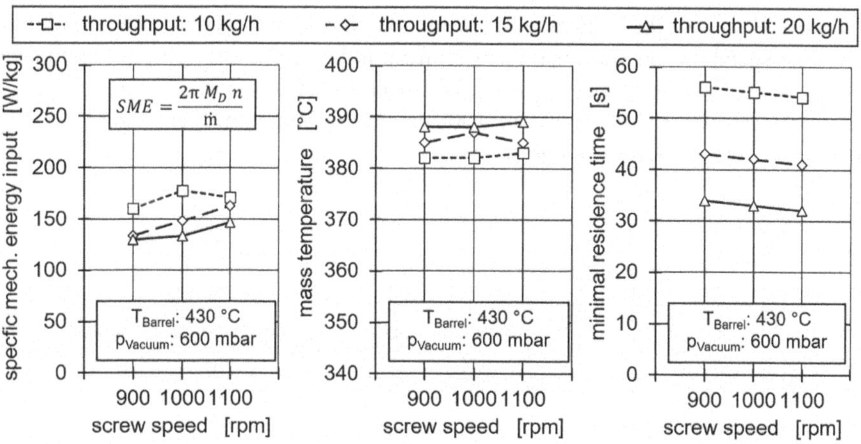

Fig. 8. Influence of throughput and screw speed on the minimum residence time of the melt, mass temperature and specific mechanical energy input

same time, the specific mechanical energy input increases slightly with a reduction in throughput. As low throughput results in a long residence time and thus a good degassing performance or a high evaporation cooling, the mass temperature decreases with reduced throughput. Due to the long residence time at low throughputs, a high degree of decomposition and a high condensate yield can be expected.

The results in Fig. 9 confirm these assumptions. Due to the long dwell time, the MFR can be increased from 51 g/10 min (1000 rpm and 15 kg/h) to 122 g/10 min (1000 rpm and 10 kg/h) by reducing the throughput to 10 kg/h. This results in a significantly higher degree of decomposition at low throughput rates and correspondingly long residence time.

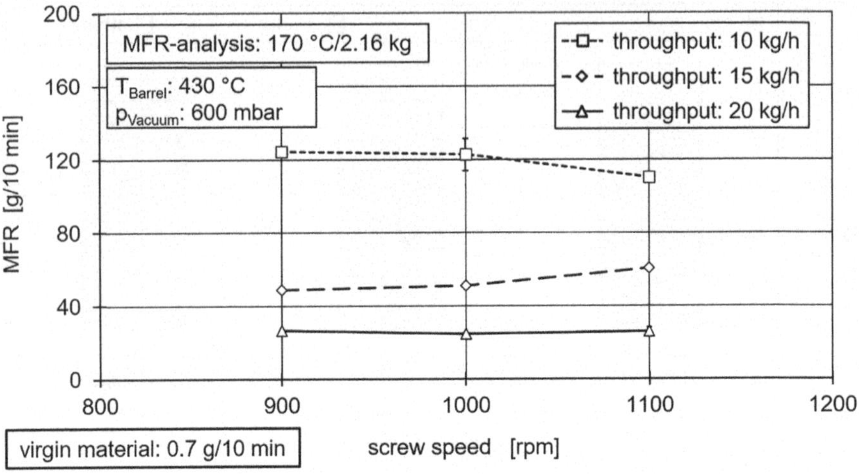

Fig. 9. Influence of throughput and screw speed on the MFR

Figure 10 shows the influence of throughput and screw speed on the condensate yield. The speed has no significant and uniform influence on the achieved condensate yields, even if the throughput is varied. However, lower flow rates lead to an increasing condensate yield. At a speed of 1000 rpm, the condensate yield increases from 13.9% (20 kg/h) to 34.4% (10 kg/h).

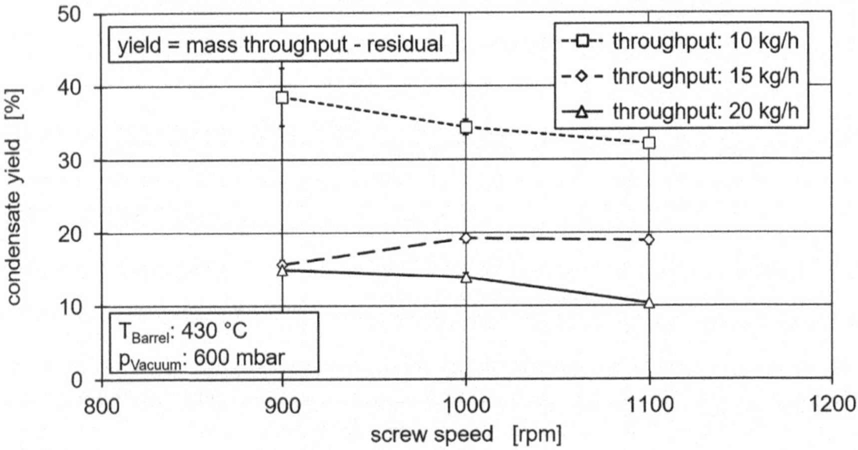

Fig. 10. Influence of throughput and screw speed on condensate yield

3.1 Influence of Different Polystyrene grades on Depolymerisation

In further investigations, other PS grades and real waste samples were investigated by means of chemical recycling. Besides the above mentioned PS grade 156F,

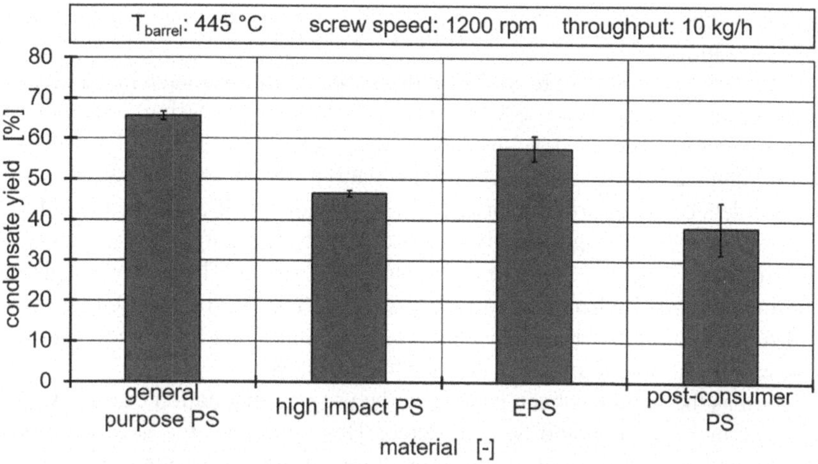

Fig. 11. Influence of the material used on the condensate yield

a high-impact PS, type 486N from INEOS Styrolution Group GmbH, Frankfurt am Main, is used. In addition, EPS (expanded polystyrene) and a sample of the PS fraction from post-consumer waste collected through municipal waste collection in Germany are used. In order to achieve a high degree of decomposition, the maximum barrel temperature was increased to 445 °C and a long residence time over a low throughput of 10 kg/h was achieved for these tests. The maximum condensate yield of 66% is achieved for PS 156F (Fig. 11). When using the PS fraction from post-consumer packaging waste, the lowest condensate yield of approx. 40% is achieved. It must be considered that the materials used are modified with fillers and contaminated with foreign polymers and dirt adherence, which reduces the amount of PS available for depolymerisation. High-impact PS is filled with an elastomer component to increase impact strength. This also reduces the level of PS available. Furthermore, side reactions can contribute to a reduction of the condensate yield. When EPS is used, condensate yields comparable to those of PS 156F are achieved. In future investigations, the influence of impurities with foreign plastics, fillers and additives on chemical recycling will be investigated.

4 Conclusion and outlook

The investigations show that the thermal degradation of PS in a twin-screw extruder is possible. High mass temperatures and long residence time lead to stronger thermal decomposition and thus to higher condensate yields. The influence of the screw speed is low, as the melt viscosity is low at the prevailing mass temperatures and therefore the energy input cannot be significantly increased by increasing the screw speed. Further investigations will analyse the content of the condensate as a function of the process control. In addition, it must be investigated to what extent the condensate, consisting of monomers, dimers and trimers and other volatile compounds, is suitable for polymerisation.

To increase the condensate yield of the process, the absolute material flow of extracted gases must be increased. For this purpose, new degassing domes that have a larger free volume will be developed. This will lead to a faster and higher degassing of the components resulting from depolymerisation. A cascade process of two twin-screw extruders offers a further approach to improve the yield of the process. The second twin-screw extruder significantly increases the process length and residence time, which results in a longer reaction time for depolymerisation. The mass throughput can thus be increased at higher relative recovery rates. The use of reaction accelerators offers a further possibility to increase the discharged material flow. These shorten the process length and time required to degrade the polystyrene.

As soon as the yield has been increased, the use of a distillation cascade with subsequent polymerisation should be considered. This eliminates the need for reheating for distillation and stabilisation required for styrene transport. In this way, several successive processes could be merged.

Acknowledgements. The investigations set out in this report received financial support from Germany's Federal Ministry of Education and Research within the initiative "Plastics in the Envi-ronment – Sources. Sinks. Solutions." of the BMBF-framework programme "Research for Sustainable Development" (No. 033R194C), to whom we extend our thanks.

References

1. PlasticsEurope Homepage: Plastics – the Facts 2018: https://www.plasticseurope.org/download_file/force/2387/319. Accessed 5 May 2019
2. Menges, G., Michaeli, W., Bittner, M.: Recycling von Kunststoffen. Hanser, München (1992)
3. Menges, G., Haberstroh, E., Michaeli, W., Schmachtenberg, E.: Werkstoffkunde Kunststoffe. Hanser, München (2014)
4. Sasse, F., Emig, G., 10: Chemical recycling of polymer materials. Chem. Eng. Technol.: Ind. Chem.-Plant Equipment-Bioprocess. Eng.-Biotechnol. **21**, 777–789 (1998)
5. Elias, H.G.: Makromoleküle: Chemische Struktur und Synthesen. Wiley, Weinheim (1999)
6. Achilias, S.: Chemical recycling of polymers. The case of poly (methyl methacrylate). In Proceedings of the International Conference on Energy & Environmental Systems, Chalkida, Greece, pp. 8–10 (2006)
7. Kaminsky, W., Predel, M., Sadiki, A.: Feedstock recycling of polymers by pyrolysis in a fluidised bed. Polym. Degrad. Stab. **85**(3), 1045–1050 (2004)
8. Yang, M., Shibasaki, Y.: Mechanisms of thermal degradation of polystyrene, polymethacrylonitrile, and their copolymers on flash pyrolysis. J. Polym. Sci., Part A: Polym. Chem. **36**(13), 2315–2330 (1998)
9. Audisio, G., Bertini, F., Beltrame, P.L., Carniti, P.: Catalytic degradation of polymers: Part III—Degradation of polystyrene. Polym. Degrad. Stab. **29**(2), 191–200 (1990)
10. Michaeli, W., K. Breyer: Chemisches Recycling von PMMA–Depolymerisation durch Extrusion. Achema Mag. **97**(53) (1997)

11. Hottinger, A.: Konstruktion, Aufbau und Erprobung eines Laborstandes zur Depolymerisation von Polymethylmethacrylat (PMMA). Institut für Kunststoffverarbeitung, RWTH Aachen, Diplomarbeit (1993)
12. Schwarz, R.: Aufbau und Inbetriebnahme einer Extrusionsanlage zur kontinuierlichen Depolymerisation von PMMA. Studienarbeit am Institut für Kunststoffverarbeitung, RWTH Aachen (1995)
13. Born, M., Carl, C., Schneider, G.: Fachkunde Gefahrstoffe. Storck Verlag, Hamburg (2017)
14. N.N.: Entgasen beim Erstellen und Aufbereiten von Kunststoffen. VDI-Verlag, Düsseldorf (1991)

Development of a Solids Conveying Throughput Model for Grooved Barrel Extruders Based on Discrete Element Simulations

Volker Schoeppner and Florian Bruening(✉)

Paderborn University, Kunststofftechnik Paderborn (KTP),
Paderborn, Germany
florian.bruening@ktp.upb.de

Abstract. Regarding the design of single screw extruders, the prediction of the throughput needs to be of high accuracy. The calculation of the solids conveying throughput is particularly important for extruders with a grooved feed section. In contrast to smooth barrel extruders the throughput of the entire plant is determined at the feed section. Various simplifications are necessary for an analytical modelling of the solids conveying, e.g. the classification into conveying cases, the assumption of solid block flow and the assumption of pressure anisotropy coefficients.

In numerical simulations with the Discrete Element Method (DEM), which has recently been successfully used to describe solids conveying in smooth barrel extruders, the simplifications mentioned above are not taken into account. Here, the pellets are approximated as spherical particles or particles composed of spheres. Based on virtual overlaps, contact models and the solution of Newton's equations of motion, a more complex consideration of the pellet flow is possible. In each iteration step of the simulation, the particle velocities, contact forces and derived quantities, e.g. the mass throughput in the screw channel and the grooves or the radial pressure build-up along the grooved barrel can be evaluated. Therefore, a DEM simulation model can be investigated by means of statistical design of experiments in order to convert the target values into a metamodel by regression. Long computation times of numerical simulations will be avoided in this matter as well as too simple assumptions of analytical approaches.

The influencing parameters to be simulated are divided into material, geometry and process parameters. Relevant material parameters are the coefficients of friction of the polymer-polymer and polymer-steel surface as well as the restitution coefficient and the particle diameter. The geometry is varied in the form of the screw diameter, the channel depth and pitch, the number of grooves and their width, depth and angle. By varying the angle, both conventional axial grooves and helical grooves are taken into account. Finally, the process parameters speed and backpressure are also considered in the simulations. In order to reduce the simulation effort, irrelevant parameters are identified in preliminary investigations.

C. Hopmann and R. Dahlmann (Eds.): *Advances in Polymer Processing 2020*,
pp. 50–62, 2020. https://doi.org/10.1007/978-3-662-60809-8_5

Keywords: Extrusion · Solids Conveying · Discrete Element Method · Grooved barrel extruder

1 Introduction

Single-screw extrusion is one of the most important processing methods for thermoplastics in terms of quantity. Regarding the process engineering the design of the screw geometry is of great importance for economical operation of the extrusion line. For extruders with grooved barrels, the design of the feed section is particularly relevant as it determines the throughput of the entire line. In recent years, the Discrete Element Method (DEM) has moved into focus for the simulation of solids conveying processes in plastics processing, because this method allows the complex movements of the solid bed to be comprehensively monitored. In addition to a short introduction to previous calculation approaches for grooved feed sections, the fundamentals of DEM as well as the procedures for the development of a new model based on DEM simulations are explained.

1.1 Process Engineering and Conventional Calculation Methods of Grooved Barrel Extruders

The use of extruders with grooved feed sections, screw concepts matched to these and a growing understanding of the process have contributed significantly to the continuous increase in throughput rates of single-screw machines. In pipe extrusion, for example, the throughput increased eight times between 1960 and 2004 [1].

The conveying principle of grooved barrel extruders is based on the fact that axially or helical grooves in the feed section of the barrel prevent the pellets from being rotated in the circumferential direction. If the design is accurate, the pellets are forced to be conveyed, also known as "nut-spindle conveying" [2]. The feed section then acts as a solids conveying pump. This means that the throughput almost is independent of the back pressure on the one hand but dependent on the process behaviour in the feed section on the other. Further disadvantages of grooved barrel extruders are an increased susceptibility to wear and the resulting increased production and maintenance costs [3]. Therefore, in the past there have been many efforts to investigate and mathematically describe the conveying behaviour of grooved feed sections. Rautenbach and Peiffer did the first complete modelling [4, 5]. This is based on the theoretical approaches of Darnell and Mol [6] for smooth barrel extruders and first transfers of these approaches to grooved barrel extruders [7–10]. All the approaches mentioned have in common that the pellets completely fill the screw channel and are conveyed as a block driven by friction or interlocking. This assumption is also maintained in further developed models [11–14] because the description of complex kinematic interactions between the pellets is analytically difficult.

For a better overview, the procedure for the analytical calculation is shown schematically in Fig. 1. N is the screw speed, φ the mean helix angle, v_a the axial speed of the solids block, v_0 the circumferential speed, v_c the speed in channel direction and v_r the relative speed between v_0 und v_c. Here, v_0 and v_r include the conveying angle α.

With the corrected bulk density $\rho_{s,c}$ and the axial cross-sectional surface of the channel A_a, the absolute throughput can then be calculated according to Eq. 1.

$$\dot{m} = \rho_{s,c} \cdot v_a \cdot A_a \tag{1}$$

with

$$v_a = v_0 \frac{tan(\varphi) \cdot tan(\alpha)}{tan(\varphi) + tan(\alpha)} \tag{2}$$

and

$$v_0 = \pi \cdot D \cdot N \tag{3}$$

The conveying angle α is calculated by a force balance on the solid block, so that this calculation is imprecise if the assumption of block flow is not suitable. For example, a conveying mechanism deviating from the block flow is described in [15] for powder and in [16] for pellets.

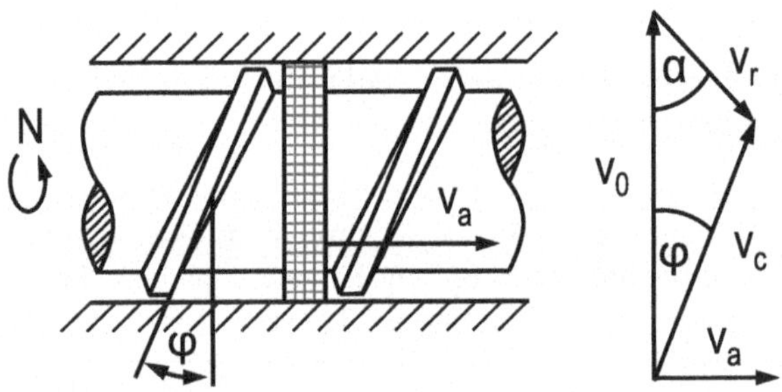

Fig. 1. Schematic representation of the conventional throughput calculation

1.2 Principles of the Discrete Element Method

Originally developed in 1979 by Cundall and Strack [17] for the simulation of granular media in geotechnics (e.g. sand and earth), the DEM has now also found application in plastics technology. After the first two-dimensional simulations of Potente and Pohl [18], further work was carried out by Moysey and Thompson [16, 19–21]. With DEM they were able to observe effects known from practice, such as the backflow of pellets from the first screw flights into the hopper and found good overall agreement with experimental results.

Also at Kunststofftechnik Paderborn (KTP) the DEM was recently successfully used for the simulation of solids conveying processes in smooth barrel extruders. Here, the focus was on the simulation of the pouring behaviour from the hopper in the screw up to high screw speeds [22] as well as the conveying behaviour under application of a back pressure [23]. Therefore, the DEM is now also used for the analysis of grooved feed sections.

In contrast to FEM, the geometries to be analysed are not meshed but modelled as an accumulation of discrete spherical particles. Since no deformations of the particles can be simulated, contact events are calculated with the help of a virtual overlap δ. The calculation model is shown schematically in Fig. 2. Starting from the initial conditions and the overlap δ resulting from an integration over the time step, normal and tangential forces are calculated using the contact model. The latter are limited by the static friction coefficient. With these contact forces the Newtonian equations of motion are then solved translatory and rotatory. After calculating the new quantities of motion, the calculation starts again in the next time step [24].

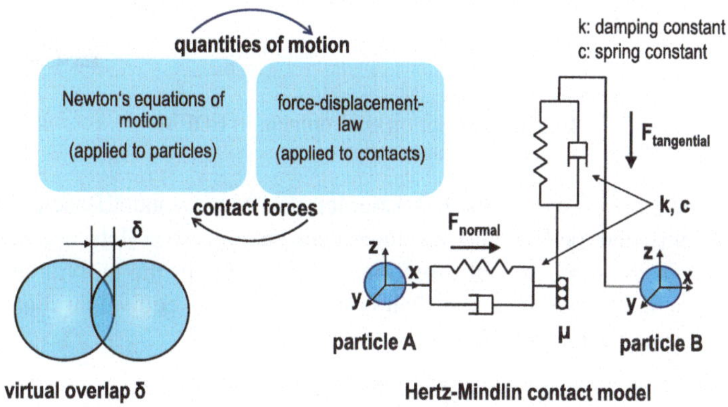

Fig. 2. Principles of the Discrete Element Method (DEM)

2 Description of the Simulation Model

The geometry of the simulation model consists of the feed section of a grooved barrel extruder with a nominal diameter $D = 50\,\text{mm}$, a length $L = 4\,\text{D}$, a channel depth $h_e = 0.09\,\text{D}$ and a pitch $t_e = 0.9\,\text{D}$. There are 10 rectangular grooves distributed around the circumference with a width of $b_G = 7.5\,\text{mm}$ and a depth of $h_G = 3\,\text{mm}$, which run out linearly from the front edge of the hopper. After transfer into the simulation software EDEM of DEM Solutions Ltd., Edinburgh, UK, the model is shown in Fig. 3. The individual segments can be seen here in order to evaluate the target values over the screw length. In order to be able to apply a counterpressure, the screw is extended at the end of the actual feed section by a cylindrical shear section in which a force field works. Each particle within this force field is loaded with a defined force F_p in negative y-direction. Since the contacts of the particles within this extended cylindrical shear section are defined as friction-free and therefore no further forces act, the axial back pressure p is obtained with the number of particles in the force field n_p and the circular ring surface A_k [25]:

$$p = \frac{n_p \cdot F_p}{A_k} \tag{4}$$

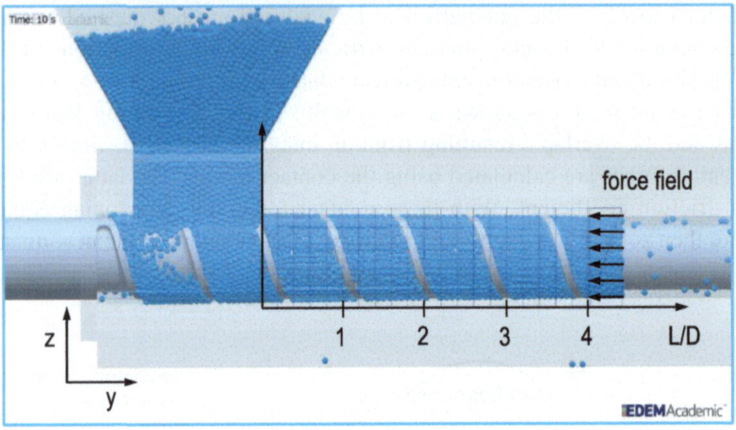

Fig. 3. Simulation environment in EDEM

The investigated material is a PE-LD Lupolen 1840 D, LyondellBasell. In order to transfer the lenticular pellets into an equivalent sphere, 50 pellets are measured in their diameter in the three spatial directions. The lengths in the individual directions are first arithmetically averaged and then converted into an equivalent diameter d_p by forming the geometric mean:

$$d_p = \sqrt[3]{\overline{d_x} \cdot \overline{d_y} \cdot \overline{d_z}} \tag{5}$$

Since the influence of the coefficient of restitution (COR) on the results of simulations in previous investigations has been found to be low, it is set to a typical value of 0.6 [23]. The further parameters of the Hertz-Mindlin contact model are taken from the data sheet and literature and are summarized in Table 1. Unless otherwise stated, the simulations are calculated with a speed $N = 175\ 1/\text{min}$ and a back pressure $p = 50$ bar. This value is equal to the pressure at the end of the grooved bush calculated with the REX 15 software (computer-aided extruder design, developed at KTP) when an experimentally determined value of 180 bar is given at the tip of the entire screw.

Table 1. Parameter for the contact model

Parameter	Unit	Value
Particle diameter d_p	mm	3.45
Solid density ρ	kg/m^3	919
Poisson's ratio ν	–	0.47
Shear Modulus G	MPa	68
Coefficient of restitution COR	–	0.6
Coefficients of friction μ	–	0.1 … 0.9
Gravity g	m/s^2	9.81

3 Preliminary Investigations

3.1 Variation of Coefficients of Friction

It is known from literature that mainly three coefficients of friction influence the conveying characteristics of the feed section in a single-screw extruder. This is the internal coefficient of friction of the plastic-plastic material combination μ_i, the coefficient of friction of the plastic-steel material combination at the barrel μ_b and at the screw μ_s. In particular, the ratio μ_b/μ_s has a large influence on the conveying angle and thus the throughput. The larger μ_b/μ_s, the larger the conveying angle [1]. Therefore, these parameters are varied in a full factorial experimental design, which is shown in Table 2. The factor levels are chosen in a wide range because the coefficients of friction in real extrusion processes strongly depend on the process parameters pressure and temperature. Concurrently, the experimental determination of these dependencies is of limited accuracy and reproducibility [26]. That is why it is examined how the DEM model reacts to changing friction coefficients.

Table 2. Variation of coefficients of friction

Number of simulation	μ_i	μ_b	μ_s
1	0.1	0.1	0.1
2	0.9	0.1	0.1
3	0.1	0.9	0.1
4	0.9	0.9	0.1
5	0.1	0.1	0.9
6	0.9	0.1	0.9
7	0.1	0.9	0.9
8	0.9	0.9	0.9

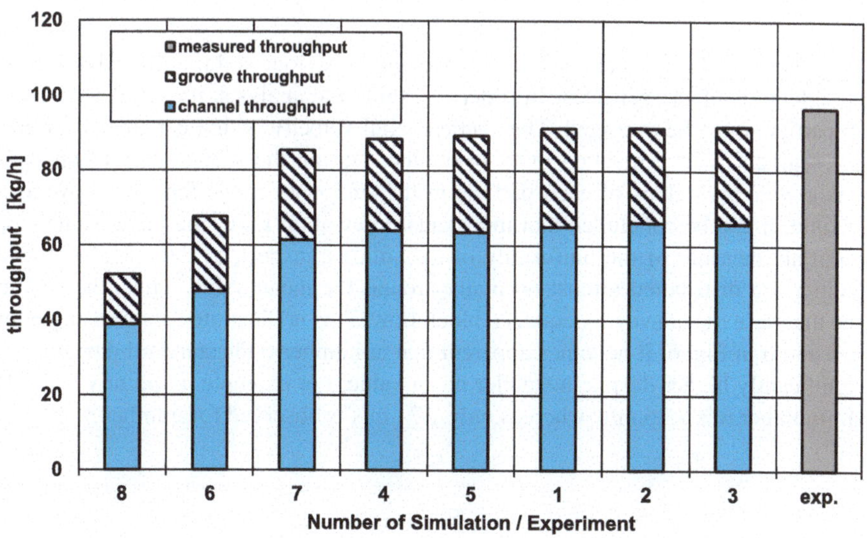

Fig. 4. Throughputs in dependence of the coefficients of friction

Figure 4 depicts the throughput rates calculated by simulation and compares them to the experimental value. Results are shown in an ascending order. Through virtual mass flow sensors, which are positioned in the channel and grooves behind the hopper, the simulated throughput is further divided into channel and grooves throughput. When interpreting the results, it should be noted that the geometric boundary conditions of the screw, the grooves, and the pellets leads to interlocking conveying in the channel and the grooves. This is a phenomenon described from an analytical point of view in case 1a by Schöppner [11]. Conveying in the beginning of the grooves was observed for each simulation. The results are in line with the analytical assumption of conveying proposed in case 1a. Further, the total flow rates, except for numbers 8 and 6, match those of the experimental value. However, a comparison of numbers 3 and 6 highlights that both the assumption of interlocking conveying and block flow are not valid for the entire feed section under the given conditions. Figure 5 uses tracer particles to illustrate that at number 3 (left side) there is a strong block flow with a conveying angle of $\alpha \approx 90°$, whereas at number 6 (right side) the particles are conveyed in the circumferential direction to a greater extent. Therefore, it can be concluded that the assumption of a block flow is not valid for these simulations.

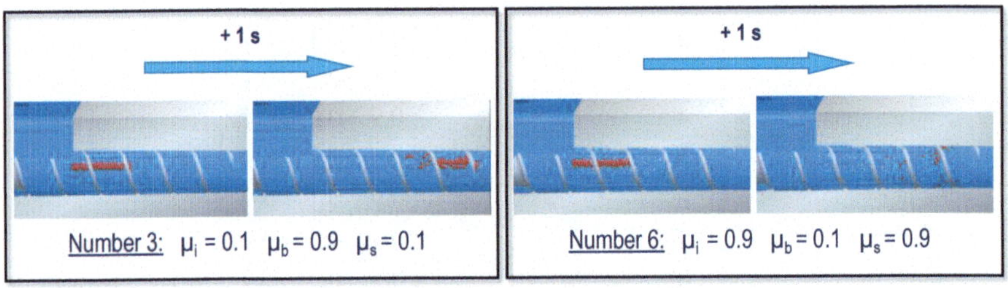

Fig. 5. Visual comparison of solids conveying depending on coefficients of friction

Another possibility to characterize conveying behaviour is through analysis of the axial velocities of the particles. In order to enable the analysis, the axial velocity of each particle must be averaged. The average axial velocity is first calculated for each screw segment, which resulted in $\overline{v_a}$ over time. During the second step of analysis, the axial velocity values of each particle in the stationary range have been averaged over time. It can be concluded that the standard deviation s_{va} of the mean value is an appropriate measure of the uniformity of the solids transport, as the calculated axial velocities are distributed almost normally around the mean value. Thus, the smaller s_{va} is, the more the conveying equals a block flow. This is illustrated showcasing numbers 3 and 6 in Fig. 6. It becomes apparent that at number 6 the standard deviation s_{va} is significantly higher despite a smaller mean value. For example, at point y = 3.5 D, s_{va} for number 6 is 7.26 m/s, whereas only 2.71 m/s is observed for number 3.

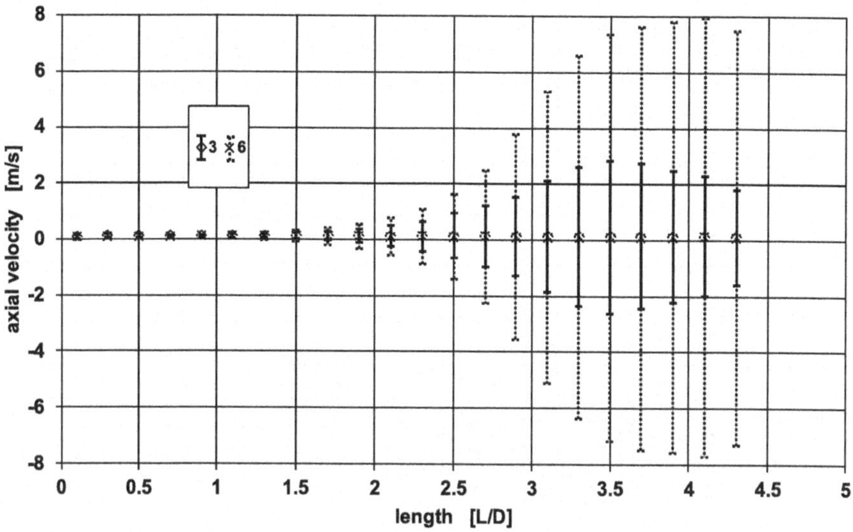

Fig. 6. Comparison of the standard deviation of the axial velocity

Furthermore, the DEM offers the possibility to evaluate the pressure build-up over the screw length. For this purpose, the normal forces of the particles are cumulated in the radial direction and related segmentally to the surface of the screw. In the following, different pressure curves will be discussed using the example of number 2 and 4, which differ only in the coefficient of friction μ_b (see Table 2). Although the mass flow rates hardly differ between 91 kg/h (number 2) and 88 kg/h (number 4), the pressure curves differ from each other, as Fig. 7 shows. The maximum pressure in simulation number 2 is 84 bar and thus 40% higher than at number 4, where only a maximum pressure of 60 bar occurs. The pressure build-up in the grooved barrel has a large influence on the dissipation. If the dissipation is too high, the pellets will melt in the grooves at the limit speed N_l and the interlocking conveying characteristic can no longer be maintained [27]. The precise prediction of N_l using DEM will be investigated in the future. For this purpose, the dissipated power is determined as a function of the influencing variables mentioned in Chap. 4 and converted into a temperature increase of the pellets by means of an energy balance and the specific enthalpy. If the temperature calculated in this way exceeds the crystalline melting temperature or the glass transition temperature, the pressure throughput behaviour is no longer independent of back pressure.

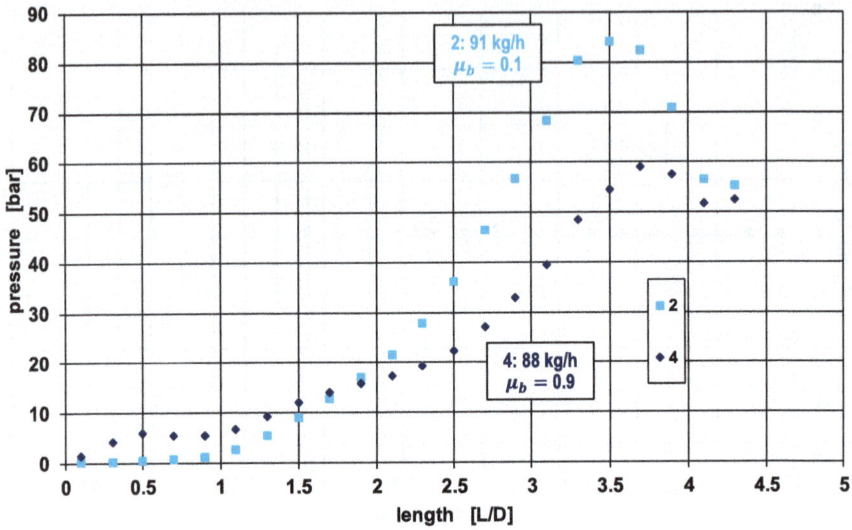

Fig. 7. Influence of barrel friction coefficient on pressure build-up

3.2 Variation of Grooves Angle

Since the groove angle will also be varied in the final regression model (see Chap. 4), the feed section from Sec. 3.1 is equipped with four helical grooves of different angles φ_G. Groove angles which are wound against the screw (25°, 60°) and those which are wound in the same direction (120°, 155°) are taken into account. The coefficients of friction are selected to $\mu_i = 0.1$, $\mu_b = 0.9$ and $\mu_s = 0.1$. This corresponds to simulation number 3 from Sec. 3.1, which has the highest throughput rate. All other parameters are kept constant. The result is shown in Fig. 8. It can be seen that all feed sections with helical grooves have a significantly lower throughput than the axial grooves (90°). A possible reason for this is that the selected combination of channel depth, groove depth and pellet diameter leads to an interlocking conveying and thus to a conveying angle that approximately corresponds to the groove angle. Nevertheless, other scenarios are also possible in which helical grooves are superior. From an analytical point of view, this is just the case if friction-locked conveying with a conveying angle of $\alpha < 90°$ prevails [1]. Then there is no relative speed between the conveying in the grooves and in the screw channel, so that the pellets are guided "as on rails" and less friction losses occur.

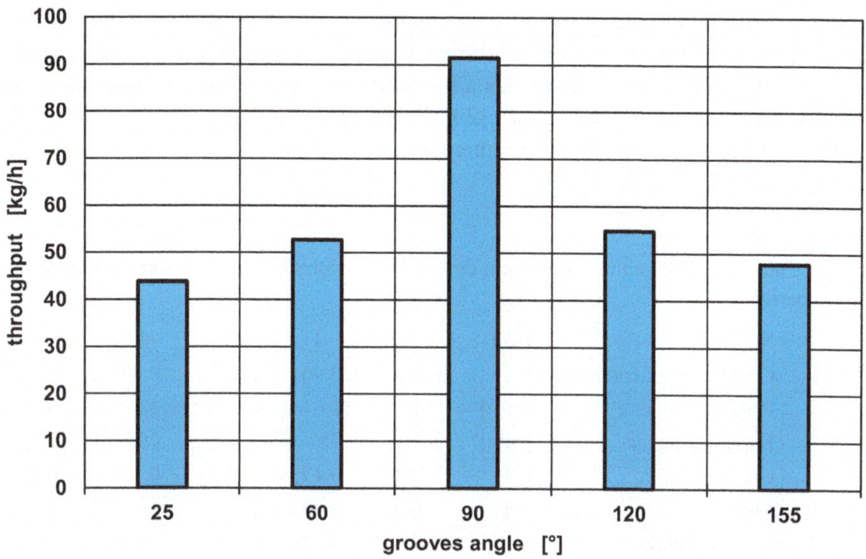

Fig. 8. Investigation of helical grooves

4 Development of a Regression Model

In order to systematically investigate the conveying behaviour of grooved feed sections by means of DEM, the relevant influencing variables must first be determined and then a suitable experimental design has to be selected. The influencing variables are divided into material, geometry and process parameters. Material parameters relevant for the conveying characteristics are the coefficients of friction μ_i, μ_b and μ_s (see Sec. 3.1) as well as the pellet diameter d_p. The number of coefficients of friction can be reduced by assuming that the two coefficients of friction of the material combination polymer-steel μ_b and μ_s are combined to an external coefficient of friction $\mu_e = \mu_b = \mu_s$. Furthermore, the inner coefficient of friction μ_i is assumed to be 0.3 greater than the outer coefficient of friction μ_e so that $\mu_e = \mu_i - 0.3$. This approach has already led to satisfactory results in [28].

The geometry parameters barrel diameter D, channel depth h_e, screw pitch t_e and grooves angle φ_G must also be taken into account. To limit the time required for the simulation, the groove geometry is fixed to a width $b_G = 8\,\text{mm}$ and a depth $h_G = 3\,\text{mm}$. In addition, the effective area of the grooves on the cylinder circumference is set to 1/3, so that the number of grooves results [1]:

$$n_G = \frac{\pi \cdot D \cdot \sin(\varphi_G)}{3 \cdot b_G} \tag{6}$$

Finally, the process parameters screw speed N and back pressure p are taken into account, resulting in a total of 8 parameters. In order to be able to describe quadratic effects with a moderate test scope, the use of a central composite design (CCD) with a fractional factorial core is a suitable approach. With the factor levels summarized in Table 3, an experimental design with 85 points results. The five levels are the axial points $\pm\alpha$, the cube points of the fractional factorial core ± 1 and the center point in the cube is the level 0. This design has a resolution of five so that the main effects are only mixed with fourfold interactions [29]. Since the values for α by definition represent the extreme values of the experimental design, it must be ensured that they are feasible when selecting them.

Table 3. Factors of the central composite design for the investigation of solids conveying in grooved barrels

Factor	Meaning	Unit	$-\alpha$	−1	0	+1	$+\alpha$
μ_i	Coefficient of friction	[−]	0.4	0.461	0.55	0.639	0.7
d_p	Particle diameter	[mm]	1	1.61	2.5	3.39	4
D	Barrel diameter	[mm]	30	75	140	205	250
h_e/D	Channel depth	[−]	0.1	0.12	0.15	0.18	0.2
t_e/D	Pitch	[−]	0.7	0.82	1	1.18	1.3
φ_G	Grooves angle	[°]	30	42.16	60	77.84	90
N	Screw speed	[1/min]	50	242.56	525	807.44	1000
p	Backpressure	[bar]	0	101.35	250	398.65	500

5 Conclusion and Outlook

The simulations carried out show that the DEM is basically suitable for modeling solids conveying in grooved barrel extruders. In order to be able to describe the effects of individual factors not only randomly, the simulation model is systematically examined with the central composite design presented in Chap. 4. After completion of the simulations, multivariate regression equations can be used to model different types of target values such as the throughput, the conveying angle, the resulting axial velocity and the limiting speed. Furthermore, the results will be validated on a solids conveying test bench, which is specially developed for the investigation of solids conveying in grooved feed sections. This allows the avoidance of uncertainties that would arise in validation tests on an extruder with downstream melting and metering sections.

Acknowledgements. The results presented here were obtained in the research project SCHO 551/36-1 which is supported by the German Research Foundation (DFG). Therefore, we would like to thank the DFG for its support.

References

1. Bornemann, M.: Erweiterung der modelltheoretischen Grundlagen zur Durchsatz- und Leistungsberechnung von Einschneckenplastifiziereinheiten. PhD thesis, Paderborn University (2011)
2. Grünschloß, E.: Schüttdichte und Massendurchsatz in Nutbuchsenextrudern. Kunststoffe **83**, 309–311 (1993)

3. Limper, A.: Verfahrenstechnik der Thermoplastextrusion. Hanser, München (2013)
4. Rautenbach, R., Peiffer, H.: Durchsatz- und Drehmomentverhalten genuteter Einzugszonen von Einschneckenextrudern. Kunststoffe **72**, 262–266 (1982)
5. Peiffer, H.: Zum Förderproblem in der genuteten Einzugszone von Einschneckenextrudern. PhD thesis, RWTH Aachen University (1981)
6. Darnell, W.H., Mol, E.A.J.: Solid transport in extruders. SPE – J. **12** (1956)
7. Goldacker, E.: Untersuchung zur inneren Reibung von Pulvern, insbesondere im Hinblick auf die Förderung in Extrudern. PhD thesis, RWTH Aachen University (1971)
8. Hegele, R.: Untersuchungen zur Verarbeitung pulverförmiger Polyolefine auf Einschnecken-Extrudern. PhD thesis, RWTH Aachen University (1972)
9. Menges, G., Hegele, R., Langecker, G.R.: Theorie der Förderung im Einschneckenextruder, Vergleich genutete und glatte Einzugsbuchse. Plastverarbeiter **23** (1972)
10. Langecker, G.R.: Untersuchungen zum Stoffverhalten von Kunststoffpulvern in der Einzugszone von Einschneckenmaschinen mit genuteten Buchsen. PhD thesis, RWTH Aachen University (1977)
11. Schöppner, V.: Simulation der Plastifiziereinheit von Einschneckenextrudern. PhD thesis, Paderborn University (1995)
12. Michels, R.: Verbesserung der Verarbeitungsbandbreite und der Leistungsfähigkeit von Einschneckenextrudern. PhD thesis, University Duisburg-Essen (2005)
13. Rahal, H.: Alternative Methoden zur Feststoffförderung und Plastifizierung in der Extrusionstechnik. PhD thesis, University Duisburg-Essen (2008)
14. Miethlinger, J.: Modellierung der Feststoffförderzone von Einschneckenextrudern unter besonderer Berücksichtigung von Wendelnutbuchsen. Kunststoffe **93**, 49–53 (2003)
15. Zhang, S., Sernas, V.: Incompressible model of solids conveying in a single-screw extruder. In: Society of Plastic Engineers (ed.) ANTEC, Orlando (2000)
16. Moysey, P.A., Thompson, M.R.: Modelling the solids inflow and solids conveying of single-screw extruders using the discrete element method. Powder Technol. **153**, 95–107 (2005)
17. Cundall, P.A., Strack, O.D.L.: A discrete numerical model for granular assemblies. Géotechnique **29**, 47–65 (1979)
18. Potente, H., Pohl, T.C.: Simulation and analyses of the polymer-pellet-flow into the first section of a single screw. In: Society of Plastic Engineers (ed.) ANTEC, Dallas, USA (2001)
19. Moysey, P.A., Thompson, M.R.: Investigation of solids transport in a single-screw extruder using a 3-D discrete particle simulation. Polym. Eng. Sci. **44**, 2203–2215 (2004)
20. Moysey, P.A., Thompson, M.R.: Determining the collision properties of semi-crystalline and amorphous thermoplastics for DEM simulations of solids transport in an extruder. Chem. Eng. Sci. **62**, 3699–3709 (2007)
21. Moysey, P.A., Thompson, M.R.: Discrete particle simulations of solids compaction and conveying in a single-screw extruder. Polym. Eng. Sci. **48**, 62–73 (2008)
22. Leßmann, J.-S., Weddige, R., Schöppner, V., Porsch, A.: Modelling the solids throughput of single screw smooth barrel extruders as a function of the feed section parameters. IPP (2012). https://doi.org/10.3139/217.2588
23. Trippe, J., Schöppner, V.: Modeling of solid conveying pressure throughput behavior of single screw smooth barrel extruders under consideration of backpressure and high screw speeds. IPP **33**, 486–496 (2018)
24. Weddige, R.: Analyse und Optimierung des Prozessverhaltens von Einschneckenextrudern im Bereich der Feststoffförderung und der druckabhängigen Viskosität. PhD thesis, Paderborn University (2015)
25. Leßmann, J.-S.: Berechnung und Simulation von Feststoffförderprozessen in Einschneckenextrudern bis in den Hochgeschwindigkeitsbereich. PhD thesis, Paderborn University (2016)

26. Rauwendaal, C.: Polymer Extrusion, 5th edn. Hanser, München (2014)
27. Potente, H., Stenzel, H., Bergedieck, J.: Output computation in the grooved-barrel extruder with regard to various conveyance techniques. Adv. Polym. Technol. **10**, 285–295 (1990)
28. Trippe, J.: Erweiterung der Modellierung zur Durchsatz- und Leistungsberechnung von Feststoffförderprozessen in der Einschneckenextrusion. PhD thesis, Paderborn University (2018)
29. Siebertz, K., van Bebber, D., Hochkirchen, T.: Statistische Versuchsplanung – Design of Experiments (DoE). Springer, Berlin (2010)

Foam Extrusion of Elastomers using Water as Physical Blowing Agent

Christian Hopmann, Sebastian Kammer(✉), Fabian Fey, and Martin Facklam

Institute for Plastics Processing (IKV) in Industry and Craft, RWTH Aachen University, 52074 Aachen, Germany
sebastian.kammer@ikv.rwth-aachen.de

Abstract. An alternative to chemical blowing agents is the physical blowing agent water which can be adsorbed by fillers and thus easily incorporated into rubber compounds. The phase transition of the water during heating of the rubber in the vulcanisation unit is used to initiate the blowing reaction. In order to achieve the required quality parameters (e.g. mechanics and cell structure), the cross-linking and blowing reaction have to be balanced. This can be achieved by optimising the compound recipe and the processing. This paper examines the influence of the extrusion and vulcanisation process on the foam structure and the mechanical properties of water-foamed rubber compounds based on ethylene propylene diene terpolymer rubber (EPDM) and nitrile butadiene rubber (NBR). In the extrusion process, the barrel temperature and screw speed are systematically varied and the foamed profiles are evaluated with regard to surface, foam structure, tension set and an optimal process point was identified. For the subsequent vulcanisation process, the effect of the hot air temperature on the tension set, the spring constant and characteristic values of the cell structure are analysed.

Keywords: Blowing agent · Elastomer · Foam extrusion

1 Introduction

1.1 Potential Regulation of Chemical Blowing Agents by REACH Directive

Due to their insulating and sealing properties, foamed elastomers are used in the automotive and construction industries as well as for a wide range of industrial applications [1–3]. The state of the art in the production of cellular elastomers is foaming with chemical blowing agents, which decompose parallel with the cross-linking reaction at a specific temperature and thus produce a foam structure in the elastomer [4, 5]. Chemical blowing agent systems such as Azodicarbonamide (ADCA) in combination with hydrazine derivatives such as p,p′ oxibis(benzenesulfonylhydrazide) (OBSH) [6] have the largest economic impact in rubber processing. Regardless of their toxicity, chemical blowing agents are widely used. Besides the blowing gases carbon dioxide

C. Hopmann and R. Dahlmann (Eds.): *Advances in Polymer Processing 2020*,
pp. 63–78, 2020. https://doi.org/10.1007/978-3-662-60809-8_6

(CO_2) and dinitrogen (N_2) ADCA decomposes into toxic products such as carbon monoxide (CO), isocyanic acid (HNCO) and ammonia (NH_3) [7, 8]. Since ADCA has been classified by the European Chemicals Agency (ECHA) as a substance of very high concern (SVHC), it is to be expected that it will be subject to regulatory approval in the European Union in the future in accordance with the regulation concerning the Registration, Evaluation, Authorisation and Restriction of Chemicals (REACH) [9, 10]. Due to the upcoming regulation of chemical blowing agents it is necessary to develop alternative non-toxic blowing agent systems for the extrusion of foamable elastomers.

1.2 Water as a Physical Blowing Agent for Elastomers

Elastomers can be foamed without chemical blowing agents by using the phase transition of fluids. The expanding fluid is then referred as the physical blowing agent.

In the past, numerous investigations were carried out with inert gases as physical blowing agents such as N_2 and CO_2 [11, 12]. However, gases are only conditionally suitable as physical blowing agents for the production of foamed elastomers. Gas-loaded rubber compounds expand instantly due to the high pressure drops and fast decrease of the partial pressure in the extrusion die. As the surface of the rubber at the die outlet is not yet cross-linked, spontaneous foaming leads to surface defects, which must be avoided for quality reasons. Furthermore, for the physical foaming with gases dosing units for the extruders are required [13]. Despite the lower volatility and the lower solubility compared to most physical blowing agents, water is attractive as a physical blowing agent from an ecological and economic point of view. Under atmospheric pressure, the volume of 1 ml water at 100 °C expands to approximately 1700 ml water vapour. In addition, water is readily available and does not require any complex manual handling and has already been incorporated into the melt in liquid form in the extruder to foam thermoplastics [14, 15].

Because of the high differences in viscosity between water and rubber compounds, water cannot be homogeneously dispersed into the rubber compound in liquid form. However, it can be very easily incorporated by mixing water loaded hygroscopic powdery substances or crystalline bound water such as hydrates into the rubber compound [16, 17]. In the subsequent extrusion process, after shaping in the extrusion die, the blowing reaction with the cell formation and cell growth takes place during the phase transition of the water and simultaneous crosslinking of the rubber in the vulcanisation unit. Cell formation requires an increase in temperature to a value above the evaporation point of water at a given pressure. If the evaporation temperature in the extruder is not exceeded, there is no spontaneous cell formation at the die outlet due to the pressure drop. The cell growth then takes place in the subsequent vulcanisation unit when heat energy is applied to the rubber profile. The cell growth depends on the viscosity. This in turn depends on the temperature and the time-dependent cross-linking state.

In preliminary investigations, the foaming of rubber with water as a physical blowing agent showed great potential both in injection moulding and in extrusion [16] and is therefore being investigated more in detail in this research. The suitability of different carbon blacks, silica and hydrates as water-carrying substances has been investigated and the compound recipe of an ethylene propylene diene terpolymer rubber (EPDM) and a nitrile butadiene rubber (NBR) compound optimised with regard to reach low foam densities. The main focus was on the synchronisation of the cross-linking and blowing reaction in order to guarantee process stability of the foamable compounds. If vulcanisation proceeds too quickly, the high degree of crosslinking makes the resistance of the matrix against cell internal pressure too high to enable the grow of bubbles. Whereas a low cross-linking rate leads to uncontrolled bubble growth and collapsing cells due to coalescence effects [17–19].

In addition to the recipe, the crosslinking and blowing reaction can also be influenced by the temperature control during processing. In the extrusion process, the temperature control of the barrel and the screw speed affect the heating of rubber [2, 20]. In the subsequent vulcanisation unit, the energy required for the cross-linking and blowing reaction can be introduced into the rubber by convection, infrared radiation (IR) or dielectric heating. The different heating methods result in locally different heating rates and thus cross-linking kinetics over the profile cross-section [18, 19, 21, 22].

Yet, it hasn't been investigated how the extrusion and vulcanisation process influence the quality characteristics of a rubber profile foamed with water as a physical blowing agent. In the following sections the compound production, the extrusion and vulcanisation process via convective heating and the methods used to determine the foam quality parameters cell structure and mechanical properties are presented. Subsequently, results of the systematic measurement are depicted and discussed how the extrusion and vulcanisation process has to be adjusted to use water as a physical blowing agent in rubber extrusion.

2 Methods

2.1 Compound Production and Incorporation of Blowing Agents

For the following investigations NBR and EPDM recipes (Tables 1 and 2) are modified. The feedstocks without blowing agents and sulphur are produced on an industrial scale internal mixer by Hexpol Compounding s.p.r.l., Eupen, Belgium.

Table 1. NBR recipe with a hydrate as blowing agent

Ingredient	Weight proportion [phr]
Perbunan 2845 F	100
Corax N550	60
Vulkanol 81	20
Zinc oxide	5
Vulkacit CZ/C	1.5
Rhenofit F	10
Sulfur 90/95	2
$CaSO_4 \cdot 2H_2O$	0.5

On the roller mill MT 8″ × 20″ of Rubicon Gummitechnik, Halle/Saale, Germany, the cross-linking agent sulphur and the blowing agent calcium sulphate dehydrate ($CaSO_4 \cdot 2H_2O$) are incorporated. The water content in the mixture is 0.62 wt.-%. As a reference, rubber compounds with 0.5 phr ADCA are produced. As an indicator for the flow behaviour of the rubber compounds, the Mooney viscosity is determined with shearing-disc viscometer according ISO 289-1:2015. The torque of the shaft is measured after 1 min heating to 100 °C and 4 min' rotation at a constant shear rate of 1.56 s^{-1} (ML (1 + 4), 100 °C). The Mooney viscosity is an instrument-specific measured parameter (torque) and is given in Mooney units (MU). The higher the MU value the higher the viscosity of the measured compound. The Mooney viscosity of the EPDM master batch is 35 MU and the NBR has a viscosity of 52 MU.

Table 2. EPDM recipe with a hydrate as blowing agent

Ingredient	Weight proportion [phr]
Keltan 6950C	100
Corax N550	80
Sunpar 2280	70
Zinc oxide	5
Stearic acid	2
Lipoxol	3
Vulkacit DM/C	1.5
Rhenocure ZAT	3.5
Rhenofit F	10
Sulfur 90/95	1.5
$CaSO_4 \cdot 2H_2O$	0.5

2.2 Extrusion and Evaluation of the Processability

The extrusion trials are carried out using a laboratory extruder from Brabender GmbH, Duisburg, Germany. A conveying screw without core progression and a diameter of 19 mm and a length of 20 D is used. The rubber compound is fed into the extruder in the form of 1 cm wide strips and drawn into the rotating screw with the help of a feeding unit. The screw speed and the barrel temperature are each varied in three stages: 30 min^{-1}, 60 min^{-1} and 90 min^{-1} as well as 40 °C, 60 °C and 80 °C. The barrel temperature is measured with a thermocouple in the barrel at the screw tip. The melt temperature was measured in the bulk in the extrusion die with a plug-in thermometer. For the production of round profiles an extrusion die with a diameter of 10 mm is used.

2.3 Continuous and Discontinuous Vulcanisation and Foaming of the Profile Samples

In order to investigate the influence of the extrusion process, the round profiles with a diameter of 10 mm are continuously vulcanised in the downstream hot-air unit type

HLK-UHF 101 of TSM GmbH, Stolberg, Germany. The air temperature is varied at three levels (180 °C, 200 °C and 220 °C).

To investigate the influence of the vulcanisation process, the extruded profiles were cooled to room temperature, cut into 25 cm long strands and afterwards cross-linked in a three-stage variation of the hot air temperature in the discontinuous vulcanisation oven type L200 from Gerlach Maschinenbau GmbH, Nettetal, Germany. The curing time is the t_{90} time derived from RPA measurements at 180 °C. The curing time for the EPDM samples is 300 s and for the NBR 180 s [18, 19].

2.4 Evaluation of the Extrudates Surface Quality

The surface quality is an important quality feature of rubber profiles. Rubber profiles are extruded and vulcanised by varying the barrel temperature and screw speed. During extrusion, the high shear stresses in the extrusion die can exceed the tear strength of the rubber compound, resulting in cracks on the surface of the extrudate. During vulcanisation, the blowing reaction can cause small bubbles in the surface layer, resulting in a rough surface. According to Table 3, the surface quality of the profile is evaluated and divided into three classes. The higher the score, the smoother and more homogeneous is the surface.

Table 3. Quality score of the surface of extruded and vulcanised samples

Quality score	Surface characteristic	Example
3	Good: smooth, no defects	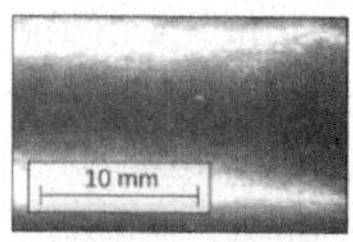
2	Poor: coarse, occasional small bubbles, no cracks	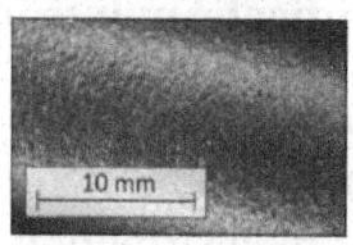
1	Very poor: very rough, uneven, cracks	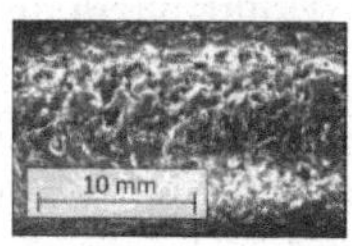

2.5 Mechanical Analysis of the Foamed Elastomers

In this paper, short-term and long term measurements are performed to characterize the mechanical properties of the foamed elastomers. For the short-term test the hardness and the static spring constant is measured. The tension set test is used to determine the long-term behavior of the rubber compounds.

The Shore A hardness of the elastomer profiles is measured with the durometer HPE II of Bareiss Prüfgerätebau GmbH, Oberdischingen, to characterise the mechanical short-term properties. The hardness value is a relative value between 0 and 100. The lower the hardness value, the softer the material. In accordance with DIN ISO 7619-1 [24], the test is carried out on six test specimens.

The static spring constant of the foamed rubber samples are measured with the Jidoka S 101 II elastomer tester from CAS GmbH & Co. KG, Reinbek. For the measurement of the static spring constant, the profiles are cut into 10 mm slices perpendicular to extrusion direction. The static spring constant (SC) is the proportionality factor of the force which the material opposes to the deflection and the resulting deformation. In order to increase measurement reliability, they are carried out on three different samples for each process point.

The tension set (TS) is determined as the characteristic value for the mechanical long-term behaviour. The tension set was performed as an alternative to the more common compression set, as specimen preparation is much easier and multiple specimens can be measured simultaneously. The deformation set after elimination of an impressed deformation corresponds to the viscous part of the deformation and correlates with the degree of cross-linking. The lower the deformation set, the higher the degree of crosslinking. The measurement is based on the standard DIN ISO 2285, method B [25]. To determine the TS, up to seven specimens with a length of 100 mm are clamped next to each other in a clamping device and then stretched by 30% to 130 mm. After storage at 70 °C for 24 h, the specimens are removed from the clamping device and the residual deformation is measured after 30 min. Due to the high experimental effort involved in specimen production and measuring time, the number of measurements in the extrusion trials is limited to those at a barrel temperature of 80 °C, since the hardness measurements showed no significant change with varying barrel temperature.

2.6 Optical Analysis of the Cell Structure

In order to draw conclusions about the influence of the varied process parameters on the foam structure, comparable information on foam characteristics such as the cell size and its distribution is required. To evaluate the foam structure resulting from the various parameter or material variations, macroscopic images of the cross-sectional area of all elastomer profiles are taken. The images are taken with the acA3800-10 gm area scan camera from Basler AG, Ahrensburg. The sensor has a maximum imaging circle of 6.17 × 4.55 mm and a resolution of 3840 px × 2748 px. The Cosmicar/Pentax Television Lens lens from Ricoh Imaging Company, Ltd., Tokyo, Japan with a luminous intensity of 1:1.4 and a fixed focal length of 25 mm is used. The camera is mounted in a bracket approximately 28 cm above the specimen. A LED light ring from CCS Inc, Kyoto, Japan with white light (5500 K) is used for uniform illumination of the sample surface from all sides. The software Ozella (Version 4.0) developed at the IKV for optical cell analysis enables the characterisation of the foam structure of elastomer profiles by means of digital image processing. In addition to the average cell diameter, the software calculates the porosity and the surface cell density as well as the sample area, the mean distance to the nearest neighbouring cell and the number of cells found in closed foam structures [26]. From this, a cell size distribution in the form of a histogram can be generated, which provides information on the homogeneity of the cell diameters in the sample cross-section. As the ratio of the cavity volume to the total volume, the porosity gives an indication of the respective density reduction of the sample [26]. The distance to the neighboring cell describes the distance between the centers of the two cells. For the analysis of the cell structure, six different macroscopic images of the cross-section per sample are analysed.

3 Results

3.1 Barrel Temperature of the Extruder Affects Surface Quality and Foam Structure

Figure 1 shows the quality score of the extrudate surface of the EPDM and NBR compounds as a function of the extrusion parameters barrel temperature and screw speed. The highest surface quality is achieved for both compounds at the barrel temperature of 40 °C. For NBR, a good surface quality is achieved even at a barrel temperature of 60 °C. For the following investigations only profiles with a very good surface quality are extruded. Therefore, in the following tests the barrel temperature is set to 40 °C and the screw speed to 60 min^{-1}.

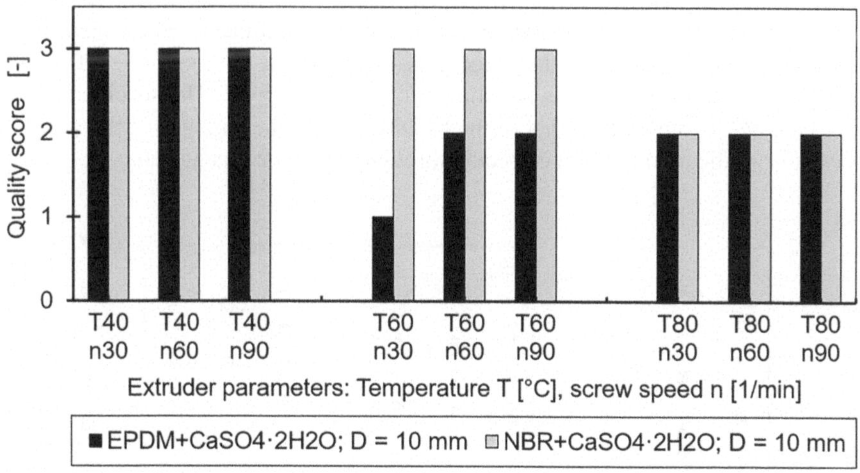

Fig. 1. Surface quality of vulcanised samples as a function of extruder parameters

In addition to the surface properties, the extrusion parameters have an effect on the resulting foam structure of the rubber profile. Figure 2 shows the cross section of one representative EPDM sample of each foamed profile as a function of the barrel temperature and the screw speed. After extrusion the EPDM profiles are vulcanised for 300 s at 200 °C. At a barrel temperature of 60 °C, a homogeneous fine but mixed open and closed cell structure is obtained independent of the screw speed.

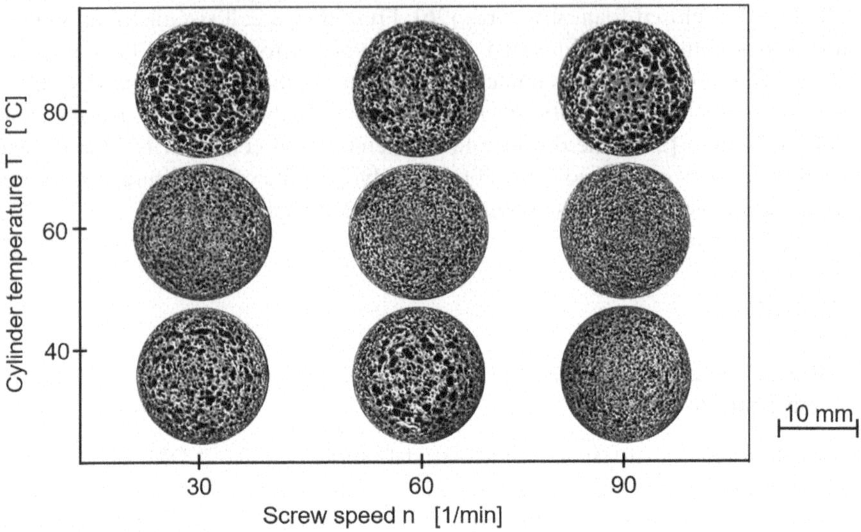

Fig. 2. Resulting foam structure of the cross-section of water foamed EPDM profiles after vulcanisation at 200 °C for 300 s as a function of extruder parameters

The actual temperature in the rubber extrudate deviates by 10 °C to 20 °C from the set barrel temperature, as shown in Fig. 3. By increasing the screw speed from 30 min^{-1} to 90 min^{-1}, the temperature of the rubber is then increased by about 20 °C. The specific throughput was determined gravimetrically at the tool outlet. The mean values from three measurements are shown for each test point. The specific throughput of the NBR is higher than that of the EPDM. Due to the variance of the measured values, no significant trend of the specific throughput can be recognised.

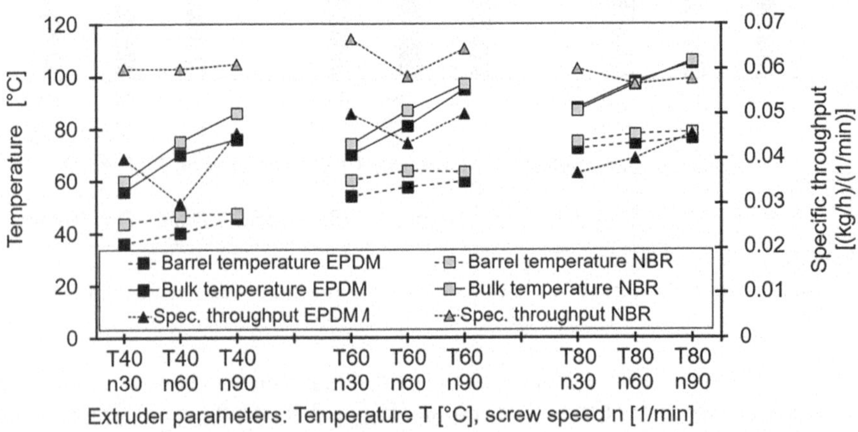

Fig. 3. Temperature development of the barrel and rubber as a function of extruder parameters

Regardless of the extrusion parameters and the resulting variations in foam structure, the Shore hardness of the EPDM compound is between 22 and 26 (Fig. 4). As a reference the hardness of the unfoamed elastomers reaches values of 50 Shore A (EPDM) and 65 Shore A (NBR). The effects of the process settings are smaller than the deviation of the measured hardness and thus not significantly different from each other. The tension set of the EPDM is in the order of 25% and about 15% lower than NBR. The effect in the measured tension set as a function of the extrusion parameters is about 3%. Since only one measured value per test point is available for the tension set and the effect is very low, its significance cannot be assessed.

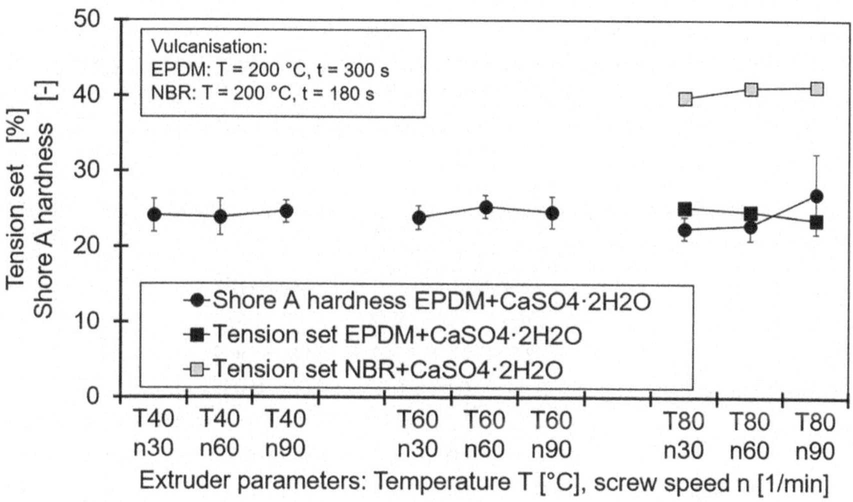

Fig. 4. Mechanical properties of foamed compounds as a function of extruder parameters

3.2 Influence of the Vulcanisation Process on the Extrudate Quality

The mean values $\bar{x}$ of the characteristic values from the cellular structure analysis with their 95% confidence interval $\Delta\bar{x}$ are summarised in Table 4 as a function of the compound and the vulcanisation temperature. With a temperature increase from 180 °C to 220 °C, the sample area of the EPDM increases significantly from 177 mm^2 to 265 mm^2 and for NBR decreases from 256 mm^2 to 244.9 mm^2. The average cell diameter does not change significantly for both compounds due to a change in temperature. The porosity also does not change significantly with the vulcanisation temperature. However, the porosity of the EPDM compound is significantly higher than the porosity of the NBR. The number of cells in EPDM and NBR increases slightly with temperature, but this effect is not statistically significant either. However, the number of cells in the NBR samples is significantly higher than the number of cells in the

EPDM foams. The temperature has for EPDM no influence on the distance between the neighbouring cells. With NBR, the distance to the neighbouring cell decreases slightly from 0.44 mm to 0.41 mm as a result of a temperature increase from 180 °C to 220 °C. The cell density does not decrease significantly with EPDM when the temperature is increased. The cell density of the NBR is significantly different for the temperatures of 180 °C and 220 °C and increases with higher temperature from 2.38 1/mm^2 to 2.70 1/mm^2.

Figure 5 shows an example of the cell size distribution of the two compounds for the crosslinking temperature of 220 °C. It can be seen by the larger number of cells smaller than 400 μm that the foam structure of the NBR is finer than that of the EPDM. It can also be seen that EPDM has more cells than NBR which are larger than 1000 μm. Since the resistance of the rubber matrix to biaxial deformation due to the growing bubbles is higher for higher viscosities, the cell growth is suppressed, so that no coalescence occurs and the cell structure becomes finer. The higher viscosity of NBR (52 MU) compared to EPDM (35 MU) leads to a finer cell structure.

Table 4. Results of the cell structure analysis of the elastomers foamed with calcium dihydrate

		$\bar{x}$	$\Delta\bar{x}$	$\bar{x}$	$\Delta\bar{x}$	$\bar{x}$	$\Delta\bar{x}$	$\bar{x}$	$\Delta\bar{x}$
Compound	–	EPDM				NBR			
Temperature	°C	180		220		180		220	
Sample area	mm^2	177.5	11.1	265.2	62.0	256.2	4.0	244.9	3.8
Average cell diameter	mm	0.39	0.03	0.47	0.06	0.039	0.01	0.37	0.01
Porosity	%	44.8	5.7	49.2	3.0	39.3	2.3	41.4	1.6
Number of cells	–	431	60	502	80	609	29	660	28
Distance to neighbour cell	mm	0.42	0.01	0.48	0.06	0.44	0.01	0.41	0.01
Cell density	1/mm^2	2.44	0.34	2.00	0.48	2.38	0.11	2.70	0.15

Table 5 shows exemplary cross-sections of the foamed profiles. It can be derived that the vulcanisation temperature has no impact on the cell structure of the EPDM. At a vulcanisation temperature of 200 °C, a very fine and homogeneous cell structure of NBR can be achieved. EPDM foamed with the chemical reference blowing agent ADCA at 200 °C has a more homogeneous and fine-pored closed-cell structure than with the hydrate as blowing agent. The foam structure achievable with water as blowing agent in NBR at 200 °C is finer than the chemically foamed reference sample with ADCA.

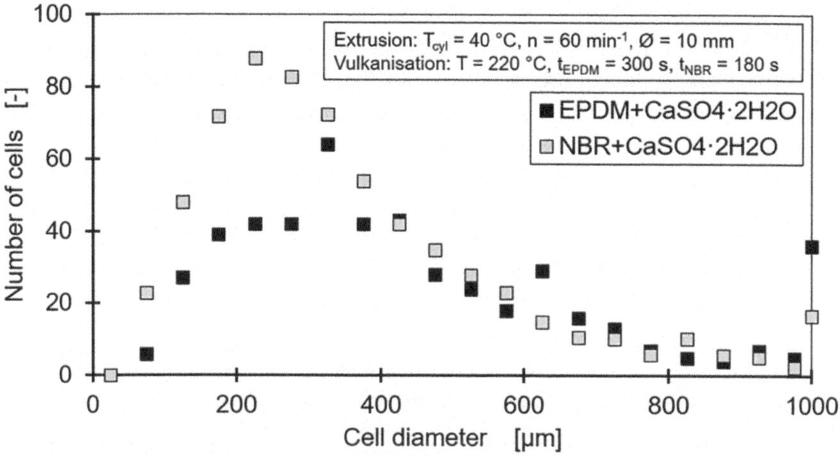

Fig. 5. Cell size distribution of water foamed EPDM and NBR compounds

Table 5. Effect of the vulcanisation process on the foam structure

Compound	180 °C	200 °C	220 °C
EPDM+$CaSO_4 \cdot 2H_2O$			
EPDM+ADCA		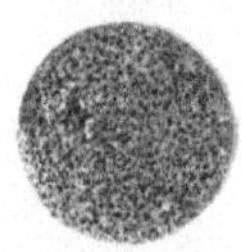	
NBR+$CaSO_4 \cdot 2H_2O$	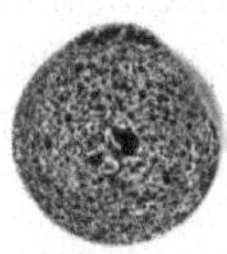	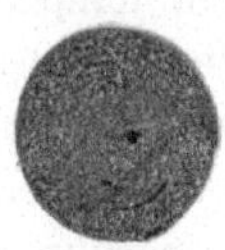	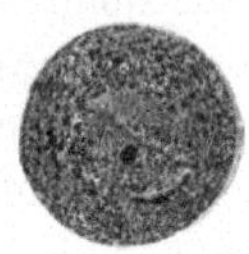
NBR+ADCA		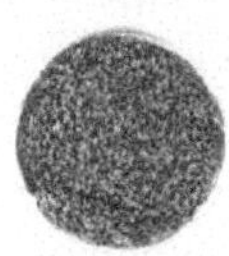	

The mechanical properties as a function of the vulcanisation temperature are shown in Fig. 6. The tension set of the EPDM remains almost constant between 20% and 25% when the temperature is increased. The tension set of the NBR decreases significantly with a temperature increase from 180 °C to 220 °C from about 60% to 20%. At the same time, the spring constant increases from 23 N/mm to 42 N/mm. In contrast, the spring constant of the EPDM foamed with water decreases with rising temperature from 20 N/mm to 12 N/mm. The spring constant of the chemically foamed EPDM is 28 N/mm at a vulcanisation temperature of 200 °C and 48 N/mm for the chemically foamed NBR and thus higher than the samples foamed with water.

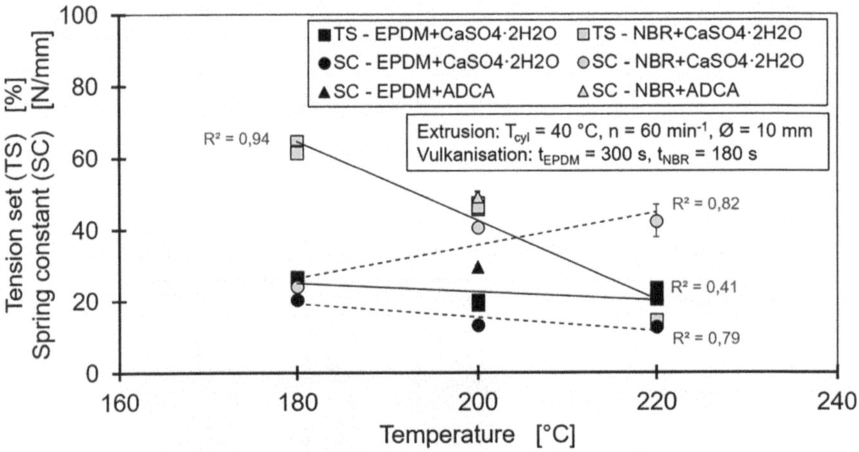

Fig. 6. Effect of the vulcanisation process on the tension set

4 Discussion

Regarding surface quality, the temperature in the extrusion process has a significant influence. This can be explained by the temperature dependency of tensile strength and viscosity. The bulk temperature is strongly influenced by the parameters of the extrusion process. It would be expected that a cold barrel wall would lead to higher specific throughputs [2, 20] resulting in lower temperatures and higher wall shear stress and thus exceeding the tensile strength of the rubber. This trend is not recognisable for the measured throughput but for the surface quality.

Surface defects with cracks occur at low temperatures. At higher die exit temperatures bubbles break through the surface and cause defects. The continuing good surface quality of the NBR compound at a set barrel temperature of 60 °C can be explained on the one hand by the higher viscosity compared to EPDM and on the other hand by the faster crosslinking kinetics [17, 18], so that bubbles cannot penetrate through the rapidly crosslinking surface due to the higher deformation resistance of the matrix. While the crosslinking rate increases strongly with temperature, the decomposition of the hydrate is linked to a fixed temperature range between 100 °C and 135 °C. This leads to a premature blowing reaction at mass temperatures above 100 °C with simultaneously low cross-linking rates. The same effect can be observed with too low heating rates in the profile cross-section if the blowing reaction takes place faster than the cross-linking.

The extrusion process shows no significant influence on the investigated mechanical properties although the foam structure is influenced by the extrusion process by the effects described above. This can be explained by the fact that the mechanical properties are dominated to a certain extent by the degree of cross-linking and are less dependent on the resulting foam structure. This also coincides with the findings regarding the interrelationships between foam structure and mechanical properties [21].

The cell structure analysis showed that the cross-section of the EPDM profile and the porosity increase significantly with increasing vulcanisation temperature. This can be linked to the initially decreasing viscosity when the profile is heated, the initially slow cross-linking reaction and the relatively early blowing reaction, resulting in excessive bubble growth. With the faster cross-linking of the NBR mixture, it can be derived that a smaller foam cross-section with many smaller cells is produced at a temperature of 220 °C due to the lower incubation time and thus faster completion of the cross-linking reaction. The faster crosslinking reaction compared to EPDM leads to the formation of many small cells due to the higher resistance of the matrix, so that coalescence is avoided. In comparison to the chemical reference blowing agent ADCA, a fine-cell foam structure comparable to ADCA can be achieved using the NBR mixture recipe, which is already well suited to the process.

The mechanical properties of the EPDM in case of the tension set do not change significantly with the increase in the vulcanisation temperature. This can be attributed to the sufficiently long crosslinking time. The slightly decreasing spring constant can be explained by the slightly coarser cell structure of the EPDM. The decreasing tension set of the NBR with increasing temperature shows that the crosslinking is only completed after 180 s at temperatures above 220 °C. The spring constant of the foamed NBR increases with temperature. This can be explained on the one hand by the finer-cell foam structure of the NBR as the temperature rises, and on the other hand by the higher degree of crosslinking. Compared to the ADCA foamed compounds, the water foamed EPDM and NBR profiles have a slightly lower spring constant. This is due both to the coarser cell structure of the EPDM and to the lower degree of crosslinking. With ADCA as the reference blowing agent, the same EPDM masterbatch achieved tension set values between 10% and 20% [21] under comparable vulcanisation parameters. The lower degree of crosslinking of water-foamed samples was also observed in the measurement of crosslinking isotherms [18, 19] and can be attributed to interactions with the accelerator system.

5 Summary and Conclusion

Systematic investigations are carried out to evaluate the influence of relevant parameters of the process on the resulting quality characteristics of water-foamed elastomers. It is derived that due to dissipative heating and heat conduction from the extruder into the rubber, the tendency to surface defects such as cracks and bubbles increase. The blowing reaction takes place in a defined temperature range between 100 °C and 135 °C and the crosslinking speed increases disproportionately with temperature. This results in a coarse cell structure at low melt temperatures of about 60 °C due to slow heating. A coarse cell structure is also induced at high bulk temperatures of up to 100 °C at the die outlet due to an early blowing reaction. Because of the higher viscosity and the lower incubation time of NBR, a finer cell structure was obtained compared to EPDM. Although the bulk temperature before entering the vulcanisation unit has an influence on the foam structure, the mechanical properties are dominated by the cross-linking process. Analogous to the results from the extrusion tests, the slower curing reaction and lower viscosity the coarser the foam structure. Compared to the NBR compound the EPDM shows a coarser foam structure at with higher curing temperatures.

The tension set does not depend on the foam structure but on the degree of crosslinking and thus on the vulcanisation temperature and duration. The spring constant correlates with the foam structure as a measure of hardness as well as with the degree of crosslinking.

It has been shown that it is possible to extrude foamed rubber profiles with water as a physical blowing agent. The process investigations have so far only been carried out on a laboratory extruder. On an industrial scale machine with a larger screw diameter, higher dissipation and lower temperature homogeneity can be expected, so that the investigation of the temperature sensitivity of chemical blowing agents and water based blowing agents should be validated.

With regard to the achievable quality parameters, it is purposeful to adapt both the compound recipe and the processing process. With the NBR formulation as shown, a fine-cell foam structure with similar properties as with the reference blowing agent can already be produced. A fine cell structure of the water-foamed EPDM can be achieved by adapting the recipe regarding the accelerator system from an inhibited sulfone amide (CBS) to a faster mercapto benzo thiazole (MBT).

Acknowledgments. The depicted research "Fundamental research on foaming of elastomers with water as physical blowing agent and description of mechanisms of foaming" (HO4776/32-1) has been funded by the Deutsche Forschungsgemeinschaft (DFG). We would like to extend our thanks to the DFG.

We would also like to thank Hexpol Compounding s. p. r. l, Evonik Ressource Efficiency GmbH, Sigma Aldrich GmbH, Avokal GmbH and H-JG Consulting for their support.

References

1. Klempner, D., Frish, K.C.: Handbook of Polymeric Foams and Foam Technology. Hanser, Munich (1998)
2. Röthemeyer, F., Sommer, F.: Kautschuk Technologie: Werkstoffe – Verarbeitung – Produkte. Hanser, Munich (2013)
3. Wahlen, L.: Chemical blowing agents – Properties and applications. In: Proceedings of the Blowing Agents and Foaming Processes, Munich (2006)
4. Haberstroh, E., Kremers, A., Epping, K.: Extrusion von physikalisch geschäumten Kautschukprofilen. Kaut. Gummi Kunst. **58**, 449–454 (2005)
5. Kremers, A., Krusche, A., Haberstroh, E.: Analysis of the production of sponge rubber profiles. Macromol. Mater. Eng. **284**, 70–75 (2000)
6. Maier, R.D., Schiller, M.: Handbuch Kunststoff-Additive. Hanser, Munich (2016)
7. Mergenhagen, T.: Chemical foaming agents in thermoplastics and thermosets. In: Proceedings on Blowing Agents and Foaming Processes, pp. 1–6, Mainz, Germany (2013)
8. N.N.: Plastic Additives: Blowing Agents. Product brochure, Lanxess AG, Leverkusen (2004)
9. N.N.: REACH – Registration, Evaluation, Authorisation and Restriction of Chemicals. (EG) Nr. 1907/2006, European Comission (2006)
10. N.N.: Revision of Annex XIV of Regulation (EG) Nr. 1907/2006, European Comission (2006)
11. Tomasko, D., Li, H., Liu, D., Han, X. et. al: A review of CO_2 applications in the processing of polymers. Ind. Eng. Chem. Res. 42(25), pp. 6431–6455 (2003)
12. Kropp, D., Michaeli, W., Herrmann, T., Schröder, O.: Foam extrusion of thermoplastic elastomers using CO_2 as blowing agent. J. Cell. Plast. **4**(34), 304–313 (1998)
13. Sitz, S.: Extrusion of physically foamed silicone rubber profiles. Dissertation, RWTH Aachen (2015). ISBN: 978-3-95886-054-4
14. Dumbauld, G.: US5070111A, Foaming Thermoplastic Elastomers. US Patent 12.03.1991
15. Sahnoune, A.: Foaming of thermoplastic elastomers with water. J. Cell. Plast. **37**, 149–159 (2001)
16. Hopmann, C., Lemke, F., Binh, Q.: Foaming of EPDM with water as blowing agent in injection molding. J. Appl. Polym. Sci. **133**(43613), 1–11 (2016)
17. Hopmann, C., Kammer, S.: Identification of suitable water carrying substances for the physical foaming of rubber. In: Rubber Meets Science 2017, Aachen (2017)
18. Hopmann, C., Kammer, S.: Analysis of processing relevant material properties of modified rubber compounds for the physical foaming with water. In: DKT2018, German Rubber Conference, Nuremberg, (2018)
19. Hopmann, C., Kammer, S.: Foam extrusion of elastomers using water as physical foaming agent. In: KHK, 13th Rubber Fall Colloquium, Hannover (2018)
20. Limper, A., Barth, P., Grajewski, F.: Technologie der Kautschukverarbeitung. Hanser, Munich (1989)
21. Hopmann, Ch., Kammer, S., Lemke, F.: Investigations on the influence of vulcanization parameters on the foam structure and mechanical properties of expanded EPDM. Kaut. Gummi Kunst. KGK **71**(11/12), 16–22 (2018)
22. Hopmann, C., Kammer, S.: Towards a digitalisation of the continuous vulcanisation process. In: Rubber Meets Science 2019, Aachen, (2019)
23. N.N.: DIN ISO 7619-1: Elastomere oder thermoplastische Elastomere – Bestimmung der Eindringhärte – Teil 1: Durometer-Verfahren (Shore-Härte). Beuth, Berlin (2012)

24. N.N.: DIN ISO 2285: Elastomere oder thermoplastische Elastomere – Bestimmung des Zugverformungsrestes unter konstanter Dehnung und des Zugverformungsrestes, der Dehnung und des Fließens unter konstanter Zugbelastung. Beuth, Berlin (2013)
25. Peters, R.: Schaumstrukturanalyse mit digitalen Bildverarbeitungsmethoden. Dissertation, RWTH Aachen (2003)
26. Hopmann, C.: Einstellung der Zellstruktur von chemisch geschäumten Kautschukprofilen durch Optimierung von Verarbeitungsparametern und Polymerstruktur. Institut für Kunststoffverarbeitung, RWTH Aachen, Schlussbericht IGF-Forschungsvorhaben Nr. 16701 N (2015)

Homogenisation of the Wall Thickness Distribution of Thermoformed Cups by using Different Pre-stretch Plugs and Process Parameter Settings to Improve Material Efficiency

Dennis Balcerowiak(✉) and Christian Hopmann

Institute for Plastics Processing in Industry and Craft, RWTH Aachen University, Aachen, Germany
Dennis.Balcerowiak@ikv.rwth-aachen.de

Abstract. In thermoforming up to 90% of the product costs are related to the material. For this reason, the efficient use of materials is essential in the production. Material efficiency of thermoformed packaging products, especially cups, can be increased by using pre-stretching plugs. These plugs affect the wall thickness distribution of the product. Thin areas, which can cause problems concerning top-load stability, can be avoided. By adjusting the shape of the pre-stretch plugs and the thermoforming process the wall thickness distribution changes significantly. Thus the distribution can be homogenised and the wall thickness can be increased more than 100% in the thinnest area. The plug design is varied in wall angle and the radius which joints the side wall and the flat top of the plug. Using DoE the effects of different geometry combinations and sheet temperatures are analysed. It can be shown, that the wall angle of the plug and the sheet temperature have a big influence on the homogenisation. Top-load stability is an important feature of packaging cups. By adjusting the wall thickness distribution, the top-load stability can be raised significantly. This high top-load stability in turn allows a reduction of the sheet's thickness, increasing material efficiency. Additionally, the required plug forces to pre-stretch the sheet are investigated, which can improve the process control or serve as a parameter for quality management.

Keywords: Material efficiency · Plug-assisted thermoforming · Wall thickness distribution

1 Introduction and Motivation

The constantly increasing demand on the sustainability of plastic products requires higher material efficiency. Especially for packaging products it becomes more important to reduce the amount of waste. Thermoforming is one of the most important technologies to produce thin-walled packaging products. The wall thickness distribution

C. Hopmann and R. Dahlmann (Eds.): *Advances in Polymer Processing 2020*, pp. 79–92, 2020. https://doi.org/10.1007/978-3-662-60809-8_7

of these products has to be adjusted to use less material. In addition to sustainability, material savings are essential for thermoformed product because up to 90% of the costs depend on the material. Beside the material savings, the entire thermoforming process can be operated more efficiently. The required energy and the cycle time decrease if less material has to be heated and cooled after forming [1–5].

Beside temperature profiling, plug assisted thermoforming is the leading approach to influence the wall thickness distribution [2, 5–8]. In negative thermoforming process pre-stretching plugs are used to draw the plastic material into areas where otherwise thin areas would occur, because of locally different stretching ratios. Therefore, thin spots must be taken into account when designing thermoformed products because they are regarded as weak points of the product [3]. Thickening those weak points, for example, leads to higher top-load stability of packaging products. Also due to volume constancy of the sheet, thicker areas of the product are less oversized because the material is distributed more homogeneously. When using pre-stretched plugs, however, the resulting wall thickness distribution depends strongly on the geometry of the plug, the process parameters and the plastic material [5–7, 9–12]. First, the design of pre-stretch plugs and the resulting wall thickness distributions of those different plugs are described. Afterwards it will be shown that different wall thickness distributions lead to different cup properties and that measured plug forces might be used as a quality control parameter.

2 Design of Pre-stretch Plugs in Thermoforming

There are numerous publications on plug-assisted thermoforming. These publications show a varity of wall thickness distributions of different plugs used in plug assisted thermoforming processes. However, the local wall thicknesses are very difficult to compare because of different parameters like plug and mould geometries, plastics materials and sheet temperatures used in these publications [3, 7, 10, 14–17]. Due to the complex process and plug/material interactions only a few guidelines for designing pre-stretch plugs exist. In addition, these guidelines are very general and no information is given on the resulting wall thickness distribution [6, 8]. Alternatively, simulation might be employed (e.g. T-Sim), but material data acquisition and determination of friction coefficients between sheet and plug are complex and demanding. For this reason, tests will continue to be carried out to demonstrate the influence of plug geometry and various process parameters on wall thickness distribution.

Different pre-stretch plugs are shown in Fig. 1. They are designed based on the guideline of Schwarzmann [6, 8]. Further information can be found in the corresponding literature of [6, 8] and the publications of [5, 18] for design specifications and the exact design procedure. The plug geometry depends on the used negative thermoforming mould and thus of the investigated cup geometry. The chosen cup has an opening diameter of 60 mm, a depth of 40 mm and a draft angle of 4°. This leads to a plug with a 4° wall angle and a 3 mm radius [6]. The radius connects the side wall and the flat bottom of the plug. Based on the plug designed according to the guideline the other plugs are changed in wall angle and the radius. The used plug material is a syntactic foam (type: Hytac W, manufacturer: CMT Europe BV, Waalwijk/Netherlands).

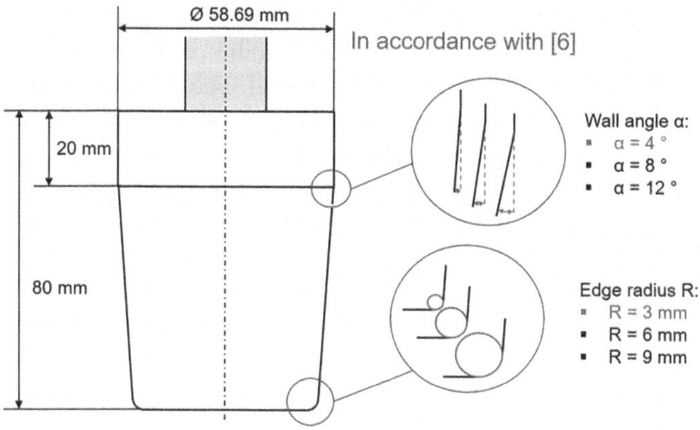

Fig. 1. Selected pre-stretch plugs to analyse the resulting wall thickness distribution [5]

3 Process Parameters and Evaluation Methodology

All tests mentioned are performed on the thermoforming machine Kiefel KD 20/25, Kiefel GmbH, Freilassing, Germany. Beside the sheet material and temperature, the pre-stretching distance (plug displacement), the plug speed, the delay in activation of compressed air and the geometry of the plug have an influence on the wall thickness distribution. The pre-stretching distance is selected according to the guideline [6]. The speed of the plug influences the stretching resistance during forming. Generally, the speed is set as high as possible to minimise cycle time and to achieve a low heat transfer into the plug [7]. With the used mould technology the activation time of the compressed air has no significant influence on the wall thickness distribution [18]. The sheet temperature and gradient are measured with a pyrometer and contact thermometer to adjust emitter settings with the objective of a homogeneous temperature distribution. The chosen constant process parameters are shown in Table 1:

Table 1. Constant processing parameters

Parameter	Value and entity
Pre-stretching distance	37.8 mm of 40 mm
Plug velocity	273 mm/s
Delay in compressed air activation	After 100% of pre-stretch distance
Forming pressure	5 bar
Time of forming pressure	3 s
Plug temperature	35 °C

The resulting wall thickness distribution of different plugs used in the thermoforming process is analysed along a measuring path over the entire cup cross-section. Average values are taken after one side is folded to the other side. For a better overview only one half of the cup is displayed as shown in Fig. 2, which shows a wall thickness distribution of a cup formed only with compressed air is displayed. Measuring location 1 (ML 1) is located in the middle of the cup base (bottom) and ML 14 on the cups top end near the edge of the wall. The thin spot at ML 6 and ML 7 is conspicuous. This thin spot must be specifically avoided and the material has to be less thinned out. Since the wall thickness distribution varies if different materials are used, polystyrene (PS) and polypropylene (PP) are investigated. Both materials are common for packaging applications. Because of different stretching properties and interactions between sheet and plug, different wall thickness distributions are to be expected [3]. The sheets have a thickness of 0.8 mm. The sheets are produced by W.u.H. Fernholz GmbH & Co. KG, Meinerzhagen, Germany. Per experiment five cups are produced.

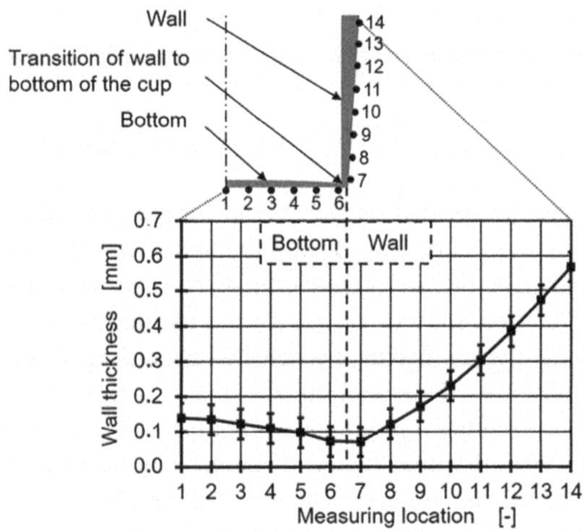

Fig. 2. Measuring locations on the cup and representation of a polystyrene cup formed only with compressed air [5]

4 Analysis of the Wall Thickness Distribution using Different Pre-stretch Plug Geometries

Since there are many different plug geometries to investigate and the sheet temperature also has a particularly large influence on the wall thickness distribution, it is difficult to determine which parameter has the biggest effect on the homogeneity of the resulting cup. In order to quantify the influence of the various geometric parameters and the sheet temperatures on homogeneity of the cups, a DoE test plan is drawn up. Since interactions of different parameters cannot be ruled out a full factorial DoE is used [19]. The test plan is shown in Table 2. Each parameter is changed from test to test, so that in total 27 parameter configurations are performed:

Table 2. Test plan of design of experiments to analyse the influence on the resulting wall thickness homogeneity

Parameter	Unit	Levels
Wall angle	[°]	4/8/12
Edge radius	[mm]	3/6/9
Sheet temperature of PP	[°C]	120/126/132

Figure 3 shows the resulting wall thickness distribution when using PS at 120 °C sheet temperature. As shown, different angles of the plug (4, 8 and 12°) with a constant edge radius of 9 mm lead to different wall thickness distributions. The wall thickness can be influenced significantly with variation of the wall angle.

The influence is conspicuous with regard to the quenching marks. Using plugs with a wall angle of 4° and 8° thicker areas occur at ML 8 and ML 10 compared to the rest of the cup. The quenching marks can be minimised by increasing the wall angle. Because of the higher angle the material no longer gets into contact with the plug during stretching. The sheet can be stretched freely because there is more free space between the plug and the mould and therefore it does not cool immediately by touching the plug. In addition, more freely stretchable areas lead to more equal local stretching and thus to increased homogeneity. Depending on the plug geometry, the minimum wall thickness can be more than doubled from 0.05 mm (Fig. 2) to 0.13 mm (Fig. 3) compared to pure compressed air forming. Changing the edge radius while keeping the wall angle constant has less influence on the wall thickness distribution [18]. For more detailed investigations of the resulting wall thickness distributions of PS, see [5].

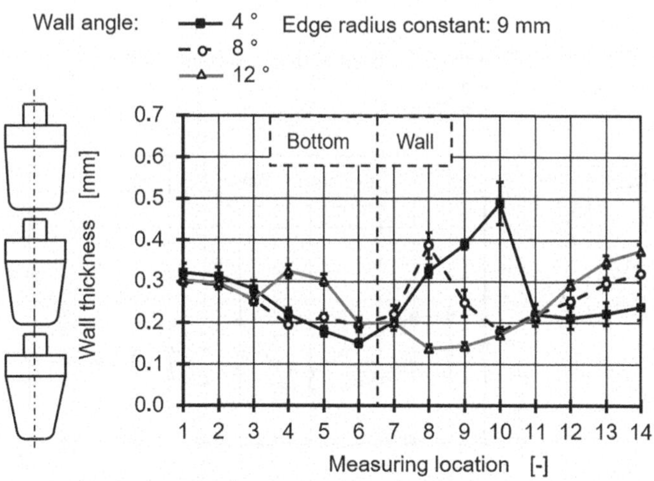

Fig. 3. Influence of the plug wall angle of flat plugs with constant edge radius when using PS [5]

If using different sheet materials, the determined wall thickness distributions of PS cannot be transferred easily to the new material. Therefore, wall thickness distributions of PP are shown.

Because of different stretching resistances compared to PS different amounts of material will be pulled off the plug, which leads to different local stretching ratios. Figures 4 and 5 show the resulting wall thickness distributions using PP at a sheet temperature of 126 °C.

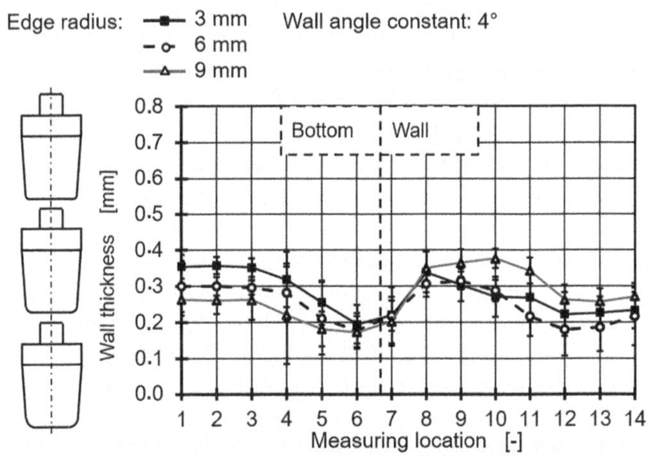

Fig. 4. Influence of the plug edge radius of flat plugs with constant wall angle when using PP

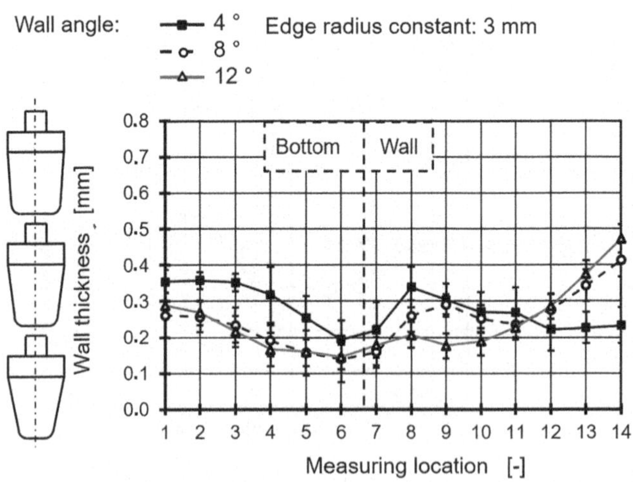

Fig. 5. Influence of the plug wall angle of flat plugs with constant radius when using PP

As determined with PS in [18], the radius has a smaller influence on the wall thickness distribution than the wall angle of the plug when using PP. The distributions do not change a lot over the different measuring locations. Generally, a bigger edge radius leads to more material pulled off the plug, because the material is easier to pull around bigger radii as the deflection of sheet around a small radius requires a higher force (Fig. 4). If 8° and 12° angles are used, the effects of the radii are less significant, because it is even more easy to pull the material off the plug with bigger wall angles. So no clear statement can be made about the influence of the edge radius, since it differs a lot dependent on the used wall angle. There are no big differences in the profile.

Variating the wall angle has a greater influence on the resulting distributions (Fig. 5). The material in the bottom is more stretched when using a wall angle of 12° compared to a wall angle of 4°. With a 12° angle the material is more easily pulled around the plug radius compared to the 4° angle even if the radius itself is constant (3 mm). The wall thickness can be more influenced by changing the wall angle than by radius because of different sizes of free stretchable areas which are not in touch with the plug. Thus different stretching resistances occurs and leads to different deflection off the plug.

In general, pre-stretch plugs allow the wall thickness distribution to be adjusted or even to be able to form products. In these tests it is not possible to produce PP cups only with compressed air. The wall thickness in the transition area of the cup is that thin that the stability is not sufficient to measure with the used Magna Mike 8600 of Olympus Deutschland GmbH, Hamburg, Germany, because the sheet tears apart during the measurement.

Altogether it can be stated that the wall thickness profile itself is influenced by the change of wall angle. Changing the wall angles lead to profiles which differ significantly or at least more than the change of the edge radius does. Using pre-stretch plugs enables to change the wall thickness distribution and makes it possible to homogenise it. The thinnest areas can be greatly increased compared to cups produced only with compressed air.

5 Analysis of Homogeneity of the Wall Thickness Distribution

To describe the homogeneity of different wall thickness distributions a mathematical criterion has to be defined. The wall thickness distributions give a good impression of the local thickness depending on the various measuring locations, but the diagrams are not suitable for conclusions about the homogeneity of the distribution. Equation 1 is one possibility to describe the homogeneity of a formed cup. Δs is described as the difference between the thickest (s_{max}) and the thinnest (s_{min}) wall thickness of the cup according to [20]. If the difference is low the wall thickness is almost the same at all measuring locations and thus does not differ in huge ranges.

$$\Delta s = s_{max} - s_{min} \quad (1)$$

The observed effects based on the DoE test plan described in chap. 4 are shown in Fig. 6. It can be stated that the change of the wall angle and the interaction of wall angle and temperature have the biggest influence on the homogeneity of the wall thickness distribution. If the wall angle is increased the difference of the thickest and thinnest wall thickness of the cup increases more than 0.05 mm. This means that the produced cup is less homogenous. Whereas changing the wall angle and sheet temperature from low to high setting leads to the best wall thickness distributions. The homogeneity can be increased up to 0.1 mm. Both of the effects are significant. The other parameters are less significant or indifferent.

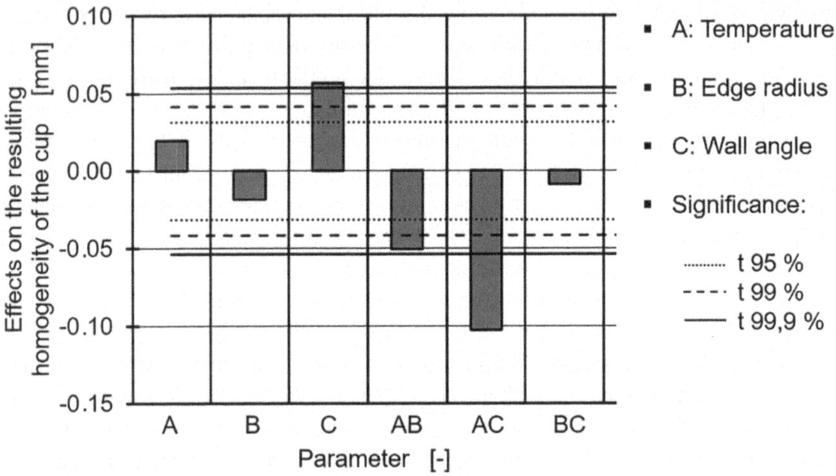

Fig. 6. Effects and interactions of the different parameters on the resulting homogeneity of the cups when using PP

In Fig. 3 increasing the wall angle of the plug leads to a more homogenous wall thickness distribution when using PS whereas it decreases when using PP. Comparing these two materials it can clearly be seen that effects cannot be based on process or geometry parameters only. The material has a great influence on the characteristics of effects. Figure 5 shows that an increase of the wall angle leads to a thicker side wall and a thin bottom. In this case the homogeneity is not as good as with a 4° wall angle. This is contrary to the explanation that more freely stretchable areas lead to homogenous wall thickness distributions due to more equal local stretching ratios (like seen when using PS). Up to now, no guideline is established, which is able to predict the influence on the wall thickness distribution using different materials.

In addition to the flat plugs more geometrically complex plugs are investigated. These plugs cannot be considered in DoE, because they are no flat plugs and thus cannot be described in the used geometrical pattern. These plug geometries are common in thermoforming and their influence on the wall thickness distribution is investigated as well [6, 8].

Pointed plugs which have no flat area at the bottom are constructed (Fig. 7). One pointed plug with a wall angle of 12° and one with 17° to be more sharp-ended. In addition, more complex plugs are investigated. These so called warts are separated into flat and convex warts (Fig. 7). They are developed based on the plug designed by the guideline, too. They own an undercut which prevents the sheet from getting in contact with the plug in this area and thus can avoid quench marks.

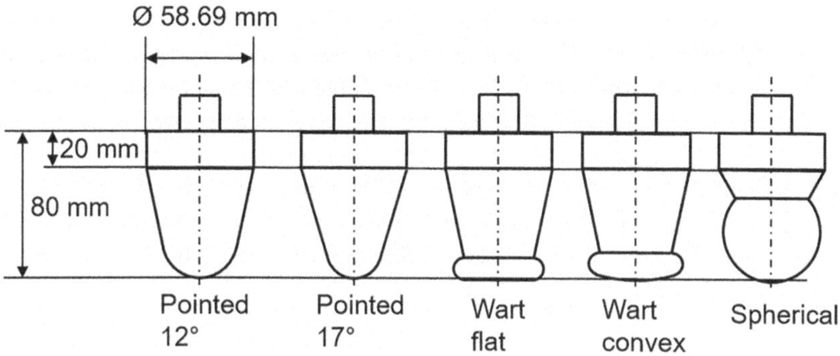

Fig. 7. Pointed plugs, warts and spherical pre-stretch plugs to analyse the influence on the wall thickness distribution

The wall thickness distributions of pointed plugs are shown in Fig. 8 at a PP-sheet temperature of 126 °C. Less material is pulled out of the bottom area if pointed plugs are used. This leads to particularly thick bottom areas. The larger freely stretchable area causes different stretching resistances. Since more material is available for free stretching (without plug contact), the local stretching resistances of the sheet decrease and less material is pulled out of the bottom. The required forces for stretching the freely areas are lower than those forces which are needed to exceed the adhesive forces between the sheet on the plug. The profiles of the wall thickness distributions are qualitatively the same when using pointed plugs as well as when using PS and PP [5, 18]. The wall thickness can be improved from not formable to at least 0.11 mm.

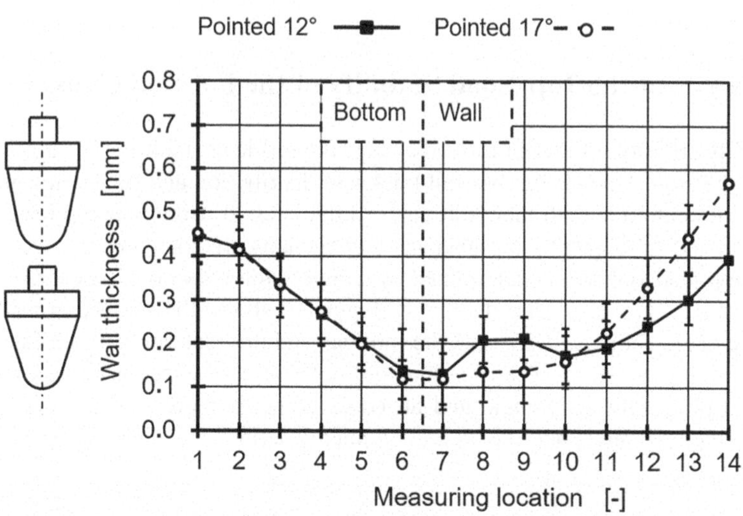

Fig. 8. Influence of pointed plugs on the wall thickness distribution when using PP [5]

Figure 9 shows the resulting wall thickness distributions of the complex plug geometries. All three geometries lead to better material distributions. The wall thicknesses change significantly to those of flat or pointed plugs due to different stretching conditions. These stretching conditions cannot be explained properly because of the complex sheet and plug interaction due to changing contact areas during forming. No clear statements can be made in comparison to flat plugs, because a defined difference between freely stretchable sheet areas and areas with plug contact does not exist. It should be shown that complex plugs can lead to approximately the same or even more homogeneous wall thickness distributions.

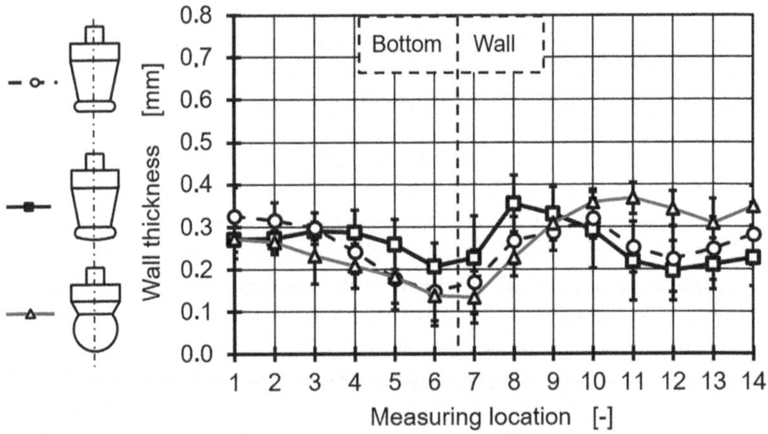

Fig. 9. Influence of complex plug geometries on the wall thickness distribution when using PP

6 Analysis of the Top-Load Stability of the Formed Cups

One quality criterion of thermoformed cups is the top-load stability. The top-load stability of the cups depends on the wall thickness distribution and the homogeneity. To test the different cups with their various wall thickness distributions top-load test are carried out and subsequently top-load and homogeneity are compared.

The top-load stability is determined by compression tests on a Zwick Z10, Zwick Roell GmbH & Co. KG, Ulm, Germany. Figure 10 shows the maximum wall thickness differences (homogeneity) and the top-load stability determined as a function of different plug geometries. The used material is PP at a sheet temperature of 126 °C. 4°R6 is a plug with a 4° angle and an edge radius of 6 mm. So the first number describes the wall angle and the second number describes the edge radius. The tested cups are those with the three best homogeneities, one with an average and another one with a less homogeneity.

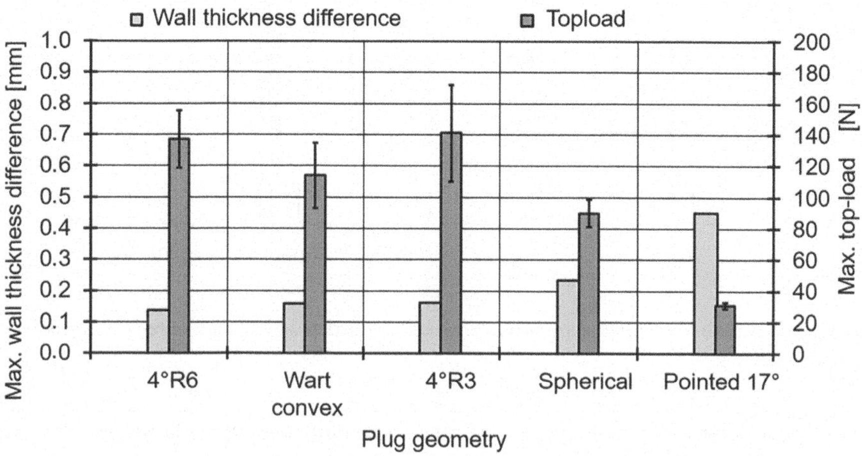

Fig. 10 Maximum wall thickness differences and top-load as a function of plug geometries [5]

Homogeneous cups can stand higher top-load forces (140 N). The higher stability results from thickening the thin areas. Cups with a very unequal wall thickness distribution produced with pointed plugs are weak in comparison, because of thinner wall thicknesses in the transition area (32 N). More homogeneous distributions lead to better mechanical properties of the cups in top-load tests. This can be confirmed with the selected cup geometry. The tendency that better top-load stability can be achieved with more homogeneous wall thickness distributions can also be confirmed with other sheet materials and sheet temperatures.

7 Analysis of Plug Force and Wall Thickness Distribution

Different process parameters can be established as control parameters in thermoforming machines. To pre-stretch the sheet the plug has to be pushed into the material. Depending on the plug geometry different plug forces are needed. In modern thermoforming machines the plug forces are measured and can be used as one control parameter [3]. To develop a further quality control criterion the plug forces are measured and compared to the resulting maximum wall thickness difference and thus homogeneity. In Fig. 11 the plug forces and homogeneity are determined as a function of different plug geometries.

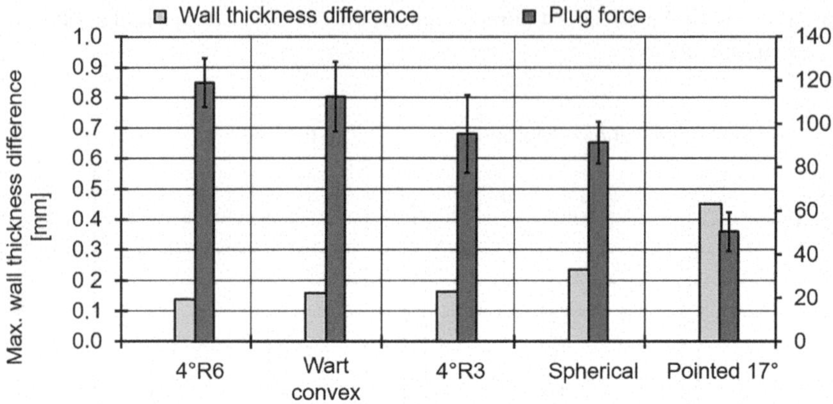

Fig. 11. Maximum wall thickness differences and needed plug forces to pre-stretch the sheet when using PP (126 °C)

Figure 11 clarifies that a high plug force is an indication for a homogeneous wall thickness distribution. The shown figure are the results of forming PP with a sheet temperature of 126 °C. The other plug forces while using a different material or a different sheet temperature show the same tendencies. As shown, the plug forces can be used as a quality control parameter in thermoforming machines if plug forces are detected and correlated with the resulting wall thickness distributions. A higher plug force is needed if the contact area of the plug and the sheet increases, thereby drawing more material towards the thin areas at the bottom wall. If the plug forces decrease when using different plug geometries, it is likely that the wall thickness distribution is inhomogeneous.

8 Conclusion and Outlook

The wall thickness distribution can be influenced by using different plug geometries and process parameters. The plug wall angle enables the adjustment and the homogenisation of the wall thickness distribution. In general, the profile can be influenced significantly. The plug edge radius has a comparatively little influence on the resulting wall thickness distribution.

The resulting wall thickness distributions are not possible to predict in detail even though the plug geometry is known before. They differ enormously depending on the sheet material and sheet temperature. The sheet temperature has a big influence because of greatly changed stretching resistances of the material. Using PS, a higher wall angle leads lower deviations of the wall thickness distributions whereas using PP a small wall angle homogenises the distribution. Pointed plugs lead to a poor wall thickness distribution for all materials.

The top-load stability can be increased for all cups, with the exception of the pointed plugs. The highest top-load stabilities are achieved with the most homogeneous wall thickness distributions. The wall thickness of the used sheet can be lowered retrospectively, if the top-load can be raised. The plug forces as well as the top-load stability correlate with the maximum wall thickness differences and thus might be used as a quality criterion or control parameter. In further trials it has to be investigated to what extent the results can be transferred to different mould geometries.

Acknowledgements. The research project (19342 N) of the *Forschungsvereinigung Kunststoffverarbeitung* has been sponsored as part of the "industrielle Gemeinschaftsforschung und –entwicklung (IGF)" by the *German Bundesministerium für Wirtschaft und Energie (BMWi)* due to an enactment of the *German Bundestag* through the AiF. We would like to extend our thanks to all organisations mentioned. Furthermore we would like to thank Kiefel GmbH, Freilassing, Marbach Werkzeugbau GmbH, Heilbronn, Kiefer Werkzeugbau GmbH, Schwaigern, and W.u.H. Fernholz GmbH & Co.KG, Meinerzhagen for the provision of equipment, mould technology and test materials.

References

1. Ederleh, L.: Simulative and experimental research on the forming behaviour of thermoplastics during thermoforming. Dissertation, RWTH Aachen, 2014, ISBN: 978-3-95886-026-1
2. Hopmann, C., Michaeli, W.: Einführung in die Kunststoffverarbeitung. Hanser, München (2015)
3. Moser, A.: Nutzung von Prozesswissen beim Thermoformen von Verpackungen. Dissertation, Universität Duisburg-Essen, 2013
4. Throne, J., Beine, J.: Thermoformen Werkstoffe – Verfahren – Anwendungen. Hanser, München (1999)
5. Hopmann, C., Balcerowiak, D.: Optimierung der Stempelvorstreckung im Thermoformen zur Steigerung der Materialeffizienz. Extrusion **25**(5), 38–42 (2019)
6. Illig, A., Schwarzmann, P.: Thermoformen in der Praxis. Hanser, München (2008)
7. Martin, P.J., Duncan, P.: The role of plug design in determining wall thickness distribution in thermoforming. Polym. Eng. Sci.: Wiley Inter Science **17**(4), 804–813 (2007)
8. Schwarzmann, P.: Typisch und unentbehrlich – Durch Vorformen die Wanddickenverteilung beim Thermoformen steuern. Kunststoffe **105**(4), 74–77 (2015)
9. Schwarzmann, P.: Wann haben Formteile Schreckmarken? – Ursachen, Einflüsse und Abhilfe bei Schreckmarken. Kunststoffe **105**(7), 74–76 (2015)
10. Bernard, C., Correia, J., Bahlouli, N., Ahzi, S.: Numerical simulation of plug-assisted thermoforming: Application to polystyrene. Key Eng. Mat. **554–557**, 1602–1610 (2013)
11. Marathe, D., Rokade, D., Azad, L., Jadhav, K., Mahajan, S., Ahmad, Z., Gupta, S., Kulkarni, S., Juvekar, V., Lele, A.: Effect of plug temperature on the strain and thickness distribution of components made by plug assist thermoforming. Int. Polym. Process. **2**(31), 166–178 (2016)
12. Collins, P., Harkin-Jones, E.M.A., Martin, P.J.: The role of tool/sheet contact in plug-assisted thermoforming. Int. Polym. Process. **17**(4), 361–369 (2002)
13. Haberstroh, E., Wirtz, J.: Helfer für die Becherformung. Kunststoffe **93**(12), 52–55 (2003)
14. Martin, P.J., Choo, H.L., O'Connor, C.P.K.: Measurement & modelling of slip during plug-assisted thermoforming. Key Eng. Mater. **504–506**, 1105–1110 (2013)
15. McCool, R., Martin, P.J.: The role of process parameters in determining wall thickness distribution in plug-assisted thermoforming. Polym. Eng. Sci. **10**(50), 1923–1934 (2010)
16. McCool, R., Martin, P.J., Harkin-Jones, E.: Process modelling for control of product wall thickness in thermoforming. Plastics, Rubber and Compos. **8**(35), 340–347 (2006)

17. O'Connor, C.P.J., Martin, P.J., Sweeney, J., Menary, G., Cato-Rose, P., Spencer, P.E.: Simulation of the plug-assisted thermoforming of polypropylene using a large strain thermally coupled constitutive model. J. Mater. Process. Technol. **213**, 1588–1600 (2013)
18. Hopmann, C., Balcerowiak, D.: Der Schreckmarke den Kampf angesagt. Kunststoffe **118**(11), 81–85 (2018)
19. Kleppmann, W.: Versuchsplanung – Produkte und Prozesse optimieren. Hanser, München (2017)
20. Hopmann, C., Martens, J., Balcerowiak, D.: Analyse der Temperaturprofilierung durch Masken zur Steigerung der Materialeffizienz im Thermoformen. Z. Kunststofftechnik/WAK **13**(2), 126–143 (2017)

Barrier Screw Design for High-Performance Plasticizing in Injection Molding

Thomas Köpplmayr[1(✉)], Dominik Altmann[2,3], and Georg Steinbichler[2]

[1] Engel Austria GmbH, Ludwig-Engel-Strasse, 1, 4311 Schwertberg, Austria
thomas.koepplmayr@engel.at
[2] Institute of Polymer Injection Moulding and Process Automation, Johannes Kepler University, Altenberger Strasse, 69, 4040 Linz, Austria
[3] Kompetenzzentrum Holz GmbH, (Wood K plus) – Biobased Composites and Processes, Altenberger Strasse, 69, 4040 Linz, Austria

Abstract. Short cycle times in the injection molding process due to fast and effective cooling of the molds extend the demands for novel screw concepts starting from double-flighted screws and leading to barrier screws with dispersive and distributive mixing elements. We present different high-performance screw concepts for various applications in the injection molding industry. Besides a comparative view of different screw designs, we focus on modelling the plasticizing process in barrier screws using our recently developed software S3 (Screw Simulation Software) to investigate both the melting and the melt conveying process in more detail. The three-dimensional helix shape of the screw channel consisting of feed, barrier and metering section can be approximated via an unwound screw channel. Melting performance, pressure build-up and temperature distribution can be calculated numerically based on physical models and finite-difference approximation schemes. The axial screw motion during the plasticizing process and the idle time during the injection and holding phase are considered as well. Our simulations show good agreement with experimental studies conducted with various screw sizes and barrier configurations.

Keywords: Injection molding · Plasticizing time · Barrier screw · Melt homogeneity

1 High-Performance Plasticizing

1.1 Comparison of Different Screw Concepts with Respect to Plasticizing Rate

The main difference between extrusion and injection molding is that in an injection molding machine, a reciprocating screw is used, and the screw does not rotate during the injection and holding phase. Therefore, we differentiate between recovery rate $\dot{m}_{plast}$ and total throughput $\dot{m}_{total}$:

C. Hopmann and R. Dahlmann (Eds.): *Advances in Polymer Processing 2020*, pp. 93–107, 2020. https://doi.org/10.1007/978-3-662-60809-8_8

$$\dot{m}_{plast} = \frac{m}{t_{plast}} \tag{1}$$

$$\dot{m}_{total} = \frac{m}{t_{cycle}} \tag{2}$$

While the recovery rate (shot weight m divided by plasticizing time t_{plast}) does not necessarily give any information about the melt homogeneity, the throughput (shot weight m divided by cycle time t_{cycle}) is used to define the residence time required to obtain a certain melt quality. Unwanted inclusions in the plasticized polymer melt can either be detected visually during a purging experiment or by our recently developed non-invasive ultrasound-based measurement system [1]. These inhomogeneities will reduce the visual and structural properties of the resulting injection-molded part.

The recovery rate can be divided by the screw speed (recovery rate given in gram per screw revolution) and plotted against the throughput. Figure 1 shows a comparison of different screw geometries with respect to conveying rate and melt homogeneity determined during a purging experiment. The recovery rate is determined by the specific screw design while the throughput is increased by increasing the screw speed. The so-called throughput limit (i.e. the minimum possible residence time for melting) is determined by the screw design as well. Typical general-purpose screws (GPS) can only be used for standard applications, where the cycle time is sufficiently long compared to the shot weight. To obtain both higher recovery rates and increasing throughput limits, more specific screw designs must be used. On the one hand, a more complex screw design increases the costs of the plasticizing system, but on the other hand, using screw designs with higher throughput limits provides the possibility of selecting a smaller screw diameter or reducing the cycle time.

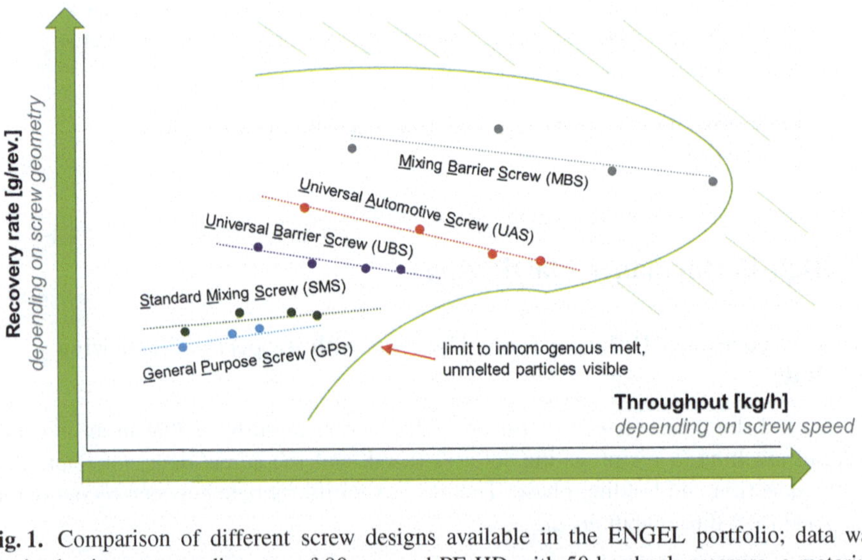

Fig. 1. Comparison of different screw designs available in the ENGEL portfolio; data was obtained using a screw diameter of 90 mm and PE-HD with 50 bar back pressure, a metering stroke of two times the screw diameter and a cycle time of two times the plasticizing time during a purging experiment.

Most of the melting occurs at the interface between the solid bed and the melt film. The melted material collects in the melt film and is dragged away to the melt pool ("drag-induced melt removal"). This mechanism was first described by Maddock [2].

The use of multiple flights can also improve melting, because the average melt film is thinner than in a single-flighted screw, see Fig. 2. However, the most stable melting process is achieved by using barrier screws. A barrier screw has two flights, a main flight and a barrier flight. The purpose of the barrier flight is to separate the solid material from the melt. Basically, there are two types of barrier sections [3]:

- In the melting zone the function of the barrier flight is to physically separate the solid bed from the melt pool. This type of barrier section is referred to as a barrier melting section.
- In the melt conveying zone the function of the barrier flight is to provide a high shear stress exposure in the narrow barrier flight gap. This type of barrier section is referred to as a barrier mixing section or a fluted mixing section.

A screw with a barrier melting section is usually referred to as a barrier screw, while a screw with a barrier mixing section is usually referred to as a mixing screw. Two examples for mixing screws are also shown in Fig. 1. The Standard Mixing Screw (SMS) is a double-flighted screw with a mixing section and provides similar flow rates, but a higher throughput limit compared to GPS. Optimizing channel height, helix angle and the total screw length can improve the melting and melt conveying capability (Universal Automotive Screw UAS).

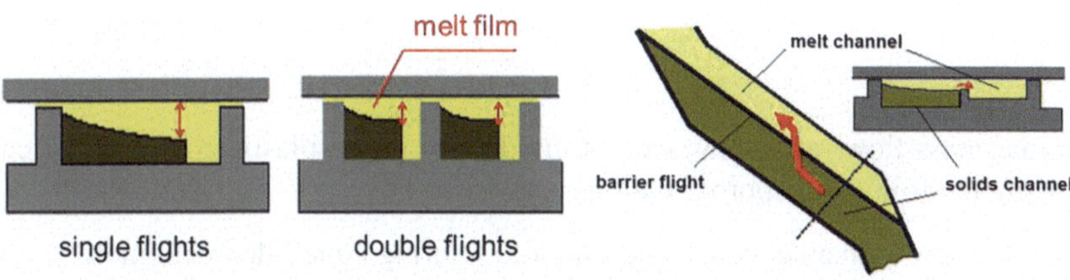

Fig. 2. Different screw concepts used in injection molding; the thickness of the melt film decreases from single-flighted screws to double-flighted screws and is minimized in barrier screws, where solids and melt are separated, which stabilizes the melting process.

1.2 Barrier Melting Sections

The inventor of the barrier screw is Charles Maillefer [4]. The main feature of a Maillefer screw is that the helix angle of the barrier flight is larger than the helix angle of the main flight. The continuous reduction in the width of the solids channel and the corresponding increase of the width of the melt channel allows a smooth and gradual change of the channel cross-section. However, the melting rate is directly determined by the width of the solid bed and the major drawback of a Maillefer screw is the lower melting rate compared to other barrier screw designs. The Universal Barrier Screw (UBS) mentioned in Fig. 1 is based on a Maillefer design without a mixing section.

Other barrier screw designs developed by Barr [5], Dray and Lawrence [6] or Kim [7] are characterized by a barrier flight which runs parallel to the main flight. The cross-section of the solids channel is then reduced by reducing the channel height while at the same time the channel height of the melt channel is increasing. This increases the solid-melt interface and improves the melting performance. However, more knowledge is required to avoid abrupt changes of the channel cross-section at the beginning and at the end of the barrier section. Deep melt channels require additional barrier mixing sections to assure good melt homogeneity. The Mixing Barrier Screw (MBS) mentioned in Fig. 1 is based on this design.

2 Simulation of Polymer Flow in Barrier Screws

2.1 Grid Generation and Modelling of the Plasticizing Process

Each zone of the barrier screw (feeding, barrier and metering zone) is defined via a specific axial length and height. The three-dimensional geometry is approximated via an unwound screw channel. With this reduction to a two-dimensional system, it is possible to reduce the computational power and therefore the simulation time. A third dimension for the timesteps Δt_{grid} has been introduced, considering the idle phase and the axial screw motion due to the metering stroke. The definition of a time grid depends on plasticizing time t_{pl} and pre-defined timesteps n (plasticizing and idle phase):

$$\Delta t_{grid} = \frac{t_{pl}}{\left(n_{pl} + n_{idle}\right) - 1} \tag{3}$$

The initial mass flow rate (show weight divided by initial plasticizing time) is calculated using the following approximations:

- unwound screw channel consisting only of metering zone filled with melt
- melt temperatures equals barrel temperature
- axial screw motion is neglected (extrusion mode)
- pressure gradient equals back pressure divided by channel length

To calculate the flow rate, a general equation can be formed, consisting of two major parts. The first part describes the volume flow rate due to drag flow and the second part due to pressure flow [8].

$$\dot{V}_{0,total} = \underbrace{\frac{1}{2} b \cdot h \cdot v_{0,z}}_{drag\,flow} - \underbrace{\frac{b \cdot h^3}{12\eta} \frac{dp}{dz}}_{pressure\,flow} \tag{4}$$

In this equation, the flow direction is z, b and h are the width and height of the unwound screw channel, $v_{0,z}$ is the velocity component in z-direction, η the melt viscosity and dp/dz the pressure gradient. Each contribution to the total flow rate can be corrected by (i) the leakage flow in the gap between the main flight and the barrel, and (ii) the influence of the screw flight for high *h*/*b* ratios, where the channel cannot be approximated by an infinite slit.

The Bird-Carreau-Yasuda model is used to describe the viscosity as a function of shear rate including all relevant regions (first plateau, shear thinning, second plateau):

$$\frac{\eta - \eta_\infty}{\eta_0 - \eta_\infty} = \frac{1}{\left[1 + (k \cdot \dot{\gamma})^a\right]^{\frac{(1-n)}{a}}} \tag{5}$$

where η_0 is the zero-shear rate viscosity, η_∞ the infinity shear rate viscosity, k a time constant, $\dot{\gamma}$ is the shear rate, a is defined as the Yasuda-parameter and n describes the power law index in the shear thinning region.

Newtonian and shear-thinning fluids have a common streamline in the flow channel at which the shear rate and hence the velocity is the same [9]. In other words, at a certain distance over the channel height (y-direction), the shear rate of a Newtonian and a shear-thinning fluid are equal. This is called the representative distance and using the corresponding shear rate, a local viscosity can be determined. This distance depends on the power law index, which can be calculated using tangents on the Bird-Carreau-Yasuda function (see Fig. 3). The temperature dependency is described by the Arrhenius shift, which mainly applies to semi-crystalline polymers [10]. To describe the specific volume as a function of pressure and temperature, the Tait model is used [11].

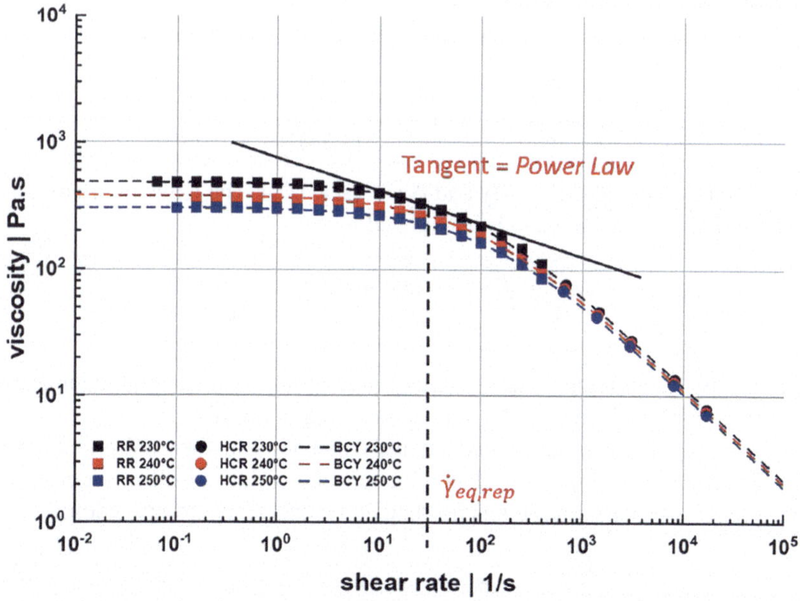

Fig. 3. Approximation of the viscosity curve by the slope of the tangent to the Bird-Carreau-Yasuda function, which is used to calculate the representative shear rate (RR … rotational rheometer, HCR … high-pressure capillary rheometer, BCY … Bird-Carreau-Yasuda model).

As a first approximation, the axial screw velocity $\dot{s}$ is calculated with the initial mass flow rate. Hence, the barrel velocity components must be adapted with the velocity components of the screw motion in x- (transverse) and z- (flow) direction depending on the helix angle φ.

$$\begin{aligned} v_{1z} &= v_0 \cdot \cos\varphi + \dot{s} \cdot \sin\varphi \\ v_{1x} &= v_0 \cdot \sin\varphi - \dot{s} \cdot \cos\varphi \\ v_1 &= \sqrt{v_{1x}^2 + v_{1z}^2} \end{aligned} \tag{6}$$

After an initial mass flow rate has been calculated, melting and viscous dissipation are also considered. Therefore, the melting process is simulated using the drag-induced melting model by Tadmor and Gogos [12]. If the barrel temperature exceeds the melting point of the polymer, a melt film is formed. Afterwards the solids bed reduces in width, whereas the melt pool increases. The pressure profile is calculated in the melt pool of the solids channel as well as in the melt channel along the flow direction.

A simple heat-source model is used in the screw channels, the melt film and the barrier gap to account for viscous dissipation. All relevant regions are shown in Fig. 4. The power consumption can be divided into following parts: drag-, transverse and pressure flow in the screw channel and leakage flow in the screw clearance. It should be mentioned that the transverse flow does not contribute to the volume flow rate, but to the energy dissipation.

An explicit time integration (3-point-formula) is applied to solve the transient heat conduction. The second derivative in y-direction (channel height) of a two-columns array (i_y, i_z) can be calculated, assuming an equidistantly spaced grid.

$$\frac{T_{i_y}^{i_z+1} - T_{i_y}^{i_z}}{\Delta z} = a \frac{T_{i_y+1}^{i_z} - 2T_{i_y}^{i_z} + T_{i_y-1}^{i_z}}{(\Delta y)^2} \tag{7}$$

with a defined as thermal diffusivity and $\Delta z \leq (\Delta y)^2/2a$. The temperature of one step forward along the z-direction can be calculated by:

$$T_{i_y}^{i_z+1} = T_{i_y}^{i_z} + \Delta z \cdot a \frac{T_{i_y+1}^{i_z} - 2T_{i_y}^{i_z} + T_{i_y-1}^{i_z}}{(\Delta y)^2} + \Delta z \cdot \eta\left(T_{i_y}^{i_z}, \dot{\gamma}\right)\dot{\gamma}^2 \tag{8}$$

In addition, the mean temperature is determined along the unwound channel length (1D in z-direction) using composite trapezoidal integration rule.

Finally, using a numerical procedure, the calculation of the total pressure differential in direction of the unwound channel is repeated with a modified mass flow rate, until the pre-defined back pressure is reached at the end of metering section.

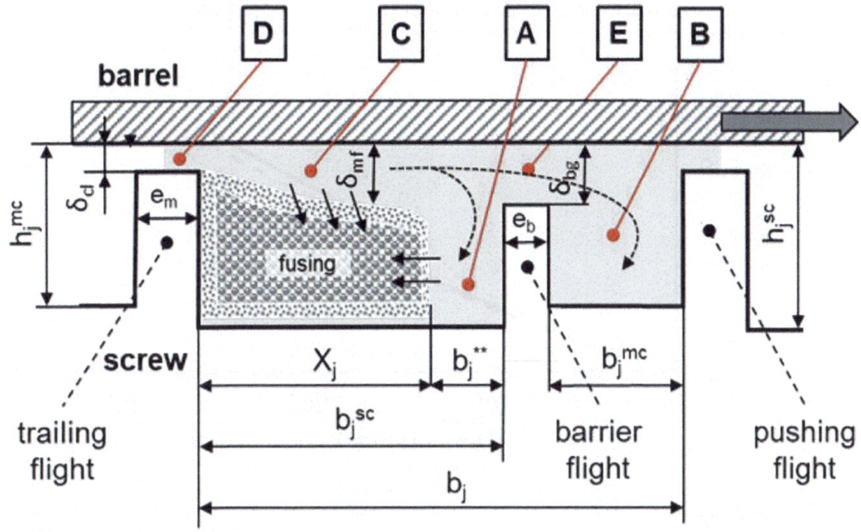

Fig. 4. Schematic overview of the different regions in a barrier zone considered in the simulation model: melt pool in the solids channel (A), melt pool in the melt channel (B), melt film between solid and barrel (C), leakage flow in the gap between main flight and barrel (D), melt flow in the barrier gap (E).

2.2 Material Characterization

For the comparison of calculation and experimental analysis, we used two types of polypropylene and one type of high-density polyethylene from Borealis (Schwechat, Austria) in this study.

The materials were rheologically characterized with a plate-plate rheometer (MCR 301, 25 × 1 mm) from Anton Paar (Graz, Austria) and with a high-pressure capillary rheometer (Rheograph 25) from Göttfert (Buchen, Germany). To capture the thermal stability all tests were carried out under nitrogen (N_2). The viscosity curves were fitted using Bird-Careau-Yasuda model and Arrhenius temperature shift with the parameters listed in Table 1. The enthalpy and heat capacity were measured with a Mettler Toledo DSC 822 (Vienna, Austria) in accordance with the standard ISO 11357-4 (saphire and blank curve). The results are shown in Fig. 5.

Table 1. Bird-Carreau-Yasuda and Arrhenius model parameters of the tested materials.

BCY parameter	Unit	PP-1	PP-2	PE-HD
η_0	Pa · s	384.4	326.2	2360.9
η_∞	Pa · s	0.0	0.0	0.0
k	s	0.01054	0.00823	0.02715
a	–	0.72898	0.40981	0.41424
n	–	0.24755	0.25000	0.33960
T_0	°C	240.0	230.0	220.0
E_0	J/mol	50566.1	42894.8	22605.4

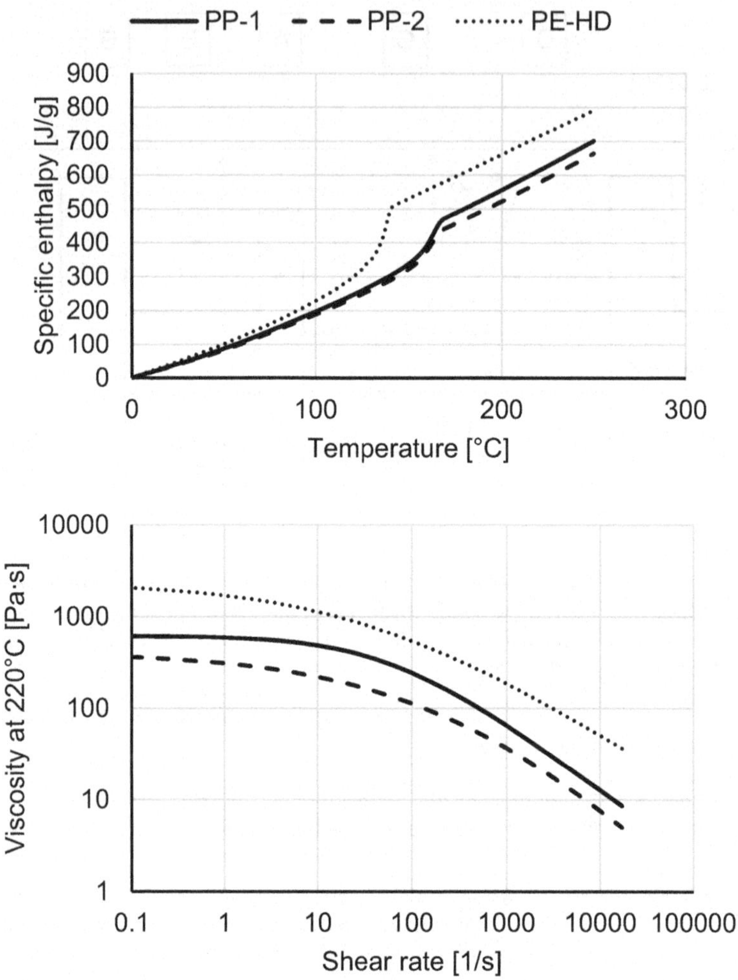

Fig. 5. Specific enthalpy and comparison of viscosity curves for all materials used in this study.

2.3 Experimental Evaluation of Simulation Results

For the experimental validation of the melting profile, so called pull-out tests are used. The traditional procedure is to stop the screw immediately and cool down the whole configuration consisting of barrel, screw and processed material. Afterwards some components of the plasticizing unit like the nozzle, the flange and heaters are disassembled. Finally, one can pull out the screw and identify the solid and melt fraction by color (masterbatch is used and the pigments disperse in the melt). This procedure is very inaccurate and inefficient, since the cooling process lasts several hours. Analyzing pressure signals at different axial positions along the plasticizing unit, the solid-bed ratio can be determined in-situ during the process [13]. Figure 6 shows a schematic relationship between the material state in the screw channel and the corresponding pressure signal.

The results using PP-1 and a barrier screw without mixing head designed for a lab machine (screw diameter 40 mm, L/D ratio 20) are shown in Fig. 7. Three rotational screw speeds (0.3 m/s, 0.5 m/s and 0.7 m/s) were tested and six pressure transducers were used to measure the melt pressure along the barrier and metering section. The error bars correspond to the pressure difference between pushing and trailing flight. Both the pressure signals and the melting curve of experiment and simulation are in good accordance. Only at the lowest screw speed, melting in our simulation is faster than during the experiment.

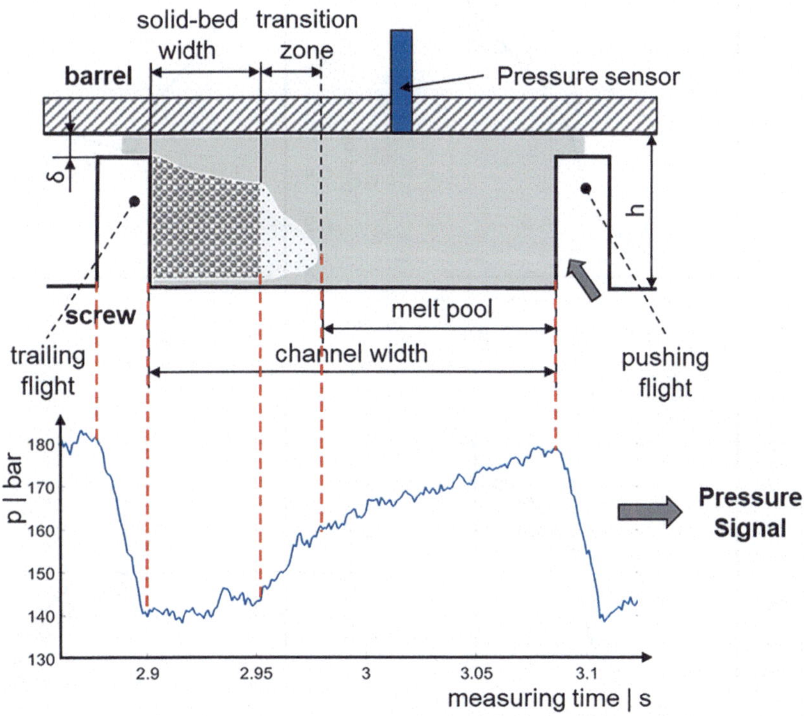

Fig. 6. Evaluation concept for determining the solid-bed width by analyzing the pressure signal.

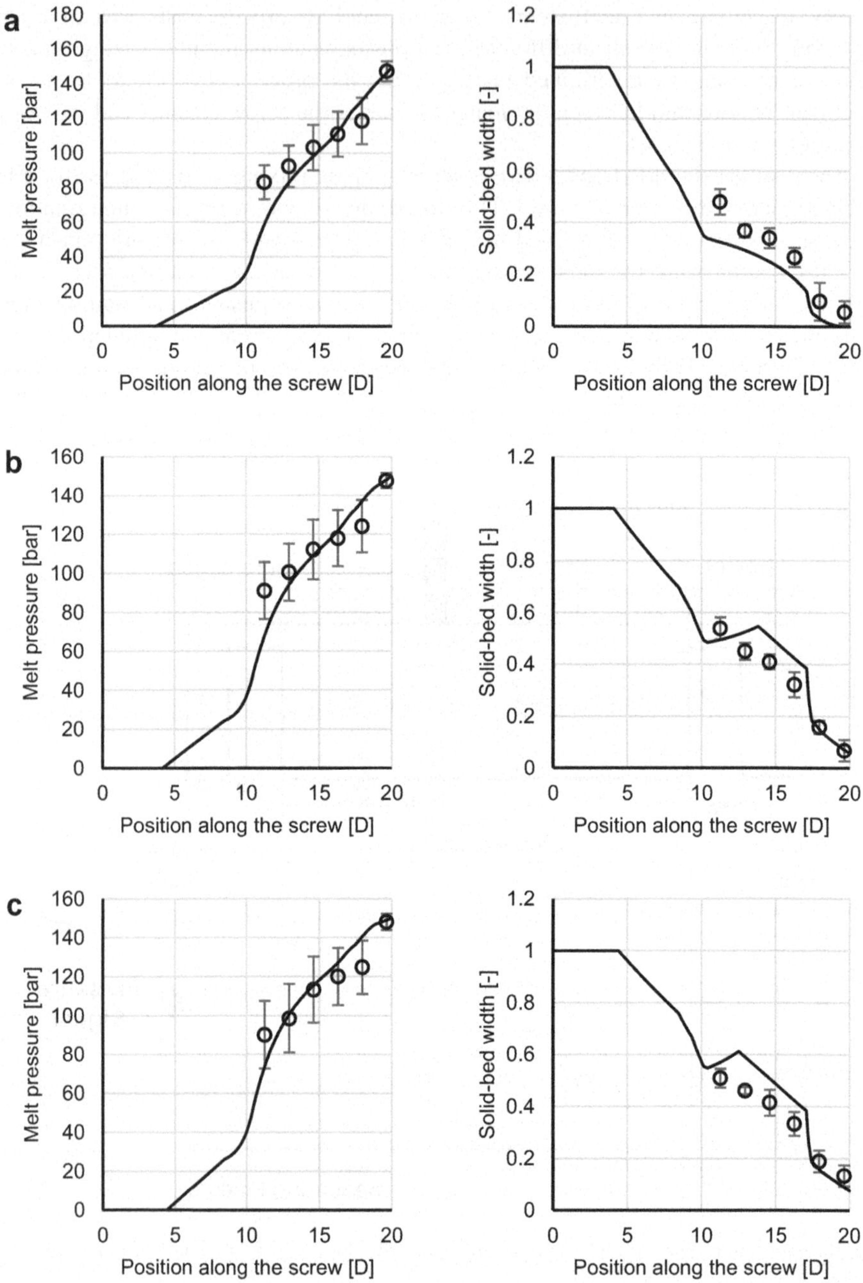

Fig. 7. Comparison of melt pressure and melting curves determined by experiment (○) and simulation (straight line) using PP-1 at (a) 0.3 m/s, (b) 0.5 m/s, (c) 0.7 m/s rotational speed; metering stroke 40 mm, cycle time 55 s, back pressure 150 bar, barrier screw Ø 40 mm.

Besides melt pressure and melting curve, the recovery rate and the corresponding plasticizing time are of great importance for the experimental verification. Figure 8 shows a comparison of experimental and simulated plasticizing times for different processing conditions. Different screw speeds (0.3 m/s, 0.5 m/s, 0.7 m/s) and back pressures (50 bar, 100 bar, 150 bar) were used in combination with the two types of polypropylene mentioned in Sect. 2.2. All results are plotted in one graph to provide an overview on the agreement of the data for one type of screw. Most of the data are located inside the corridor of ±10% deviation from the experiment. For 100 bar back pressure and the highest screw speed in combination with PP-2 (lower viscosity, see Fig. 5) the simulated plasticizing time is about 30% below the experiment. Leakage for this process condition will be higher, which results in a longer plasticizing time than expected.

Scale-up is essential for any further screw development. Therefore, two of our UBS (screw diameter 60 mm, L/D ratio 24; screw diameter 90 mm, L/D ratio 22) were used under real process conditions outside the lab. With the smaller screw, we used a test mold which gives us the possibility to vary the shot volume. All materials listed in Sect. 2.2 were tested with this screw. Metering strokes of 60 mm and 120 mm were used under different processing conditions (rotational speed 0.2 m/s, 0.6 m/s and back pressure 50 bar, 150 bar). The results for the plasticizing time are shown in Fig. 9. For the deviation from the experiment a corridor of ±20% can be defined. Most of the results are located inside this corridor. Values calculated for the low screw speed in combination with the high back pressure are on the borderline. Barrier screws are usually designed for high screw speeds and low back pressures.

With the 90 mm screw, we used a test molds for a flower tray and compared both PE-HD and PP-1 (cycle time 40 s, back pressure 40 bar). Again, we used three different screw speeds for comparison (0.26 m/s, 0.45 m/s, 0.52 m/s). Figure 10 shows the results of plasticizing time and melt temperature. For the plasticizing time, a corridor of ±10% deviation from the experiment can be defined, for the melt temperature it is ±5%.

In general, the agreement is fairly well due to the fact that there are several simplifications in the involved models: (i) an unwound screw channel with rectangular shape is used instead of the spiral channel, (ii) the screw is assumed to stand still while the barrel is rotating, (iii) solids conveying in the feeding zone is neglected, (iv) an idealized melting model is used which neglects the transition zone in Fig. 6, and (v) the melting point of the polymer is used as the flow temperature. The employed viscosity model is valid in the liquid state and the melting of the polymer starts when the local temperature exceeds the predefined melting point.

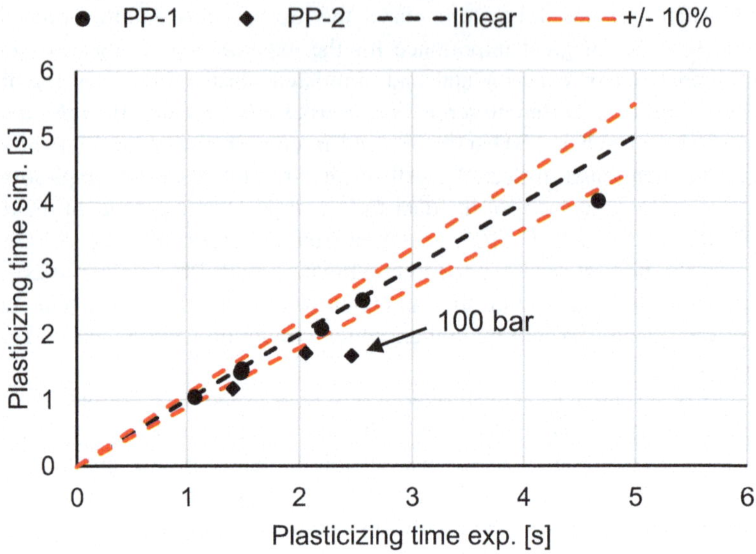

Fig. 8. Comparison of plasticizing time for experiment and simulation using different processing conditions (rotational speed 0.3 m/s, 0.5 m/s, 0.7 m/s and back pressure 50 bar, 100 bar, 150 bar; screw configuration UBS Ø 40 mm).

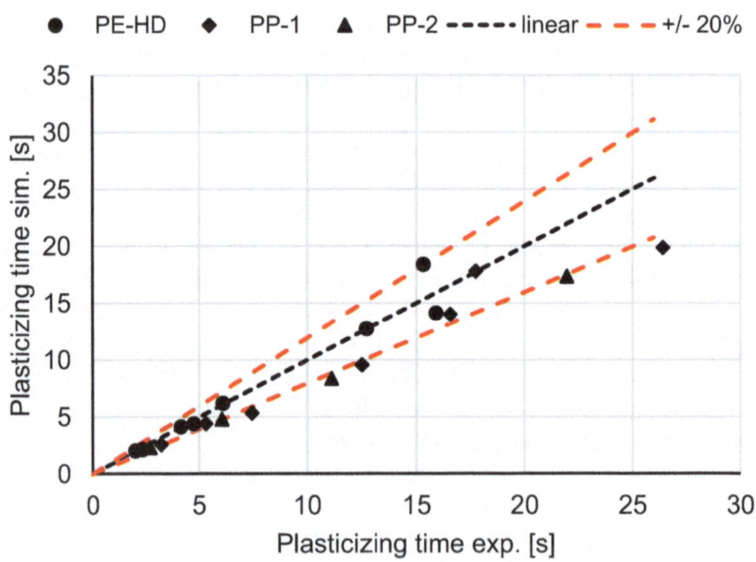

Fig. 9. Comparison of plasticizing time for experiment and simulation using different processing conditions (rotational speed 0.2 m/s, 0.6 m/s; back pressure 50 bar, 150 bar; metering stroke 60 mm, 120 mm; screw configuration UBS Ø 60 mm).

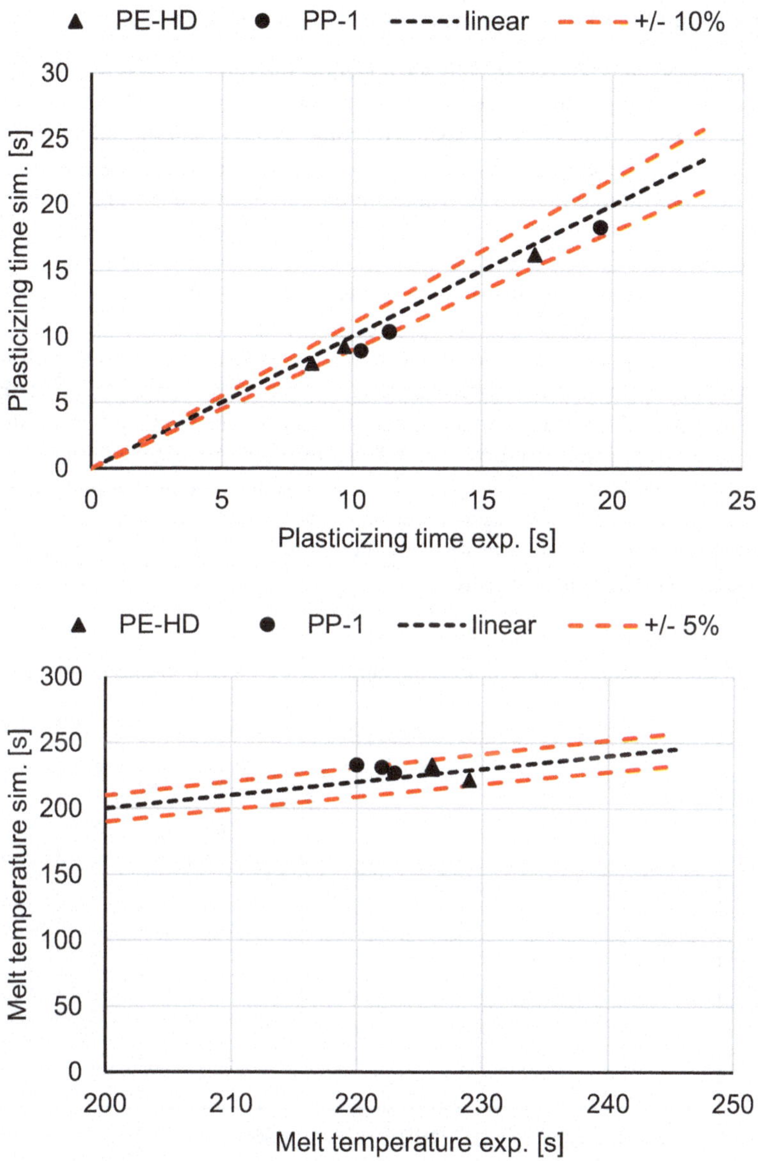

Fig. 10. Comparison of plasticizing time and melt temperature for experiment and simulation using different rotational speeds 0.26 m/s, 0.45 m/s, 0.52 m/s; metering stroke 180 mm, cycle time 40 s, back pressure 40 bar, screw configuration UBS Ø 90 mm.

2.4 Increasing the Plasticizing Performance with Barrier Screws

Higher recovery rates are needed in an injection molding process when the plasticizing time otherwise increases the total cycle time. One the one hand, higher screw speeds can be used to reduce the plasticizing time, but on the other hand melt homogeneity also must be considered. Figure 11 shows a comparison of a barrier screw with and without mixing head. The employed mixing head is a spiral Maddock which can be considered as a barrier mixing section. As soon as the maximum throughput of the barrier screw is reached (here approx. 140 kg/h), small solid particles pass the barrier gap and cannot be melted in the residual metering zone. The mixing head acts as another barrier to improve the homogeneity of the melt.

Moreover, the specific recovery rate (gram per revolution) decreases with increasing throughput due to the limited melting length in the barrier zone. At a certain point, less melt can pass the barrier gap and plugging can be observed. When on the other hand the L/D ratio is increased by e.g. 3D, the available melting length increases as well, which reduced this effect. This results in a more horizontal gradient, see Fig. 11. Therefore, barrier screws for high-performance plasticizing are characterized by an additional barrier mixing section towards the screw tip and a longer barrier section (i.e. an increased total screw length).

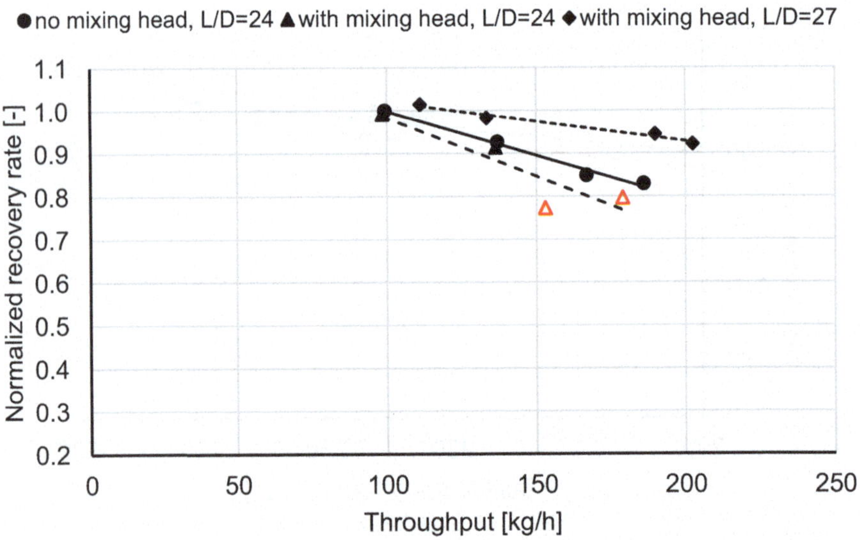

Fig. 11. Comparison of recovery rates of a barrier screw with and without mixing head (data marked by Δ indicate inhomogeneities in the melt) as well as the effect of increasing the L/D ratio on the plasticizing performance; screw diameter 70 mm, PE-HD.

2.5 Summary and Outlook

With our Screw Simulation Software S3, melting performance, pressure build-up and temperature distribution can be calculated numerically based on physical models and finite-difference approximation schemes. In contrast to conventional scale-up tools, rheological and thermo-dynamical material characterization is used to fit these models, which can be employed to simulate the material flow in the unwound screw channel. Measured and calculated plasticizing times show good accordance if the back pressure is low (e.g. 50 bar) and the screw speed is high (e.g. 0.3 m/s or higher), which usually can be assumed when barrier screws are involved. Both PE-HD and PP were considered in our study. The lower the viscosity, the more difficult is pressure build-up and results show better correlation when PP-1 instead of PP-2 is used.

In the future, we are going to improve our simulation routine by (i) focusing on the power consumption with respect to screw torque and (ii) including barrier mixing sections to be able to reproduce results shown in Sect. 2.4 and make predictions for the expected recovery rates of barrier screws with mixing heads. In addition, we developed a test setup in our plasticizing lab, which can be used to investigate the screw torque at different positions along the feeding zone. Energy consumption in the feeding zone depends on solids conveying with frictional parameters often not known in detail and on the formation of a melt film in the so-called delay zone before the drag-induced melting process starts. We believe that these modifications will further improve the simulation results.

References

1. Praher, B., Goldmann, M., Steinbichler, G.: Inline melt homogeneity measurement in injection molding. AIP Conf. Proc. **2055**(1), 120006 (2019)
2. Maddock, B.H.: Visual analysis of flow and mixing in extruder screws. SPE J. **15**(5), 383–389 (1959)
3. Rauwendaal, C.: Extruder screws with barrier sections. Polym. Eng. Sci. **26**(18), 1245–1253 (1986)
4. Maillefer, C.: Screw extruder. Swiss Patent 363149 (1959)
5. Barr, R.A.: Extruder, or extruder-like melting apparatus. US Patent 3698541 (1971)
6. Dray, R.F., Lawrence, D.L.: Apparatus for extruding plastic material. US Patent 3650652 (1970)
7. Kim, H.T.: Apparatus for extruding polymeric material. US Patent 3867079 (1972)
8. Rauwendaal, C.: Polymer Extrusion, 5th edn. Hanser, Munich (2014)
9. Osswald, T.A., Menges, G.: Material Science of Polymers for Engineers, 3rd edn. Hanser, Munich (2012)
10. Arrhenius, S.: Über die Dissociationswärme und den Einfluss der Temperatur auf den Dissociationsgrad der Elektrolyte. Z. Phys. Chem. **4**(1), 96–116 (1889)
11. Hayward, A.T.J.: Compressibility equations for liquids: a comparative study. Br. J. Appl. Phys. **18**(7), 965–977 (1967)
12. Tadmor, Z., Gogos, C.G.: Principles of Polymer Processing, 2nd edn. Wiley, Hoboken (2006)
13. Altmann, D., Praher, B., Steinbichler, G.: Simulation of the melting behavior in an injection molding plasticizing unit as measured by pressure and ultrasound measurement technology. AIP Conf. Proc. **2055**(1), 040003 (2019)

Development of IoT Device for Temperature and Cavity Pressure Measurements

Dragan Kusić[1](✉) and Matej Slapšak[2]

[1] TECOS Slovenian Tool and Die Development Centre, Kidričeva, 25, 3000 Celje, Slovenia
dragan.kusic@tecos.si

[2] L-TEK Elektronika d.o.o., Obrtna cesta, 18, 8310 Šentjernej, Slovenia

Abstract. This paper presents the obtained software and hardware results of developed Internet of Things (IoT) measurement device that is intended for real-time temperature and cavity pressure measurements in injection molding process. We set-up an appropriate IT-architecture for easier transportation of obtained measurement data and prepared a web-based GUI along with appropriate robust and precise miniature electronic PCB boards that can be easily adopted and used for monitoring and/or control applications in plastics processing industry.

As polymer materials are not good thermal conductors, the elimination of large quantities of heat needed to solidify a part also creates major problems if rapid production needs to be maintained. Therefore, a proper IoT measurement system needs to be established that is capable to precisely measure the temperature and cavity pressure behavior during different processing stages.

Furthermore, a specially constructed prototype mold was selected for testing purposes which has two cavities that are used for testing shrinkage behavior of various thermoplastic materials. The measurement results practically confirmed that the developed measurement device system fulfils the requirements for precise and robust wireless IoT based measurements of temperatures on cylinders, molds and tempering units, and cavity pressures inside the prototype mold according to Industry 4.0 requirements.

Keywords: Industry 4.0 · Injection molding · Wireless measurement · Temperature · Cavity pressure

1 Introduction

Injection molding is one of the most important processing techniques used in plastics processing industry. Certain factors need to be considered before thermoplastic materials can be adequately processed. These factors include the granule or pellet characteristics, the hygroscopic behavior of the thermoplastic material (whether it picks up water), flow properties, thermal properties (such as thermal stability and heat transfer), shrinkage degree, crystallization behavior, and molecular orientation.

C. Hopmann and R. Dahlmann (Eds.): *Advances in Polymer Processing 2020*, pp. 108–115, 2020. https://doi.org/10.1007/978-3-662-60809-8_9

Thermoplastic materials need large heat inputs to increase their temperatures to those required for melt processing and they also vary greatly in the amount of heat energy required to bring them up to processing temperatures. Such differences are not merely due to the varied processing temperatures required but are also due to the fact that varied thermoplastic materials have varying specific heats.

Different quantities of heat are required by various thermoplastic materials to increase their temperature to a fixed temperature degree. For example, when processing a semi-crystalline thermoplastic material, additional heat needs to be supplied to melt the crystal structures. In the case of an amorphous resin, this additional heat input is not required.

Anyway, both types of plastic materials will need a large quantity of heat to be quickly added into it. This leads to some issues because polymers are poor conductors of heat and can have limited thermal stability at the used processing temperatures [1, 2]. As polymer materials are poor thermal conductors, the elimination of large quantities of heat needed to solidify a part also creates major problems if rapid production needs to be maintained. Therefore, a proper monitoring/control system needs to be established that is capable to precisely measure the temperature behavior during different processing stages.

Currently, there are different monitoring systems and monitoring approaches, already available on the market and/or offered by injection molding machine suppliers which are basic ones (wiring solution) and not directly intended for IoT applications (for example solutions from Optris).

In case of plastic processing industry such monitoring/control system needs to have a wireless communication option that is in accordance with Internet of Things (IoT) and follows the guidelines of 4th Industrial revolution [3]. For this reason, we went for this approach and developed a low-cost solution that meets the requirements of plastic production world and is described in next chapter.

2 Experimental Setup

For measuring the cavity pressure and temperatures of different units in plastic production sector we have developed all the technical segments of the monitoring/control system that is based on different sensors in order to monitor the state of the mold during processing stage in real time, which will be as close as possible to the critical sites determined by the experts in the field of plastic injection.

A two-cavity prototype mold was selected for testing purposes (plates of different sizes) used to test the shrinkage behavior of various thermoplastic materials, as shown in Fig. 1. Both partners TECOS and L-TEK defined and designed together the locations of sensor connectors, while at the same time we dedicated ourselves to the method of integrating the measuring-communication electronics, because due to the high temperature on the mold itself, it should be thermally insulated, and at the same time we must ensure the transfer of a wireless switch in an environment full of metal parts, near power lines and converters.

Another potential problem is related to movable parts of the prototype mold where the wires can be potentially damaged during the mold closing or opening stage so all wires need to be mounted around the mold in order to prevent possible accidents which can stop the production for a certain amount of time [4]. Potentially, also the thermoplastic material inside the cylinder can degrade because of this which leads to additional economical loss [5, 6].

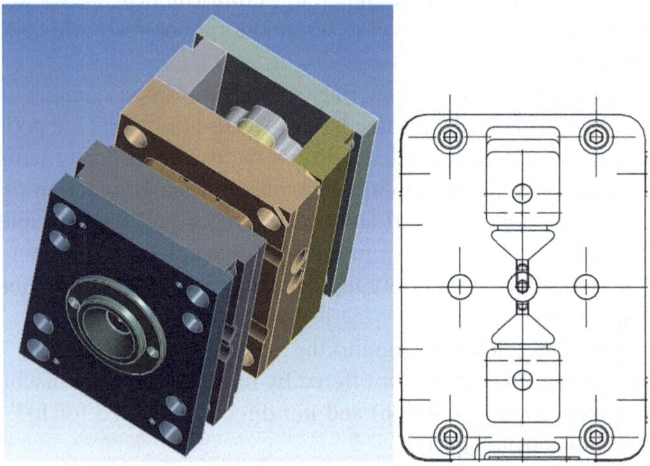

Fig. 1. CAD model of the used prototype mold.

Development work was directed in such a way that the planned electronics units could meet the most modular construction, which in the future would allow for greater flexibility in support of individual specific applications in plastic industry production environments. The basic architecture of the proposed IoT system (IoT) system and their connectivity is presented in the following Fig. 2. It enables local independent operation, device connectivity, levels, data exchange, storage and remote management of devices.

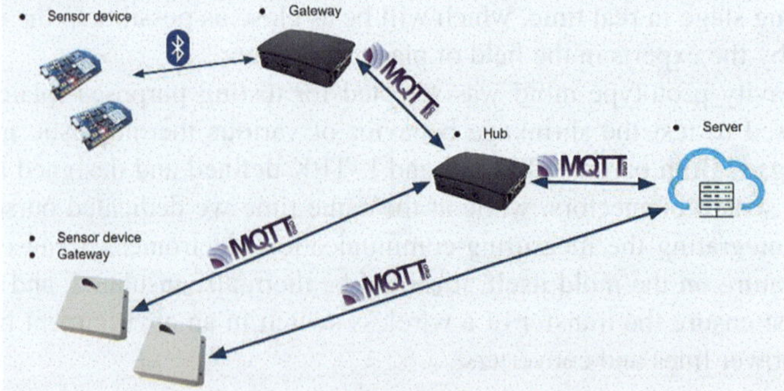

Fig. 2. Architecture of connecting IoT building blocks.

For the needs of the first prototype, we decided to use the combination of Bluetooth (BLE) and Wi-Fi wireless protocols and the Message Queuing Telemetry Transport (MQTT) transmission form [7–12]. For the needs of company TECOS testing, a code for communication has been developed that enables simultaneous local and cloud connectivity, and thus bidirectional communication between sensors, devices and control systems. The latter was in the testing phase successfully tested.

We started with the design and development of the prototypes for the purpose of testing different types of sensors in the production environment on and around the injection molds. The purpose of using the building blocks is in particular testing the robustness of the operation and the SMD (surface mount devices) measurements accuracy dependency on the type of sensor and the whole circuit of which such a sensor is part. In addition to the hardware part, software was developed for the specific measurements of pressure, temperature and humidity.

Because the latter sensor works on the battery, the software has been customized and optimized in such a way that energy consumption is minimal. This ensures a longer life duration. The software has also been developed in such a way that the frequency of reading the data can be changed through the API (Application Programming Interface), which allows us to adapt to the application of the building blocks in various end applications on the injection molding machines. Figure 3 shows from side view the developed test prototype measuring device of the control/monitoring system on the working table for measuring for example the humidity and temperatures in/on the mold and in the surroundings of the injection molding machine.

In order to measure the temperature of the production environment, on prototype testing mold and the tempering device, we decided to use the classical thermocouples type J (Fe-CuNi), which are coated and equipped with connection cables and insulated from the housing. The cavity pressure voltage signal can be measured directly on the inputs of the developed measuring device.

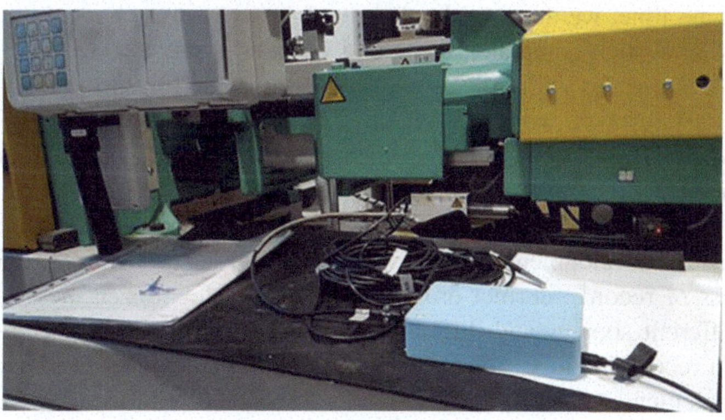

Fig. 3. Side view of the prototype test measuring device on the working table of injection molding machine.

Partners decided to install on the prototype testing mold: up to 7 temperature sensors (4 on the surface of the mold, 1 in the interior, 2 on the tempering device), an internal sensor will be inserted into a hole with a diameter of 1.5 mm and a depth of

35 mm, measuring range between 20–300 °C, accuracy of the digital temperature sensors ±0,5 °C (in the range up to 85 °C), thermocouples ±1 °C (type J), 2 cavity pressure sensors (1 in each mold cavity) with a diameter of 2.5 mm, 1-Wire protocol for testing the acquisition of temperatures, cable length for 1-Wire protocol approx. 1,5 m, sampling frequency of 2–3 Hz, external power supply of circuits, data acquisition locally and further processing. In particular, during the injection molding process, we decided to measure the temperature with the help of 4 digital 12-bit temperature sensors, to observe the temperature change in the mold during the injection process. The sensors operate on the 1-Wire protocol, which allows us to connect even a larger number of sensors. To measure the temperature conditions in the mold itself, a thermocouple type J was used, as already mentioned. For the purpose of reading data from sensors and storing data, a programmable module was developed, which is generic and allows the addition of both additional digital sensors and thermocouples of type J.

3 Results

For the purposes of developing cloud-based solution to connect the mold to the cloud, the partners identified the appropriate communication protocol and hardware, which in our case is based on TCP/IP (Transmission Control Protocol/Internet Protocol) communication, specific to the MQTT protocol and Web services technologies. This enabled us to ensure appropriate connectivity both with data systems, as well as with the level of graphic user interfaces and various data services. Developed system can, in the first place, perform all key functions locally, independently without interaction with other systems. This ensures the first degree of independence, and at the same time a systematic step towards system robustness. From security point of view all measured data are directly encrypted and stored in local cloud/server which is protected with appropriate certificate against hacking and intrusion.

At the same time, we developed software algorithms in such a way that connecting with similar system components or cloud services is possible and that it would be necessary to achieve this with little as possible inputs of settings from the user side.

A basic test infrastructure was set up for testing purposes. The latter required research and selection of the database where we selected noSQL (non-relational database) InfluxDB (open-source time series database), and for the meta data level a classic mySQL (open-source relational database management system) database. The emphasis was on development and configuration in order to achieve efficiency (tests of correctness of records, correct order, speed up to a few dozen messages per second from different sources and different forms), writing flexibility according to the required data records from various sensor BIoT (Bluetooth Internet of Things) modules, and implemented development activities to achieve both local as well as cloud layouts interchangeability between them. We also tested the stability over time with several million stored records, where it has so far proved to be an extremely reliable solution. At the same time, a meta-data structure has been developed that carries information about BIoT devices, types of records, sensor types, detail settings – allowing for correct interpretation of data, while simultaneously dynamically setting up of remote BIoT modules. In the same context, we used the data for the development of an automated procedure for the dynamic construction of user interfaces in the Grafana (open source software) environment.

The server part was installed on the Linux platform. Basic program modules were developed that enable TCP/IP – MQTT communication, decoding of data and their database records, and data generation services for the needs of the user interface developed in the Grafana environment.

A developed server environment was set up on a local server to perform the first tests. At the same time, we tested the first implementation in the cloud environment (public cloud). The measurement results in the Grafana environment are shown in the figures below, Figs. 4 and 5.

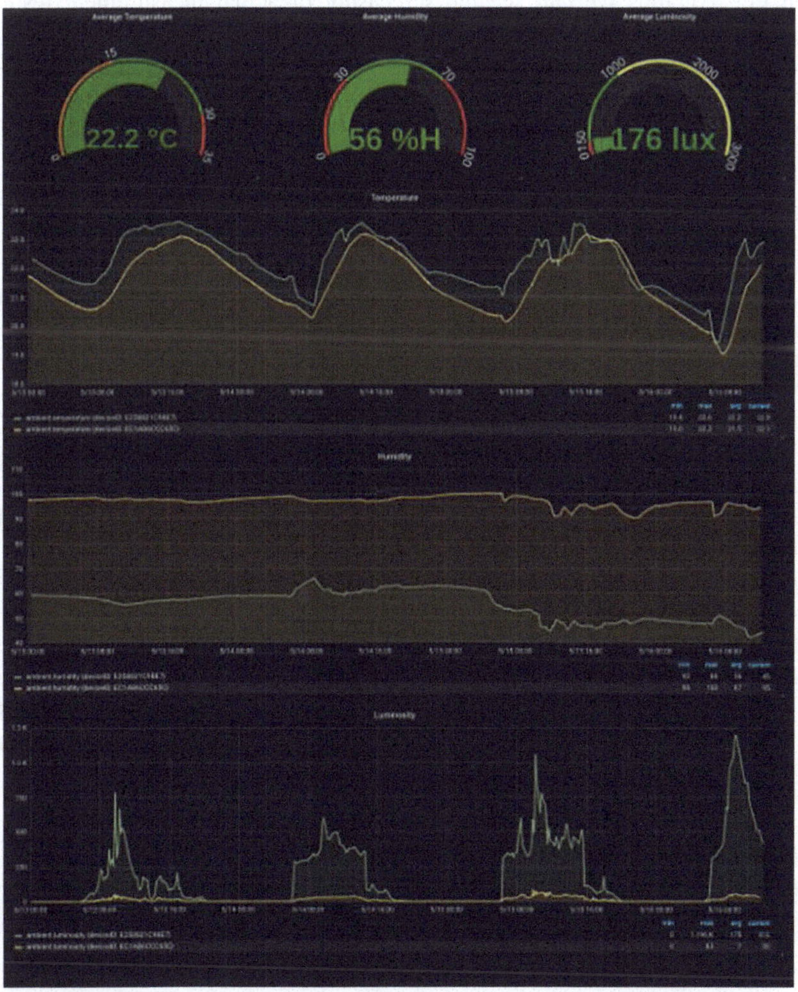

Fig. 4. Sensor measurement in Grafana environment – Example 1.

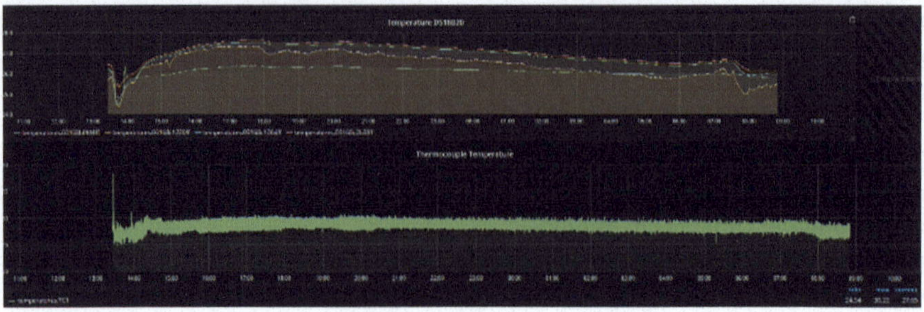

Fig. 5. Sensor measurement in Grafana environment – Example 2.

On Fig. 4 graphs we can see the screenshot of actual measurement results of temperature, humidity and luminosity. Additionally, we can measure also the cavity pressure in real-time. On Fig. 5 we can see (on another page) more in detail real-time temperature measurement results conducted with digital temperature sensor and thermocouple inside Grafana environment.

4 Conclusion

We presented in this article the development of most important technical segments of the monitoring/control system that is intended for monitoring and controlling the state of various temperature dependent units of plastic processing sector in real time.

One of the main goals was reached from economical point of view since this system is a low-cost solution that is easy to use in various plastic processing sectors. The final results confirmed practically that the developed monitoring/control system fulfils the requirements for precise wireless measurement of temperatures on cylinders, molds and tempering units in real time. Anyway, still there is a room for improvement of overall system in terms of accuracy, range of data transmission etc. which will be a subject in the forthcoming development stages.

Practically, monitoring system can be used directly once the installation of temperature and cavity pressure sensors, and its connection to the monitoring system is finished. Also, the private cloud needs to be installed and started so that all measurement results can be securely stored from the monitoring system, which is one of key advantages of our low-cost solution compared to other similar solutions in plastics processing fields.

Acknowledgment. We gratefully acknowledge the financial support of the European Union – European Regional Development Fund and the Republic of Slovenia – Ministry of Education, Science and Sport for the financial support of the GOSTOP programme in framework of which the presented work was carried out.

References

1. Raz, K., Zahalka, M.: Analysis of temperature influence on injection molding process. Proc. Manufact. Syst. **11**(2), 95–100 (2016)
2. Tang, S.H., et al.: Design and thermal analysis of plastic injection mold. J. Mater. Process. Technol. **171**(2), 259–267 (2006)
3. Official website of the GOSTOP project. www.gostop.si
4. Moayyedian, M., Abhary, K., Marian, R.: New design feature of mold in injection molding for scrap reduction. Proced. Manufact. 2, 241–245 (2015)
5. Zheng, R., Tanner, R., Fan, X.: Injection Molding: Integration of Theory and Modeling Methods. Springer, Berlin (2011)
6. Crawford, R.J.: Plastic Engineering. Butterworth-Heinemann, Oxford (2001)
7. Gomez, C., Oller, J., Paradells, J.: Overview and evaluation of Bluetooth low energy: an emerging low-power wireless technology. Sensors **12**, 11734–11753 (2012)
8. Gopalakrishnan, A., Biswal, A.C.: Applications of emerging communication trends in automation. In: 2016 IEEE 6th International Conference on Power Systems (ICPS), pp. 1–6 (2016)
9. Jung, C., Kim, K., Seo, J., Silva, B.N., Han, K.: Topology configuration and multihop routing protocol for bluetooth low energy networks. Access IEEE **5**, 9587–9598 (2017)
10. Jeon, W.S., Jeong, D.G.: Enhanced channel access for connection state of bluetooth low energy networks. IEEE Trans. Veh. Technol. **66**, 8469–8481 (2017)
11. Kim, J., Kang, S.-K., Park, J.: Bluetooth-based tree topology network for wireless industrial applications. In: 2015 15th International Conference on Control, Automation and Systems (ICCAS), pp. 1305–1308 (2015)
12. Chen, B.-R., Cheng, S.-M., Lin, J.-J.: Energy-efficient BLE device discovery for internet of things. In: Fifth International Symposium on Computing and Networking (CANDAR), pp. 75–79 (2017)

Targeted Manipulation of Fibre Orientation Through Relative Movement in an Injection Mould

Philipp Land(✉) and Thorsten Krumpholz

University of Applied Sciences Osnabrück, 49076 Osnabrück, Germany
{Philipp.Land,t.krumpholz}@hs-osnabrueck.de

Abstract. The injection moulding of fibre-reinforced plastics leads to a flow related characteristic microstructure with fibres aligned in flow direction in the shell layers and fibres aligned transversely to the flow direction in the core layer. The combination of a design related gate location and the flow related microstructure can lead to an unfavourable fibre orientation. For example, the flow direction for rotationally symmetric parts under internal pressure is often in axial direction and therefore the main fibre orientation is transversely aligned to the critical tangential tension. This leads to high wall thicknesses, increase in weight and material inefficiency. This paper shows first results of the targeted manipulation of the fibre orientation for long and short fibre reinforced thermoplastics through relative movement. The relative movement of opposing mould surfaces is realised through a rotating core in the injection mould and allows the reorientation of the fibres in tangential direction through shearing of the melt. The evaluation of this process is done by mechanical tests, microscopic investigations and computed tomography scans and shows a significant increase of transversely aligned fibres.

Keywords: Injection moulding · Fibre-reinforced plastics · Fibre-orientation · Rotating core

1 Introduction

Due to good weight-specific mechanical properties and cost-efficient injection moulding more and more metallic components are being replaced by fibre-reinforced plastics (FRPs). Thus, necessary weight reductions can be achieved for example in the automotive sector. Examples of applications include media-carrying components in the engine compartment, for which the plastics must not only fulfil the high temperature but also the mechanical requirements like internal pressure [1, 2].

For the highest possible material efficiency of fibre-reinforced plastic parts, the majority of the fibres should be oriented along the load path. The fibre orientation of short and long fibre-reinforced plastics in the injection moulding process, however, depends on the flow conditions and the gate location. The wall adhesion of the polymer melt causes a parabolic velocity profile during the injection process (see Fig. 1).

C. Hopmann and R. Dahlmann (Eds.): *Advances in Polymer Processing 2020*,
pp. 116–127, 2020. https://doi.org/10.1007/978-3-662-60809-8_10

The resulting strain of the melt in tangential direction and shearing in radial direction cause a symmetrical layer structure of the fibres [3, 4]. In this structure the fibres in the core layer orient themselves perpendicular to the flow direction and in the outer layers parallel to the flow direction. The characteristics of the layers depend, among other things, on the wall thickness, the fibre content and the process parameters [5].

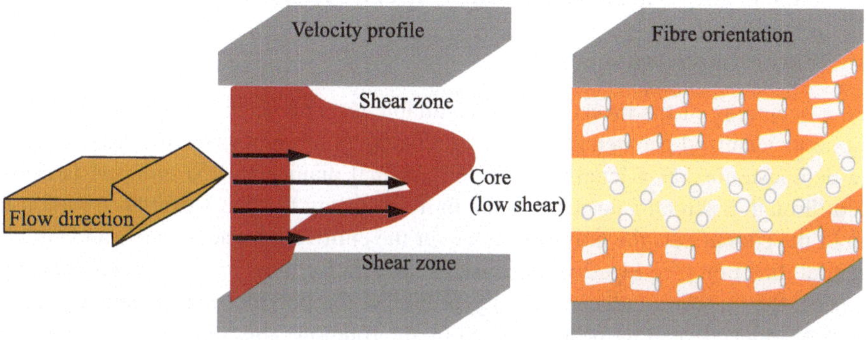

Fig. 1. Velocity profile and resulting fibre orientation over the cavity cross section according to [3]

For rotationally symmetrical components under internal pressure, the tangential stress σ_t is twice as much as the axial stress σ_a (see Eq. 1).

$$\sigma_t = 2 \cdot \sigma_a \tag{1}$$

In order to realize an optimum use of the material, the majority of the fibres in fibre-reinforced components under internal pressure should therefore be oriented in the tangential direction. Due to their demouldability and for prevention of weld lines, such components are often injected in axial direction. As a result, the majority of the fibres are oriented in the same direction. According to Eq. 1, this results in oversizing of components in axial direction if they have to withstand a certain internal pressure.

A rotating tool core can be inserted into rotationally symmetrical components to orient a large part of the fibres in the loading direction. The rotary motion causes a relative movement of the opposite cavity surfaces, which leads to a shearing of the polymer melt. The superposition of this relative movement with the shear flow of the melt during the injection process allows the reorientation of fibres into the tangential direction.

[6–9] show first investigations to this technology. [6] shows the possible reinforcing effect on unreinforced polyolefins. [7, 8] prove that the bursting pressure strength of a rotationally symmetrical cylindrical component made of glass fibre-reinforced PA and PPS can be increased by the rotating core. [9] shows the practical application of this method on a carburettor housing, where the burst pressure strength and the impact strength in the area of a weld line could be improved.

This technology itself was not further scientifically investigated or practically implemented by [6–9] in the following. The investigations shown in this paper have taken up this technology again, investigated it scientifically and furthermore applied it to long-fibre-reinforced materials.

2 Process Implementation and Experimental Procedure

2.1 Process Implementation

At the beginning of the investigations, a suitable test specimen was developed which, despite its idealized geometry, has a certain practical relevance (see Fig. 2a). The cup-like demonstrator has a length of 100 mm, a final diameter of 57 mm and is evenly filled through a screen gate. Due to the uniform filling the shear exerted by the rotation of the core affects the polymer melt over the entire component length and circumferential direction (see Fig. 2b). Because of the relative movement between the cavity surfaces, the injection-induced velocity profile of the polymer melt is superposed on the shear induced velocity profile caused by the rotation. This superposition leads to a higher degree of orientation of fibres in tangential direction.

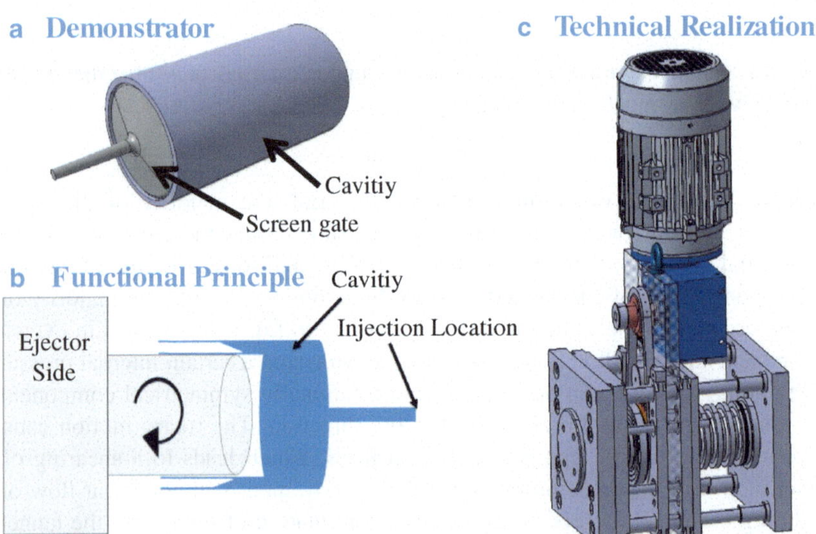

Fig. 2. Demonstrator part, functional principle and technical realization of the injection moulding with a rotating core (source: University of Applied Sciences Osnabrück)

The technical implementation is shown in Fig. 2c. An electric motor mounted on the tool transmits the unidirectional rotational movement via a chain gear to the mounted tool core. A frequency converter, whereby different rotation speeds can be realized, controls the motor. The signal for turning is transmitted to the frequency converter by the injection moulding machine Arburg Allrounder 270 c golden edition, Arburg GmbH + Co KG, Loßburg, Germany. The rotation starts 1 s before the injection process, so that the desired speed is reached at the beginning of injection. The rotation time afterwards is variable.

2.2 Experimental Procedure

The described tool concept was used to investigate a short glass fibre-reinforced polypropylene (PP) with a fibre weight content of 30%, a short glass fibre-reinforced polyamide (PA) with a fibre weight content of 50% and a long glass fibre-reinforced PA

Table 1. Overview of the examined materials

Short name	Trade name	Manufacturer
PP GF30	Polyfort FPP 30 GFC K1079	A. Schulman Inc., Fairlawn, USA
PA GF50	Grivory GV-5H EF black 9915	EMS-Chemie AG, Domat/EMS, Switzerland
PA LGF60	Grivory GV FE 16127 black	EMS-Chemie AG, Domat/EMS, Switzerland

with a fibre content of 60% (see Table 1).

The process parameters in the injection moulding process were based on the manufacturer's data from the data sheets. Subsequently the influence of the rotational speed and the influence of the absolute rotational duration on the mechanical properties of the PP GF30 were investigated. The results of this study, especially the optimal settings for the core rotation, are transferred afterwards to the PA GF50 and the PA LGF60.

3 Experimental Results

The influence of the rotary motion on the fibre orientation can be shown in different ways. On the one hand, the reorientation can be quantified indirectly by the mechanical properties of the component in the tangential direction. On the other hand, microscope images of fracture surfaces show a qualitative change in the microstructure. Furthermore, a direct measurement of the fibre orientation with a computer tomograph is possible.

3.1 Mechanical Tests

Two different mechanical tests were carried out as part of these investigations. First, ring tension tests were performed according to ASTM 2995 [10], in which the influence of the rotation parameters was investigated. In addition, burst pressure tests are carried out, confirming the results for a practical load case.

Ring Tensile Test

For the ring tensile test, specimens from the demonstrator at three different positions have been prepared and then tested in accordance with ASTM 2995. The removal position of the specimens and the test setup can be seen in Fig. 3.

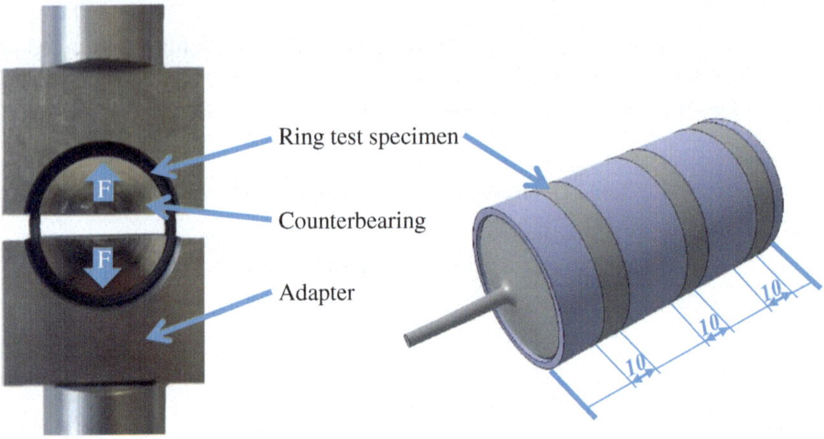

Fig. 3. Test setup for the ring tensile test and sampling position of the test specimens (source: University of Applied Sciences Osnabrück)

Since flexural stresses act to a small extent in addition to tensile stress in this test setup, only the apparent tensile stress $\sigma_{apparent}$ can be determined. This stress results from the ratio of the force F to double the cross-sectional area A of the test specimen (see Eq. 2)

$$\sigma_{apparent} = F / (2 \cdot A) \tag{2}$$

First investigations on the influence of the turning parameters on the apparent tensile strength have been carried out on PP GF30 on test rings from the middle section (see Fig. 4). The influence of the rotational speed of the tool core on the one hand and of the rotational duration on the other hand has been investigated. For the reorientation of the fibres along the entire part, the minimum turning time is equal to the injection time of 2 s. It shows that the apparent tensile strength can be increased by 32% from 45.0 MPa to 59.5 MPa with increasing rotation speed. 1.6 Hz is the maximum rotation frequency with the integrated chain gear, however, based on the increase it can be assumed that further increase in strength are possible with higher speeds.

Furthermore, the influence of the rotation time at a rotation frequency of 1.1 Hz is shown. Here it can be seen that no further increase in strength is achieved with rotation durations of more than 3 s. In Addition, when rotating during the holding pressure phase, a significant increase in standard deviation occurs.

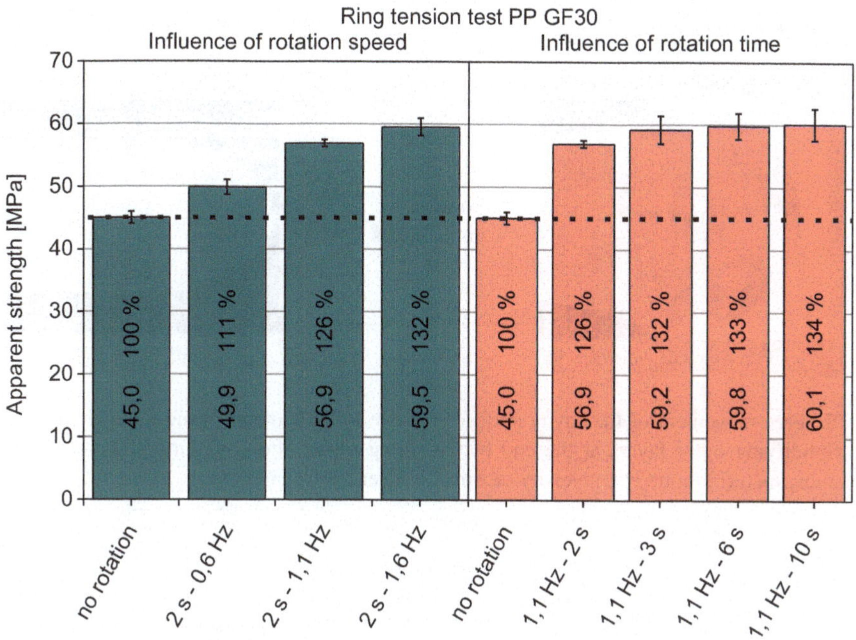

Fig. 4. Apparent strength of PP GF30 – Influence of rotation speed and rotation time (source: University of Applied Sciences Osnabrück)

The influence of the turning time is partly contradictory to the findings from [8]. There it is recommended that the rotation only takes place during the holding pressure phase. With the geometry used, a rotation during the holding pressure phase causes no or only a slight reorientation of the fibres. In this phase, the melt front has already reached the end of the flow path. The solidifying edge layers that cool down on the cold mould wall, touch and enclose the remaining melt (see Fig. 5b). Prior to this contact (see Fig. 5a), the cavity surfaces can be moved freely and the melt can be sheared. After the contact of the solidifying outer layers, the demonstrator rotates on the mould core and no further shearing of the remaining melt takes place. This behaviour prevents damage to the microstructure such as delamination at higher speeds or longer rotation times, because the melt is no longer sheared after a certain rotational resistance.

These findings indicate that the influence of the turning time depends strongly on the geometry. If the sheared area via a rotating core of a component is at the end of the flow path, no reorientation is possible during the holding pressure phase. However, if the remaining cavity, which is not affected by additional relative movements, follows after this area, a further reorientation during the holding pressure phase is conceivable.

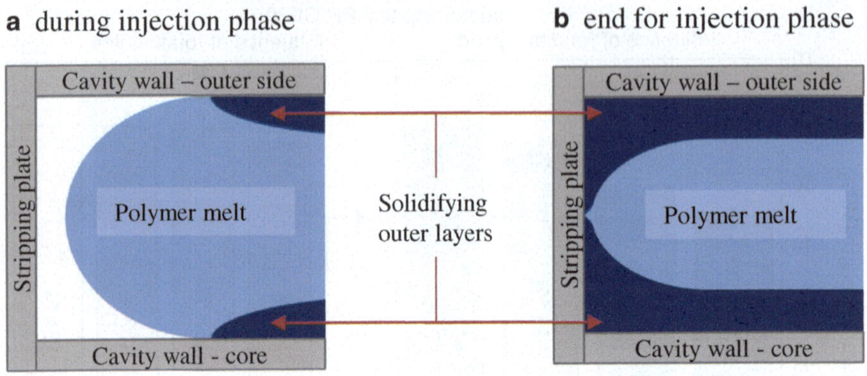

Fig. 5. Free movement of the cavity surfaces relative to each other during the injection phase (a). Solidifying outer layers at the end of the filling phase allow no further shearing of the remaining melt (b) (source: University of Applied Sciences Osnabrück)

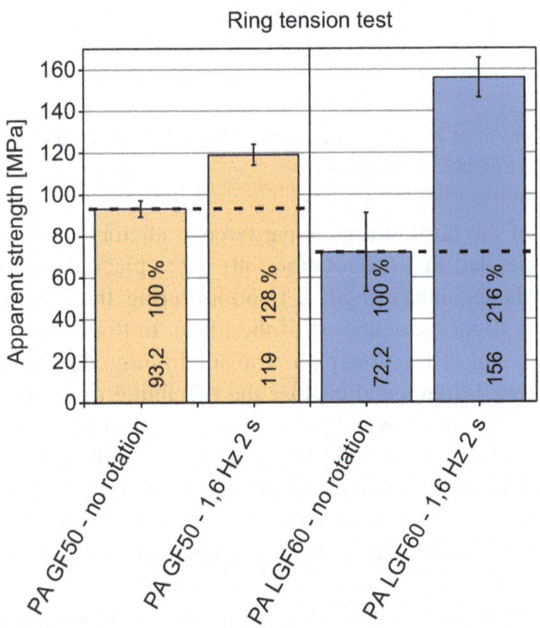

Fig. 6. Apparent strength of non-rotated and rotated test specimens of PA GF50 and PA LGF60 (source: University of Applied Sciences Osnabrück)

A similar increase in strength is possible for PA GF50 (see Fig. 6). Here the apparent tensile strength can be increased by 28% from 93.2 MPa to 119 MPa. The most significant increase in strength can be achieved with long glass fibre-reinforced polyamide. Here, the strength increase achieved is 116%. The high increase in strength of the long fibres can be explained by a rather disordered fibre structure in the non-rotated specimens and a highly oriented fibre network in the rotated specimens. Further investigations regarding the microstructure can be found in Sects. 3.2 and 3.3.

Burst Pressure Test

For the burst pressure tests, a testing device was constructed in which clamping and edge area effects are minimized. In addition, only a defined test area should be loaded during these tests. The schematic realization of this test device is shown in Fig. 7. The test medium can flow into the inner as well as into the outer area of the test specimen. Seals are fitted on the outside at the edge to the test area. This means that out of the test area the pressure applies from outside and inside and thus does not load the specimen. In the test area, the pressure applies only on the inside, which leads to an internal pressure load and ultimately to bursting in this area.

For PA LGF60, the burst pressure strength can be increased from 80.9 bar to 219.5 bar, which corresponds to an increase of 171%. In this test setup, the highly oriented fibre network due to the rotation has an even greater effect on the mechanical properties than in the ring tensile test. The burst pressure test shows that the targeted manipulation of the fibre orientation by relative movements has potential not only for uniaxial loads, but also for practical load conditions.

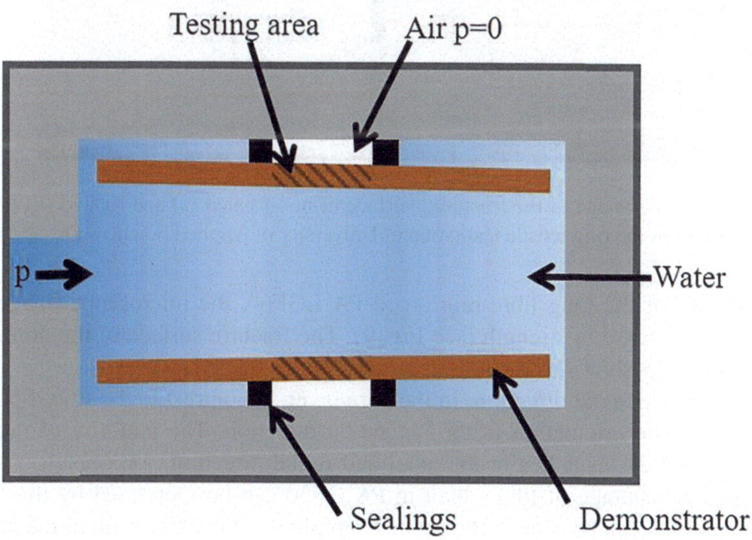

Fig. 7. Schematic realization of the burst pressure testing device (source: University of Applied Sciences Osnabrück)

3.2 Microscopic Investigations

The following microscopic images were taken on fracture surfaces of specimens tested in the ring tensile test. The filling direction in these sections is from right to left. Fibres, which are oriented in axial direction and thus in flow direction, as it is the case in the outer layers, are recognizable as elongated fibres or ellipses. The fibres oriented in tangential direction point out of the image and are therefore only visible as points.

Figure 8 shows the difference between a non-rotated and a rotated sample for the short fibre-reinforced PAGF50. The 3-layer structure known from the literature is clearly visible here (cf. Fig. 1). There are strongly noticeable outer layers and a small core layer no the non-rotated sample. The rotation of the tool core at 1.6 Hz shows a strong change in the microstructure. The outer layers are considerably thinner and do no longer show such a uniform orientation. The amount of fibres that are tangentially oriented and show out of the image has been significantly increased, which is also reflected in the remarkably wider core layer.

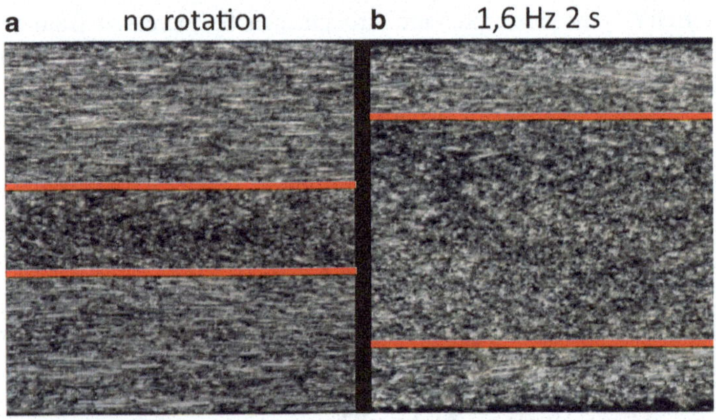

Fig. 8. Microscopic image of the fractured surface of non-rotated (a) and rotated (b) PA GF50 specimens tested in the ring tensile test (source: University of Applied Sciences Osnabrück)

In the case of the long fibre-reinforced PA LGF60, the microscope images confirm the high increase in strength (see Fig. 9). The fracture surface of the non-rotated specimen (a) also shows a layered structure. In the outer layers a large part of the fibres is oriented in axial direction. In the core layer, in contrast to the PA GF50, there is no or only little orientation in the tangential direction. The majority of the fibres in this unorientated layer lies in the axial and radial direction. As a result, a significantly larger percentage of fibres than in PA GF50 can be reoriented by the relative movement, which can be seen in the rotated sample (b). In comparison to the previous microscope images, no clear layer structure can be seen. Both in the outer areas and in the core, fibres with tangential orientation are predominantly present.

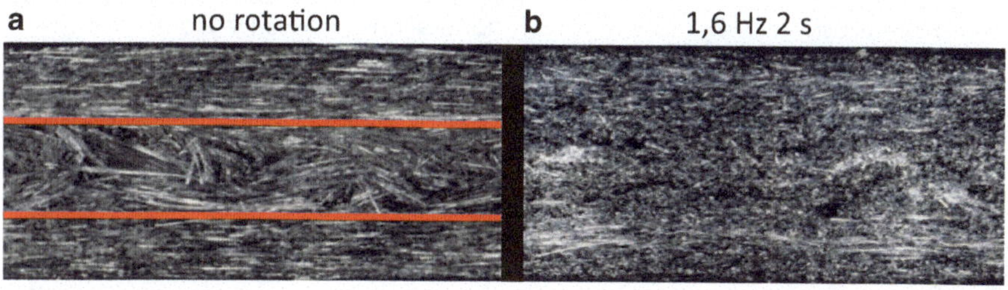

Fig. 9. Microscopic image of the fractured surface of non-rotated (a) and rotated (b) PA LGF60 specimens tested in the ring tensile test (source: University of Applied Sciences Osnabrück)

For the PA LGF60, the core layer with a high axial and radial orientation content in non-rotated specimens means that the strength in the mechanical tests is lower than in the case of the short glass fibre-reinforced PA GF50. At the same time, the process technology with the rotating core can unleash its full potential here and allows a significantly higher increase in mechanical properties.

3.3 Computed Tomography Scan

The mechanical results can also be verified using computer tomographic scans, as Fig. 10 shows. The diagram shows the degree of orientation in axial, tangential and radial direction over the normalized wall thickness for PA LGF60. The non-rotated specimen shows the layer structure known from the literature, which is also partly visible in the microscope images. In the boundary layers, about 80% of the fibres are aligned in the flow direction (axial direction) and only a small amount in the tangential and radial direction. In the core layer, the axial content decreases and the radial and tangential content increases. Here, the tangential component predominates with about 55%, but the orientation is much less pronounced here than in the outer layers. The evaluation of the rotated sample confirms the results of the mechanical and microscopic investigations. There is no noticeable boundary layer left. In the outer areas there are about the same number of fibres orientated in tangential and axial direction. The relative movement results in a highly oriented core layer, which is also significantly wider than that of the non-rotated specimen.

The integral consideration of the fibre orientation shows that the tangential orientation fraction could be increased from 28 to 64%. The axial portion was reduced from 58 to 28%.

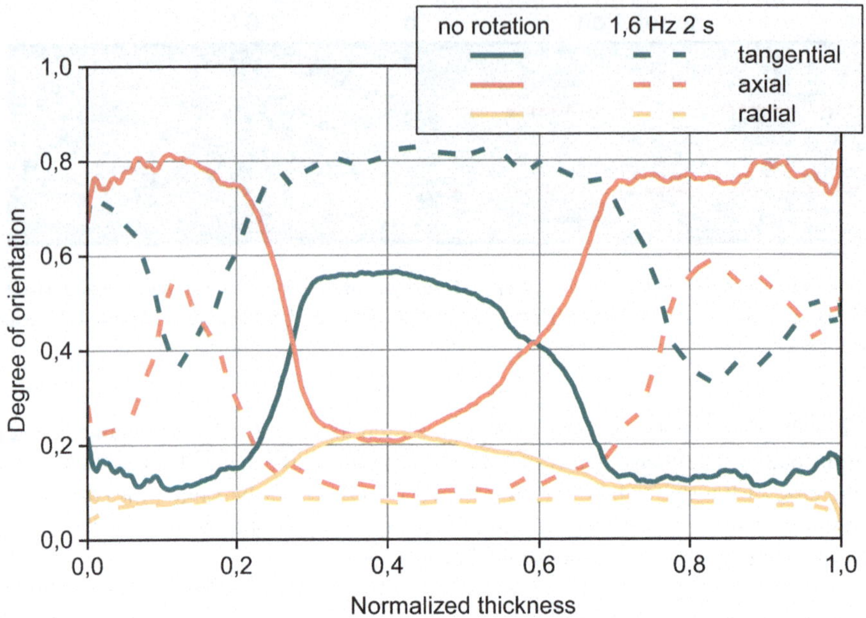

Fig. 10. Degree of fibre orientation over the normalized thickness for non-rotated and rotated PA LGF60 specimens analysed with a computer tomographic scan (source: University of Applied Sciences Osnabrück)

4 Conclusion and Outlook

Within the scope of the investigations, the fibre orientation of reinforced thermoplastics could be manipulated by a concept of a mould system with a rotating core.

Here, the rotational speed has a significant influence on the strength. At higher frequencies, more fibres could be reoriented and the greatest increases in strength could be achieved. It also shows that the maximum has not yet been reached and that higher speeds probably allow a further increase.

Furthermore, it could be shown that with turning times that are significantly longer than the injection time, no further reorientation of the fibres is caused.

Mechanical tests have shown that for PP GF30 and PA GF50 an increase in strength of about 30% is possible. For long fibre-reinforced PA LGF60, the ring tensile strength could be increased by 116% and the burst pressure strength by 171%. These results were verified by microscopic examinations and computer tomography scans.

In the future, the process will be further analysed. Investigations at different wall thicknesses and higher speeds are planned. This results will also be used to develop a design methodology for components that are to be manufactured with a rotating core. These findings will then be transferred to a practical demonstrator component as part of a research project.

Acknowledgments. The authors acknowledge the financial support by the Federal Ministry for Economic Affairs and Energy of Germany in the project *Werkzeugsystem mit drehendem Kern* (project number ZF4153401LL5). We would like to thank H. Sundermeier GmbH, Hüllhorst, Germany, for their active work as cooperation partner in this project. Further thanks goes to the companies Arburg GmbH & Co KG, Loßburg, Germany, for the provision of an injection moulding machine and EMS-Chemie AG, Domat/EMS, Switzerland and A. Schulman, Inc., Fairlawn, USA for support with material. At last, we would like to thank Mr. Uwe Becker, Managing Director of MKS-Kunststoffspritzguss GmbH, Iserlohn, Germany, for the advisory support in the project.

References

1. Flepp, A.: Optimal für den Motor. Kunststoffe **8**(2015), 86–89 (2015)
2. Baleno, B., Holtzberg, M.: Der kunststoffintensive Motor. Kunststoffe **3**(2016), 28–32 (2016)
3. Schmachtenberg, E., Brandt, M., Menning, G., et al.: Faserverstärkung richtig simulieren. Kunststoffe **5**(2004), 94–99 (2004)
4. Johannaber, F., Michaeli, W.: Handbuch Spritzgießen, 2nd edn. Hanser, München (2014)
5. Schoßig, Marcus: Schädigungsmechanismen in faserverstärkten Kunststoffen, 1st edn. Vieweg+Teubner, Wiesbaden (2011)
6. Dehennau, C., Leo, V., Cuvelliez, C.: Process for moulding a thermoplastic material by injection onto a rotating core. US Patent 5,798,072, publication date 1998/08/25
7. Dehennau, C., Leo, V., Cuvelliez, C.: Process for moulding a thermoplastic material by injection onto a rotating core. US Patent 5,824,254, publication date 1998/10/20
8. Dehennau, C., Leo, V., Cuvelliez, C.: Process for moulding a thermoplastic material by injection onto a rotating core. DE Patent 69613283T2, publication date 2002/04/18
9. Warkoski, G.: Das Spritzgießen von verstärkten Polymeren mit rotierendem Kern. Gummi Fasern Kunststoffe GAK **7**(2006), 439–443 (2006)
10. ASTM 2290: Standard Test Method for Apparent Hoop Tensile Strength of Plastic or Reinforced Plastic Pipe by Split Disk Method. ASTM International, West Conshohocken, USA (2003)

Data Driven Injection Moulding

Curdin Wick(✉), Frank Ehrig, and Guido Schuster

University of applied science Rapperswil, Rapperswil SG, Switzerland
{curdin.wick,frank.ehrig,guido.schuster}@hsr.ch

Abstract. The injection moulding process for the production of plastic parts is a very complex process. Therefore, a lot of experience and expert knowledge is necessary to produce parts with high quality. Changes in granule-batches, environmental influences and wear of the machine and the mould can strongly affect the quality of the produced parts. For this reason an injection moulding machine needs an experienced operator, who reacts properly to changing input variables and sets appropriate countermeasures. Modern injection moulding machines are able to record all countermeasures and have access to a wealth of internal machine data. Consequently, an adequate machine learning (ML) method should be able to observe, to learn the proper countermeasures and to evaluate their effectiveness. With deep learning (DL), a state of the art technology in ML, it will be possible to predictively detect process anomalies for the first time, based only on the knowledge about the internal machine data. If an operator changes the setting parameters of the injection moulding machine, the correlation between the adjustment and the anomaly is being learnt. The aim is to get process adjustment recommendations from the machine learning system.

This is a fundamentally new approach for process management in injection moulding, as the machine learning system detects problems long before they can be seen by an operator. Furthermore, the system provides process adjustment recommendations, based on the supervised and automatically generalized actions from different operators using different injection moulding machines, moulds and materials.

Keywords: Injection moulding · Machine learning · Process anomalies

1 State of the Art

1.1 Use of Machine Learning in the Field of Injection Moulding

To our knowledge today there is no machine learning or deep learning (ML/DL) based process management system for injection moulding on the market. An adaption of ML/DL by industry hasn't taken place yet, because DL is still a relatively new approach for ML, which requires a lot of data and computing power. To get more attraction from the industry for this technology, it needs to be more robust, which means the available samples for the analysis need to be increased.

C. Hopmann and R. Dahlmann (Eds.): *Advances in Polymer Processing 2020*,
pp. 128–136, 2020. https://doi.org/10.1007/978-3-662-60809-8_11

For a long time injection moulding machines have been able to record process data and to export them manually after the end of manufacturing. However only averaged values or maxima and just a small amount of curve data (often only injection pressure) can usually be exported. Modern injection moulding machines are able to record all internal machine data with a very high frequency, which increase the chances of success for a machine learning based system. But only recently it is possible to export such data from the machine, e.g. with the DataXplorer from Krauss Maffei. A further reason, why there is currently no ML based process management system for injection moulding on the market. Therefore now is the right time to invest in this technology, because theory, software, hardware and data quality are finally good enough for the development of an auspicious system.

"Added value in practice: Intelligent use of process and machine data from injection moulding production" [1] describes the state of the art. In this presentation process anomalies were detected by manually fitted features, meaning without any use of a DL approach. With such an approach detectors are learnt automatically and don't need to be programmed manually by an operator. The idea of deriving process adjustment recommendations for the operator is mentioned in the outlook.

At the University of Applied Science in Rapperswil a feasibility study [7] was performed, where with internal machine data from the DataXplorer (40 signals, 45s cycle time with a frequency of 200 Hz = 360 k data points for each part and cycle) the weight and the dimensions of an injection moulded part were predicted. The relative accuracy (standard deviation/average) of these predictions for a test set that wasn't included in the training set is better than 1%. For example, the tool temperature is intentionally increased (but no information about the tool is made available to the ML algorithm) to see if the method can still predict the resulting changes – it can, within the above relative accuracy. Although, as mentioned above, no information about the tool is available to the process, so no cavity pressure curve and/or any tool temperatures. Since these quality data could be estimated with a high relative accuracy, it is possible to perform physical quality checks less often, or at least if the quality prediction will fall outside a defined tolerance band. Whenever a physical quality check is performed, this data is fed back into the ML algorithm to further learn and improve future predictions.

1.2 Goals and Risks

Because of the good results of the feasibility study and the well-known state of the art in the detection of anomalies [2–5], it can be assumed that an automatic detection of anomalies in the field of injection moulding can be achieved. The highest risk is that not enough data samples could be generated, due to the high effort for the injection moulding trials. If a process variation occurs, the whole project slows down. This is a big difference to the ML, which Internet companies perform, because digital data is constantly generated without a physical process having to run. In other words, with real physical processes it may be difficult to exploit the power of DL because data collection is expensive and time consuming.

The next step is to learn the ML system the operator's countermeasures as a result of the detected anomaly, an "imitation learning" (IL). The paper "Bridging the gap between imitation learning and inverse reinforcement learning" together with the references in this publication define the state of the art [6]. This is a very active field of research and only a few of these ideas have been incorporated into a real product. Based on the preliminary project, however, this represents a real opportunity, since it is a clearly defined environment and the IL is therefore highly focused.

The aim is to take a step towards complete automation of the injection moulding process ("lights-out manufacturing"). Anomalies have to be detected independently by the ML system, without the help of an experienced operator, but only based on internal machine data. Injection moulding data is well suited for this purpose, as the process is inherently cyclic and a lot of very similar data is quickly available. In the last few years, current ML systems have clearly shown that they can detect such anomalies independently. During the learning phase the anomalies are detected and the following adjustment actions of the experienced operator are observed. This creates a "Supervised Learning" situation in which an appropriate action can be learned for each detected anomaly. Modern ML systems can then generalise on the basis of such data, which means that with sufficient training data and a suitable ML system, even reasonable adjustment actions for anomalies never seen before can be suggested. This is a fundamental step forward in the automation of such a complex process and represents a significant innovation in process management. By generalising the process adjustment actions to new anomalies never seen before, the ML system can retrieve more than just known actions. Thus it can propose actions that may have never been used as such before, but are still useful for this never-before-observed anomaly.

2 Investigations

2.1 Feasibility Study for the Usage of ML: Quality Data Prediction

In an already mentioned feasibility study [7] by the University of Applied Science in Rapperswil first investigations for the usage of ML in the field of injection moulding were performed. Therefore a Krauss Maffei PX120-380 injection moulding machine with an integrated DataXplorer was used. The DataXplorer records almost all machine data during one cycle, like temperatures, pressure, power, triggers and positions. More than 300 k data points per cycle could be used for interpretation and generation of the ML models. The investigations were performed on a simple ice scraper, displayed in Fig. 1. Starting from an optimised operating point, specific disturbances, which can also occur in a real production, were introduced:

- Trial series 1 (72 parts) - reference trial series → ice scrapers lie within the tolerances
- Trial series 2 (56 parts) - calcified cooling channels → temperature of the mould temperature medium increased
- Trial series 3 (57 parts) - batch fluctuation → cylinder temperature increased
- Trial series 4 (52 parts) - wrong material → 10% foreign material added

The material for all trials was polypropylene HF955MO. For each series the corresponding quality data were measured. As quality data the length, width, weight and lip distance (for the follow-up assembly of a sealing lip) were used, shown in Fig. 1.

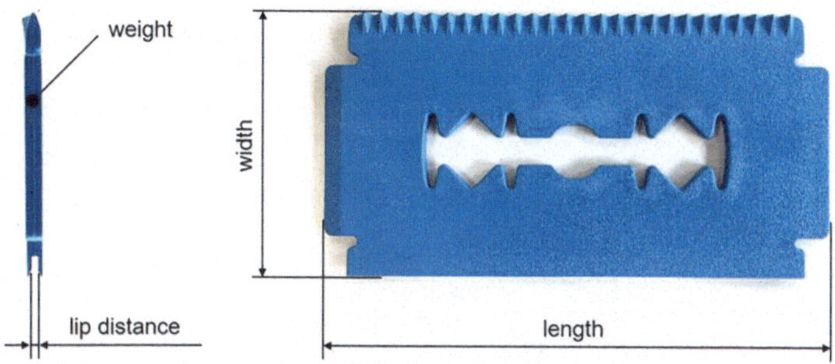

Fig. 1. Evaluated quality data of the ice scraper

The models were derived and trained by measuring the quality data of the manufactured parts and comparing them with the recorded machine data. The first step was to analyse which and how many features need to be used to classify the individual trial series. To find out which features have the biggest influence on the classification a forward stepwise selection was executed. As a result this procedure provides a ranked list of suitable features. The data from the trial series V1-V4 can be completely classified with just two of these important features. An example of features that allow this separation is shown in Fig. 2. A linear discriminant analysis (LDA) was used for the actual classification of the trial series.

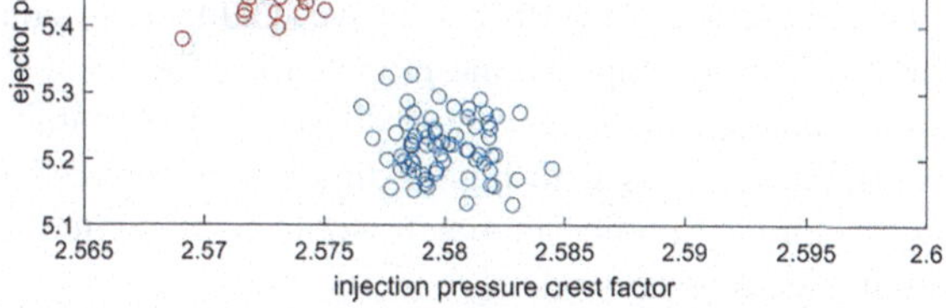

Fig. 2. Results after classification with LDA V1-V4

Test set (random data from the trial series that were not used for the development of the models) were used to test the models by predicting the quality data of the already moulded parts. The results are shown in Fig. 3.

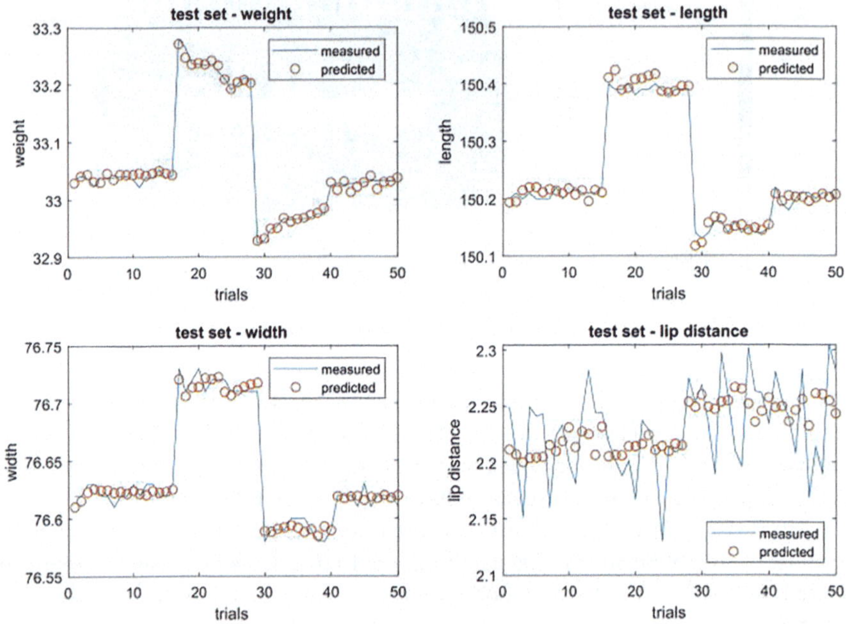

Fig. 3. Quality data prediction by the ML system

The relative accuracy (standard deviation/average) of these predictions is better than 1% for components that were not in the training set. For example, in this preliminary project 10% foreign material is intentionally added to simulate a batch variation. No batch information was made available to the algorithm and yet this method can still achieve the above relative accuracy.

Basically, the models become better for new unknown test sets if they are trained with more data. The easiest way to further improve quality is to add more data and there is still potential to further improve the models too. The results show that there could be some high-dimensional dependencies between the features used and the quality characteristics. An analysis with higher-dimensional regression models could also lead to an improvement. Another approach would be to switch to other methods of linear regression like local regression.

Objective for the next steps in prediction of quality data is to transfer the findings on other parts, because the ice scraper is a relatively simple part. Through the analysis of other parts (with different wall thicknesses, hot runner, multiple cavities) and the use of other materials more effects can be detected and the feature selection can be optimized.

2.2 Anomaly Detection: Capabilities for Predictive Maintenance

In a second investigation the focus was on the detection of process anomalies with an ML based process management system. One well-known anomaly in the injection moulding process is the wear of non-return valve. The process drifts slowly and continuously, thus after a certain time the setting parameters of the machine need to be adjusted, to guarantee proper parts. In order to simulate this behaviour, several non-return valves were machined, meaning a quantifiable damage was artificially introduced.

Starting from an optimised operating point the dosage volume was artificially increased, that even with a worn non-return valve still a sufficient melt cushion is available and the same process can be run through all trials. The material for all trials was ASA Luran S KR2858G3.

For a first set of trials non-return valves with notches with a depth of 0.33, 0.67 and 1.00 mm were machined. Afterwards the injection moulding trials with the different valves and same setting parameters were performed on the Krauss Maffei machine with DataXplorer and all internal machine data were recorded. The difference between the damaged series and the intact non-return valve can clearly be identified when considering the screw position as a function of time, shown in Fig. 4.

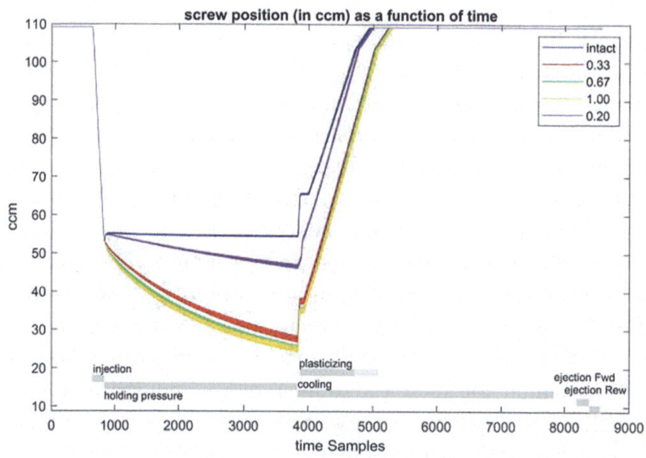

Fig. 4. Screw position as a function of time

These observations correlate with existing knowledge in the injection moulding industry which expects a decrease of the melt cushion the higher the wear of the non-return valve. A classification of single process curves to the various damages to the valve is more difficult. More process curves of the recorded internal machine data are needed for this, even then a manually assignment is very difficult or almost impossible.

By using all the recorded internal machine data and the developed ML algorithms with several features the possibilities are getting extended. Again a forward stepwise selection was executed to get a ranked list of suitable features to classify the various trials. A worn non-return valve can be clearly detected by selecting the appropriate features. Two of the best features were the first and second principal component score resulting from a principal component analysis (PCA) performed with all the calculated features. With this the various trials can be clearly classified, shown in the left diagram of Fig. 5.

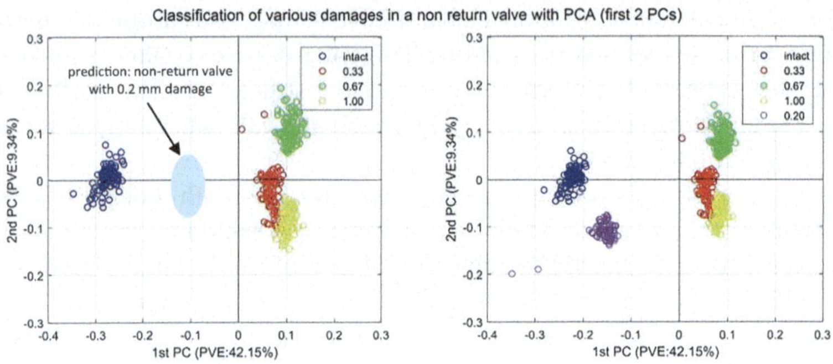

Fig. 5. Classification of the various damages with PCA

Furthermore the first principle component score seems to be a good indicator for the height or the progress of the damage on the non-return valve, probably the damage can even be reliably estimated. As the machined damage seems quite high, a valve with a smaller damage of about 0.2 mm should be located somewhere in the marked region in Fig. 5 on the left. To check this expectation another ring of a non-return valve with a notch with a depth of 0.2 mm was machined and the same injection moulding trials were done. The first principle component score is still a quite good indicator for the height or the progress of the damage on the non-return valve. Next trials with an unhardened non-return valve in combination with a highly reinforced material should simulate a continuous wear to check the correlation to existing results.

Coming back to an ML based process management system, these investigations clearly show the potential for such a system. Process anomalies should be detected probably long before they can be seen by an operator if the appropriate features are selected. A predictive maintenance system could be established too.

In a next step longer trial series will be done to detect different kind of process anomalies. If the produced parts no longer correspond to the required quality and the operator changes the setting parameters of the injection moulding machine, the correlation between the adjustment and the anomaly should be learnt in the future. The aim is to get process adjustment recommendations from the ML system, based on the supervised and automatically generalized actions from different operators using different injection moulding machines, moulds and materials.

3 Conclusion and Outlook

The biggest success factor of a ML based process management system is to record and export enough and especially the right data from the injection moulding machine. On the one hand internal machine data (process data) are needed, but on the other hand also setting parameters and quality data of the moulded parts are required to build up a multifunctional system. As the quality data of the produced parts are often not available, at least the reason for the process adjustment should be put on record. The recorded data should also be handled and interpreted with care, because of the new possibilities even software updates can be detected in internal machine data.

But one big issue is that only a few injection moulding machines, like a Krauss Maffei equipped with a DataXplorer, currently allow to collect all internal machine data and most important to easily export them. Therefore there is a lot of ongoing development on the market to export process data from the injection moulding machines and peripheral devices through uniform interfaces, like the several Euromap interfaces. Euromap 77 is one solution for a faster, easier and standardized exchange of data, but the sample frequency is limited to 10 Hz (to our knowledge), which is probably too poor for the use in a ML based process management system for injection moulding. Furthermore just the most modern machines are equipped with this interface. Euromap 63 is an alternative, but is less standardized and the sample frequency is still limited.

Other solutions needs to be found and therefore an attempt was made to collect internal machine data from a Engel e-victory injection moulding machine, with the same sample frequency as the DataXplorer provides its data. In cooperation with Siemens and Engel a first possible workflow was established, where an Industrial Computer (IPC) is connected to the injection moulding machine and gathers process data packages directly from the machine controller and uploads them into Siemens Mindsphere. Another advantage of having all the data samples in the cloud is the ability to use it live for the ML based process management system, even from several different machines.

References

1. Brexeler, I., Kruppa, S. et al.: Mehrwert in der Praxis: Intelligente Nutzung von Prozess- und Maschinendaten aus der Spritzgiessproduktion. In: VDI-Jahrestagung Spritzgiessen, Baden-Baden, 20./21.02.2018
2. Baomarand, H., Bentley, P.J.: An intelligent autopilot system that learns piloting skills from human pilots by imitation. In: 2016 International Conference on Unmanned Aircraft Systems (ICUAS), Arlington, VA, USA, pp. 1023–1031 (2013)
3. Baomarand, H., Bentley, P.J.: An intelligent autopilot system that learns flight emergency procedures by imitating human pilots. In: 2016 IEEE Symposium Series on Computational Intelligence (SSCI), Athens, pp. 1–9 (2016)
4. Baomarand, H., Bentley, P.J.: Autonomous navigation and landing of airliners using artificial neural networks and learning by imitation. In: 2017 IEEE Symposium Series on Computational Intelligence (SSCI), Hawaii, USA, (2017) (accepted)

5. Baomarm H., Bentley, P.J.: Autonomous landing and go-around of airliners under severe weather conditions using artificial neural networks. In: The 2017 International Workshop on Research, Education and Development on Unmanned Aerial Systems (RED-AUS), Linköping, Sweden (2017) (accepted)
6. Piot, B., Geist, M., Pietquin, O.: Bridging the gap between imitation learning and inverse reinforcement learning. IEEE Trans. Neural Networks Learn. Systems. **28**(8), 1814–1826 (2017)
7. Ehrig, F., Schuster, G.: Machine Learning zur Erkennung von Veränderungen beim Spritzgiessprozess. In: VDI-Jahrestagung Spritzgiessen, Baden-Baden, 19./20.02.2019

A Holistic Approach to Part Quality Prediction in Injection Molding Based on Machine Learning

Alexander Schulze Struchtrup(✉), Dimitri Kvaktun, and Reinhard Schiffers

Universität Duisburg-Essen, Lotharstraße, 1, 47057 Duisburg, Germany
{alexander.schulze-struchtrup,dimitri.kvaktun, reinhard.schiffers}@uni-due.de

Abstract. All plastics processing companies have to fulfill the objectives of time, cost and quality. Against this background, those producing in high wage countries are especially challenged, because superior part quality is often the only possibility to prevail in competition. Since this leads to high expenses on quality assurance, for some time already efforts have been made to predict the quality of injection molded parts from process data using machine learning algorithms. However, these did not yet prevail in industry, mainly for two reasons: First, because of the inevitable learning effort that is required to set up a quality prediction model and second, because of the complexity in the application. Current research in the field of transfer learning aiming to shorten learning phases addresses the first challenge. In this paper, we present a holistic approach for the data analysis steps that are necessary once process and quality data have been generated, aiming to minimize the application effort for the operator. This includes the development and application of suitable algorithms for automatic selection of data, process features as well as machine learning algorithms including hyper-parameter optimization and model adaption. Combining the two approaches could bring quality prediction one significant step forward to successful industry application. Beyond this, the presented approach is universally applicable and can therefore be used for other plastics processing methods as well.

Keywords: Injection molding · Quality prediction · Machine learning

1 Introduction

Thermoplastics injection molding is a discontinuous process that allows the automatic, highly reproducible production of molded parts with complex geometry [1–3]. Over the last couple of decades, injection molding machines have been improved with regard to both, mechanical precision and control techniques [4]. Anyway, internal and

C. Hopmann and R. Dahlmann (Eds.): *Advances in Polymer Processing 2020*,
pp. 137–149, 2020. https://doi.org/10.1007/978-3-662-60809-8_12

external perturbations, such as viscosity fluctuations of the melt, may negatively affect the quality of the molded parts.

Consequently, plastics processing companies are spending large efforts on quality assurance. Nevertheless, scrap production is usually detected only with delay in a sample based quality inspection, while bad parts produced between two samples may stay completely unnoticed. To overcome these drawbacks, research efforts have been made, to predict the quality of the molded parts directly from machine and process data using machine learning algorithms [5–11].

Despite good results, these approaches did not prevail in industry to date, although corresponding products [12] are available. In the authors' perception, this is mainly due to two drawbacks: First, because of the obligatory learning phase during which a quality prediction is not yet possible. Second, since the process of robustly building a good quality model requires many steps. These include data generation and selection, feature extraction, construction and selection as well as learning and adapting suitable models including hyperparameter-optimization, which, in the past approaches had to be carried out mainly manually and resulted in a lot of effort.

The first issue is addressed in recent research [13–17] dealing with the transfer of relationships learned from simulation data as well as other molded parts, aiming to shorten the learning phase of the new model. Still, there is no holistic approach that holistically analyzes, combines and automatically carries out the previously named data analysis steps (cf. Fig. 1) in the context of injection molding quality prediction. Therefore, we present such an approach for injection molding quality prediction and share some results of our research in the first three named areas.

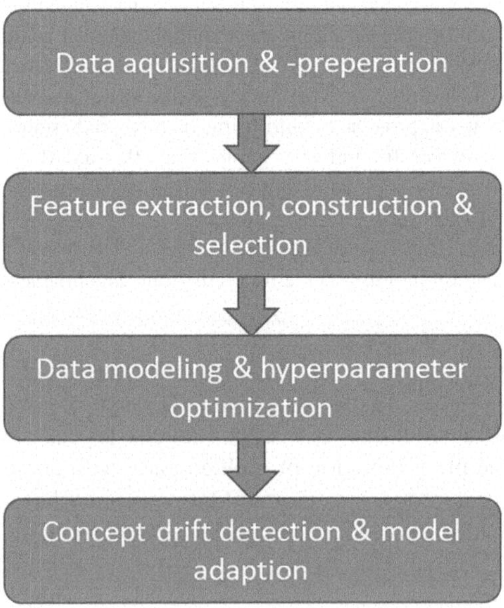

Fig. 1. General framework for holistic quality prediction.

2 Methods

2.1 Data Acquisition and Preparation

The experiments were carried out on a KraussMaffei 120-380 PX fully electric injection molding machine (IMM) in a production cell with a linear robot, conveyor belt and 100% inline quality monitoring. The machine has standard sensor technology with two additional cavity pressure sensors in the mold, which are directly connected to the machine data processing system. Six different experiments were conducted: stable process, start-up process, downtime process[1], process with re-grind material,[2] process with re-grind material and adaptive process control (APC) from KraussMaffei and a design of experiment (DOE). Therein, injection velocity, the holding pressure, the holding pressure time, cooling time and the barrel temperature at the nozzle were varied (cf. Table 1), creating 43 different combinations.

Table 1. Process setup for DOE. For all other experiments, the machine setting parameters were set to the central point of the DOE.

	Lower axial point	Lower cube point	Center point	Upper cube point	Upper axial point
Injection velocity [mm/s]	72	80	100	120	125
Holding pressure [bar]	429	450	500	550	571
Holding time [s]	2,9	3,5	5,0	6,5	7,1
Cooling time [s]	9	10	12	14	15
Nozzle temperature [°C]	226	230	240	250	254

Each experiment consists of 1000 injection molding cycles creating 1000 data samples, except the DOE with 860 cycles and 860 data samples, respectively. The weight and length of the molded parts (rectangular plate specimen, cf. Fig. 2) were measured directly after every completed injection cycle. The process and quality data were interfaced and evaluated with Matlab 2019a. In total, 48 machine and process parameters and two corresponding quality criteria were logged during each cycle.

After the data acquisition, the data needs to be prepared for the future steps like feature selection and data modeling. In general, the data is split into two parts. The first part is used for training the model and adjusting the (hyper-) parameters. The second part, the validation set, is used to estimate the generalization error of the model. The objective is a good prediction with a low generalization error. This method is called cross-validation [21]. In this work 80% of the data is used for training and 20% for validation, which is a subgroup of cross-validation called holdout method [22].

[1] In total 9 downtimes of 5, 15 and 25 min (3 times each), taking place every 100 cycles.

[2] Variation of re-grind material fraction from 0 to 100% in steps of 25%, 200 cycles each. Used material: Polypropylene LyondellBasell Moplen HP501H.

Fig. 2. Plate specimen used for the data generation experiments.

2.2 Feature Selection

The quality of the prediction model depends on the amount and quality of the data as well as the input features used for the modeling [18]. Most often, parameters are chosen through trial and error or expert knowledge [19, 20]. Since this work pursues a holistic approach for quality prediction, the parameters are chosen automatically using state of the art feature selection methods, while the resulting model quality is compared. Feature selection algorithms can be divided into three types. The first method is called embedded method where the feature selection is part of the learning process. Wrappers are the second method where the predictor which is used as a black box is tested with different subsets of features, trying to improve the overall prediction performance. Filter methods are the last approach of feature selection which are independent of the predictor. The selection is done directly by some performance evaluation metrics (PEM). Filter methods are usually less computationally expensive than embedded or wrapper methods [18], which is why they are mainly used in this work.

Search Strategies. Even when using computationally efficient feature selection methods such as filters, it may still be not feasible to evaluate every possible feature subset. Therefore, search strategies are applied, which yield still good results while minimizing the required computational resources.

In a forward selection (FS) the algorithm starts with an empty feature set and continuously adds features trying to improve the PEM. In a backward elimination (BS) the procedure starts with all features and progressively deletes the feature, which is least useful regarding the PEM [18]. Although they are computationally very efficient, both suffer from the "nesting effect". It describes the case that features which are selected through the FS, cannot be discarded later while features which are discarded in the BS cannot be re-selected [23].

A solution for this problem are floating search methods. The sequential floating forward selection (SFFS) starts with an empty feature set. In the first step the normal FS algorithm is applied and one feature is added to the feature set. In the second step one feature is conditionally excluded applying the normal BS. If this new subset is the best so far, the conditionally excluded feature is removed from the feature set and the algorithm starts with step 2 again. If the subset is not the best so far, the conditionally excluded feature is returned to the feature set and the algorithm continues with step 1 [24]. The sequential floating backward selection (SFBS) is the opposite of the SFFS and starts with all features in the feature set. In the first step the normal BS algorithm is applied and the least significant feature is excluded from the feature set.

In the second step one discarded feature is temporarily added to the feature set applying the normal FS algorithm. If the new subset gives the best PEM, the temporarily included feature is added to the feature set and the algorithm continues with step 2. If the subset is not the best so far, the feature is not added and the algorithm continues with step 1 [23].

Performance Evaluation Metric. With the performance evaluation metric, the significance of a feature is evaluated. In this work the Correlation-based Feature Selection (CFS) according to HALL [25] is selected:

$$M_s = \frac{k\bar{r}_{cf}}{\sqrt{k + k(k-1)\bar{r}_{ff}}} \quad (1)$$

where k is the number of features in the subset, $\bar{r}_{cf}$ is the average of the correlations (relevance criterion) between the features and the class (quality criterion), $\bar{r}_{ff}$ is the average feature-feature inter-correlation and M_S is the resulting PEM merit [25]. According to HALL "a good feature set is one that contains features highly correlated with the class, yet uncorrelated with each other". Other PEMs like Relief [26], minimum redundancy – maximum relevance [27] or mutual information [28] are beyond the scope of this paper. Figure 3 shows the feature selection process.

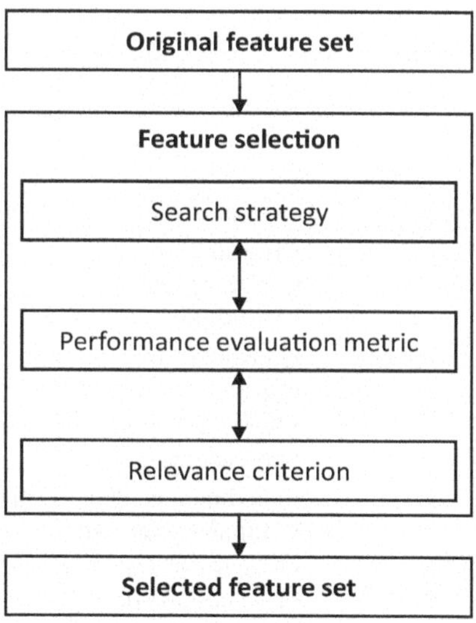

Fig. 3. Feature selection with interactions of search strategy, PEM and relevance criterion.

2.3 Data Modeling and Hyperparameter-Optimization

Machine learning methods can be distinguished in three main classes: supervised, unsupervised and reinforcement learning. All machine learning methods used in this work are supervised machine learning methods. In supervised learning the predictor learns the relation between the inputs and outputs [29]. Furthermore, supervised machine learning can be separated in two classes depending on the output data type. If the output data is discrete the problem is called classification, if the output data is continuous the problem is called regression [30]. Since the weight and the length of the component are continuous, the machine learning algorithms used in this work are those suitable for regression problems. The following six machine learning algorithms are used: Artificial neural networks (ANN) [31], support-vector machines [32], binary decision trees [33], k-nearest-neighbors (kNN) [34], ensemble methods (LSBoost [35] & random forest [36]) and Gaussian process regression [37]. Furthermore, normal multiple linear regression [38] is added to the analysis to compare classical statistical methods with machine learning.

Every machine learning method has so called hyperparameters that need to be set by the user to maximize the effectiveness of the machine learning method. They are used to define numerous configurations of the algorithm affecting both learning process and the resulting model structure. Examples are the number of neurons in the hidden layer of an ANN or the number of neighbors in the kNN-method. Most frequently, hyperparameters are set via rules-of thumb, by testing sets on a predefined grid or by the default configuration of the software-provider. In this paper the hyperparameter-optimization is done by Bayesian optimization, which proved to be a very efficient method with good performance [39]. Table 2 provides an overview over the hyperparameters chosen for optimization.

Table 2. All machine learning methods tested in this work with their hyperparameters. In total 22 predictors were learnt for every experiment and both quality key figures.

Algorithm	Hyperparameter
Multiple linear regression	–
Artificial neural networks	• Number of neurons in the hidden layer • Transfer function • Learning rate • Momentum [40]
Support-vector machine	• Penalty factor • Epsilon • Kernel function • Polynomial order [41]
Binary decision trees	• Maximum number of splits • Minimum leaf size [35, 42]
k-nearest neighbors	• Number of neighbors • Distance metric [43]
Ensemble-method	• Method • Number of learning cycles [35]
Gaussian process regression	• Kernel function • Kernel scale • Sigma [37]

3 Evaluation

3.1 Data Generation

One objective of this study is to evaluate the six different experiments, which represent possible process states occurring in real-world injection molding production. Figure 4 shows the best possible result of the 22 different predictors for the six experiments in regard of the two quality parameters. The coefficient of determination is used to evaluate the models' prediction quality on the validation dataset.

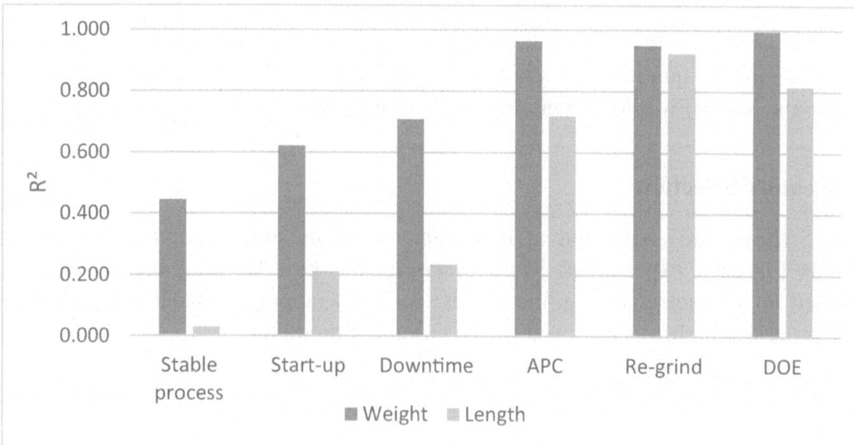

Fig. 4. Comparison of the prediction quality for the different experiments.

It can be seen that the DOE, the process with re-grind material and APC and the process with only re-grind material are easier to predict than the stable process, start-up and downtime. The best prediction result provides the DOE for the weight with a R^2 of 0.995, i.e. 99.5% of the weight variance can be explained by the model. In general, it can be observed that the weight is easier to predict than the length. An explanation could be that the measuring accuracy of the weight measurement compared to the systematical variation of the quality criterion is higher than that of the length measurement.[3] A greater measuring effort is likely to improve the model quality for the length prediction as well. It is also apparent that process situations with a low variance in the process parameters e.g. stable process are harder to predict than process situation with high variance e.g. DOE (cf. Table 3). This might be one reason, why APC, re-grind and DOE yield better results than the stable process having the smallest standard deviations.

[3] The parts were weighed with a Sartorius Entris 153I-1S balance with 1 mg reproducibility, part length were extracted algorithmically from images taken with a Canon Eos 5D Mark III DSLR with EF 70-200mm f/4L USM objective.

Table 3. Standard deviations (Mean values for weight and length are approximately 19.7 g and 182.7 mm respectively) of weight and length from the different experiments

Experiment	Weight [g]	Length [mm]
Stable process	0.005	0.023
Start-up	0.010	0.027
Downtime	0.015	0.025
APC	0.028	0.048
Re-grind	0.029	0.094
DOE	0.277	0.060

In the coming sections, only the results for the weight will be shown, as the length prediction shows qualitatively comparable behavior and a complete presentation (weight and length) would exceed the scope of this paper.

3.2 Feature Selection

Figure 5 shows the coefficient of determination of the individual feature selection methods for each experiment. More precisely, the best learning method with the respected R^2 was selected. E.g. the best R^2 for the wrapper approach is 0.334 for the stable process which was achieved by the ensemble predictor.

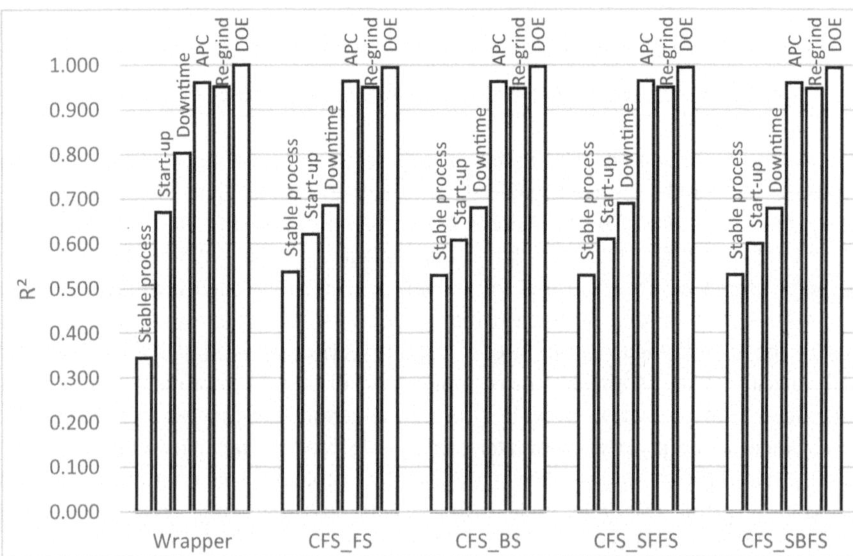

Fig. 5. Comparison of the predictive performance for the different feature selection algorithms in terms of weight.

From the comparison of feature selection methods, one can generally derive two conclusions: First, the results of the wrapper approach using multiple linear regression differs from the results of the CFS filters and second, the different search strategies only slightly affect the filters' performances. While the wrapper yields better results on the start-up and downtime datasets, the CFS performs better on the stable process data. On the other three datasets, no significant differences occur.

The wrapper was only tested with linear regression as predictor, using the FS as search strategy. The selected features with this method were used for the other predictors acting as a filter method [18]. Figure 5 also shows that the experiments with a high variance (cf. Table 3) are easier to predict that the experiments with low variance. While the features, which are selected, vary, but the overall performance is within a similar range.

Figure 6 exemplarily shows the prediction performance of the wrapper depending on the number of features selected for the re-grind experiment. The R^2 for the training dataset is most of the time higher than for the validation dataset, which was to be expected since the training data is known to the predictor while the validation set is not. The highest R^2 is reached for 13 features with a value of 0.9165. For higher feature numbers, the training R^2 continues to increase, while the test R^2 is decreasing due to overfitting.

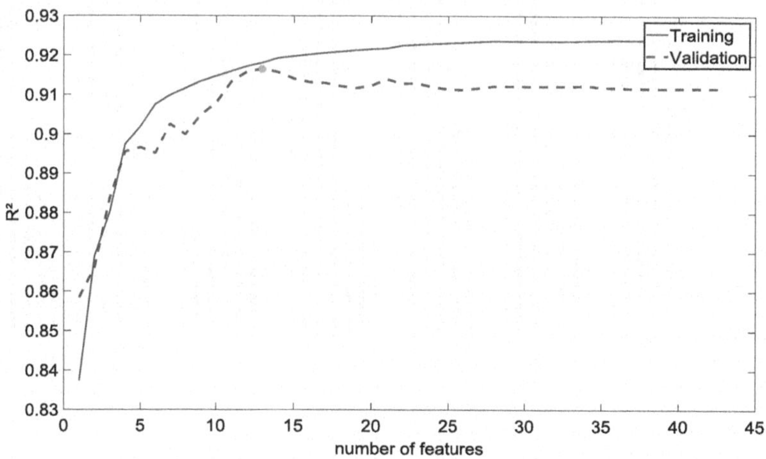

Fig. 6. Prediction quality for the weight using the wrapper approach. Coefficient of determination in terms of the number of selected features for the training and validation re-grind dataset.

3.3 Learning Algorithms

After evaluating the influence of process states and feature selection algorithms on model quality, we now want to compare the learning algorithms themselves.

As it is possible to see from Fig. 7 the Gaussian process regression is the best predictor for the prediction of the weight in every experiment. The overall highest coefficient of determination is 0.995 for the DOE with the Gaussian process regression. Like the results shown in Fig. 5 the DOE, the process with re-grind material and the process with re-grind material and APC are good to predict. Furthermore, the Gaussian process regression stands out on the stable-process data, a dataset with small variance all other algorithms have trouble to predict. It might also surprise, that multiple linear regression does not perform worst at all on all datasets: despite its simple model structure, especially on the start-up and downtime data it exceeds expectations and yield above average results. Besides kNN, which yields mainly below average results, the other algorithms (ANNs, SVM, decision trees and the ensemble) have a generally comparable predictive quality.

In general, it becomes obvious, that the process state used for data generation is much more important than the learning machine, since the algorithm can only extract correlations that are present in the data.

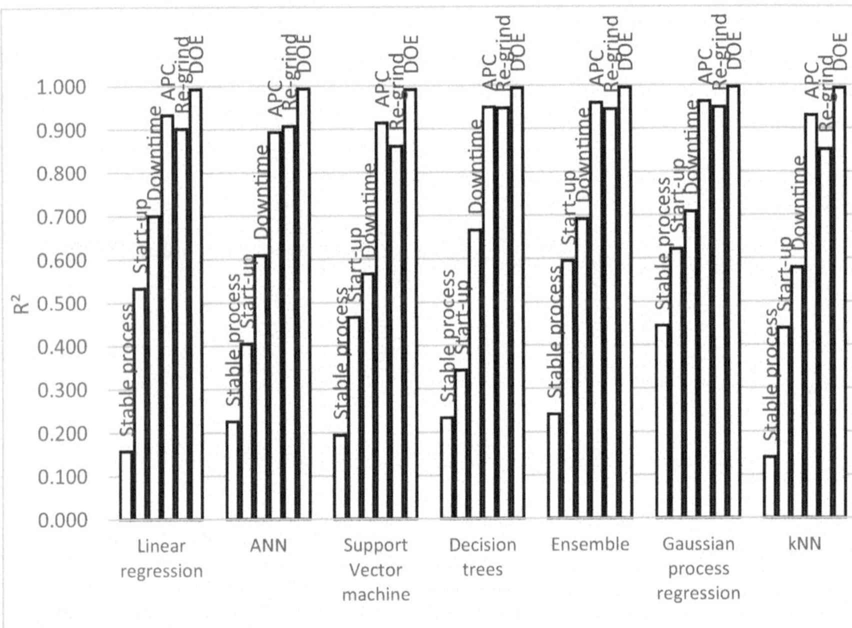

Fig. 7. Coefficient of determination for the different learning algorithms. The weight should be predicted.

4 Conclusion and Outlook

In this study six different experiments were done using a KraussMaffei 120-380 PX injection molding machine. The data include 48 machine and process parameters as well as the weight and the length of the molded parts as quality criteria. 1000 molding cycles (860 for the DOE) were carried out. The pre-processing of the data included cross-validation using 80% of the data for training and 20% for validation of the models. In the first step feature selection was executed, comparing a wrapper approach with four filter methods. The filter methods contained FS, BS, SFFS and SBFS as search strategies. The PEM was CFS according to HALL with the Pearson correlation coefficient. In total 22 predictor models were built and their hyperparameters were optimized using the Bayesian optimization. Six machine learning methods, including ANN, Support-vector machine, Decision trees, Ensemble, Gaussian process regression, kNN and normal linear regression were compared. The prediction performance of the different models was calculated through the coefficient of determination.

The results show, that process states with a high variance of the quality criteria, such as those based on the variation of the re-grind material fraction and the DOE provide the best base for learning good quality prediction models. The weight is better predictable then the length, with the highest R^2 of 0.995 for the DOE learned by a Gaussian process regression, which yielded the best results on the other datasets as well. Regarding the evaluated feature selection methods, their influence on the model quality was rather small when comparing the different search strategies. However, the presented wrapper's and filters' performances differed significantly on three out of six datasets. Still, it is hard to judge which approach is better, since there is no method outperforming the others on all datasets. Additionally, other feature selection methods might perform different, so in future work different PEMs should be addressed like mutual information or Relief. Furthermore, the framework should be expanded to other machine learning methods in particular methods for classification. Also, a holistic approach has to deal with the detection and reaction to concept drift, which might negatively affect the predictive quality.

References

1. Hopmann, C., Michaeli, W.: Einführung in die Kunststoffverarbeitung, 7th edn. Hanser, Munich (2015)
2. Hopmann, C., Michaeli, W., Greiff, H., et al.: Technologie des Spritzgießens, 4th edn. Hanser, Munich (2017)
3. Standard DIN 24450: Maschinen zum Verarbeiten von Kunststoffen und Kautschuk. Beuth, Berlin (1987)
4. Schiffers, R.: Verbesserung der Prozessfähigkeit beim Spritzgießen durch Nutzung von Prozessdaten und eine neuartige Schneckenhubführung. PhD thesis (2009)
5. Gierth, M.: Methoden und Hilfsmittel zur prozessnahen Qualitätssicherung beim Spritzgießen von Thermoplasten. PhD thesis (1992)
6. Hanning, D.: Continuous Process Control. Qualitätssicherung im Kunststoffverarbeitungsprozess auf Basis statistischer Prozessmodelle. PhD thesis (1994)
7. Häußler, J.: Eine Qualitätssicherungsstrategie für die Kunststoffverarbeitung auf der Basis künstlicher Neuronaler Netzwerke. PhD thesis (1994)
8. Vaculik, R.: Regelung der Formteilqualität beim Spritzgießen auf Basis statistischer Prozessmodelle. PhD thesis (1996)
9. Al-Haj Mustafa, M.: Modellbasierte Ansätze zur Qualitätsregelung beim Kunststoffspritzgießen. PhD thesis (2000)

10. Schnerr, O.: Automatisierung der Online-Qualitätsüberwachung beim Kunststoffspritzgießen. PhD thesis (2000)
11. Walter, A.: Methoden des prozessnahen Qualitätsmanagements in der Kunststoffverarbeitung. PhD thesis (2000)
12. Liedl, P., Haag, G., Müller, H., et al.: Spitzenqualität mit kurzen Zyklen. Kunststoffe **2**, 38–40 (2010)
13. Hopmann, C., Theunissen, M., Heinisch, J.: Von der Simulation in die Maschine – objektivierte Prozesseinrichtung durch maschinelles Lernen. In: VDI Jahrestagung Spritzgießen, Baden-Baden (2018)
14. Hopmann, C., Theunissen, M., Wipperfürth, J., et al.: Prozesseinrichtung durch maschinelles Lernen. Kunststoffe **6**, 36–41 (2018)
15. Hopmann, C., Wahle, J., Theunissen, M., et al.: Flexibilisierung der Spritzgießfertigung durch Digitalisierung. In: Kunststoffindustrie 4.0 – 29. Internationales Kolloquium Kunststofftechnik, pp. 76–88 (2018)
16. Tercan, H., Guajardo, A., Heinisch, J., et al.: Tranfer-learning: bridging the gap between real and simulation data for machine learning in injection molding. Procedia CIRP **72**, 185–190 (2018)
17. Hopmann, C., Bibow, P., Heinisch, J.: Internet of Plastics Processing. IPC Madison, USA (2019)
18. Guyon, I., Elisseeff, A.: An introduction to variable and feature selection. J. Mach. Learn. Res. **3**(3), 1157–1182 (2003)
19. Charest, M., Finn, R.; Dubay, R.: Integration of artificial intelligence in an injection molding process for on-line process parameter adjustment. In: Annual IEEE International Systems Conference (SysCon), pp. 1–6. IEEE, Vancouver, Canada (2018)
20. Gao, H., Zhang, Y., Zhou, X., Li, D.: Intelligent methods for the process parameter determination of plastic injection molding. Front. Mech. Eng. **13**(1), 85–95 (2018)
21. Duda, R., Hart, P., Stork, D.: Pattern Classification, 2nd edn. Wiley, New York (2001)
22. Arlot, S., Celisse, A.: A survey of cross-validation procedures for model selection. Stat. Surv. **4**, 40–79 (2010)
23. Pudil, P., Novovičová, J., Kittler, J.: Floating search methods in feature selection. Pattern Recogn. Lett. **15**(11), 1119–1125 (1994)
24. Chandrashekar, G., Sahin, F.: A survey on feature selection methods. Comput. Electr. Eng. **40**(1), 16–28 (2014)
25. Hall, M.A.: Correlation-based feature selection for machine learning. PhD thesis (1999)
26. Kira, K., Rendell, L.A.: The feature selection problem: traditional methods and a new Algorithm. In: AAAI'92 Proceedings of the Tenth National Conference on artificial Intelligence, pp. 129–134. AAAI, San Jose, California, USA (1992)
27. Ding, C., Peng, H.: Minimum redundancy feature selection from microarray gene expression data. In: IEEE Computer Society Bioinformatics Conference, pp. 523–528, IEEE, Stanford, USA (2003)
28. Hall, M. A., Smith, L. A.: Practical feature subset selection for machine learning. In: ACSC'98 Proceedings of the 21st Australasian Computer Science Conference, pp. 181–191. ACSC, Perth, Australia (1998)
29. Russell, S.J., Norvig, P.: Artificial intelligence, 2nd edn. Prentice Hall, Pearson Education, Upper Saddle River (2003)
30. Alpaydin, E.: Introduction to machine learning, 2nd edn. MIT Press, Cambridge (2010)
31. Hagan, M.T., Demuth, H.B., Beale, M.H.: Neural Network Design, 1st edn. PWS, Boston (1996)
32. Smola, A.J., Schölkopf, B.: A tutorial on support vector regression. Stat. Comput. **14**(3), 199–222 (2004)

33. Breiman, L., Friedman, J., Stone, C.J., Olshen, R.A.: Classification and Regression Trees, 1st edn. CRC Press, Boca Raton (1984)
34. Biau, G., Devroye, L., Dujmović, V., Krzyżak, A.: An affine invariant k-nearest neighbor regression estimate. J. Multivar. Anal. **112**, 24–34 (2012)
35. Hastie, T., Tibshirani, R., Friedman, J.H.: The Elements of Statistical Learning. Data Mining, Inference, and Prediction, 2nd edn. Springer, New York (2017)
36. Breiman, L.: Random forests. Mach. Learn. **45**(1), 5–32 (2001)
37. Rasmussen, C.E., Williams, C.K.I.: Gaussian Processes for Machine Learning, 3rd edn. MIT Press, Cambridge (2008)
38. Urban, D., Mayerl, J.: Angewandte Regressionsanalyse: Theorie, Technik und Praxis, 5th edn. Springer VS, Wiesbaden (2018)
39. Claesen, M., De Moor, B.: Hyperparameter search in machine learning. In: MIC 2015: The XI Metaheuristics International Conference, pp. 1–5, MIC, Agadir, Morocco (2015)
40. Bengio, Y.: Practical recommendations for gradient-based training of deep architectures. Lecture Notes in Computer Science 7700 LECTURE NO, pp. 437–478 (2012)
41. Ito, K., Nakano, R.: Optimizing Support Vector regression hyperparameters based on cross-validation. In: Proceedings of the International Joint Conference on Neural Networks, pp. 2077–2082. IEEE, Portland, USA (2003)
42. Matignon, R.: Data Mining using SAS Enterprise Miner, 1st edn. Wiley-Interscience, Hoboken (2007)
43. Wilson, D.R., Martinez, T.R.: Improved heterogeneous distance functions. J. Artif. Intell. Res. **6**, 1–34 (1997)

Automated 3D Skeleton Winding Process for Continuous-Fiber-Reinforcements in Structural Thermoplastic Components

Björn Beck[1(✉)], H. Tawfik[1], J. Haas[1], Y.-B. Park[2], and F. Henning[1,3]

[1] Fraunhofer Institute for Chemical Technology ICT, 76327 Pfinztal, Germany
bjoern.beck@ict.fraunhofer.de
[2] Ulsan National Institute of Science and Technology (UNIST), Ulsan, Republic of Korea
[3] Karlsruhe Institute of Technology, Institute for Vehicle System Technology, Karlsruhe, Germany

Abstract. Continuous fiber reinforcements can significantly improve the mechanical properties of thermoplastic injection-molded components compared to short and long fiber reinforcements. By aligning the continuous fibers in the main load paths, the reinforcement can be optimally exploited. The 3D skeleton winding process (3DSW) is a robot-based filament winding technique in which defined load application points are connected with wound closed loop structures. The introduction of thermoplastic commingled yarns in the winding process allows an efficient fiber impregnation to produce fiber skeletons that can be overmolded in an injection molding process to locally reinforce the final component. The combination of a robot-based winding process and injection molding as a process for large-scale production enables the use of thermoplastic materials for complex structural applications in higher quantities.

This paper presents the 3DSW manufacturing process and introduces a simple loop and a generic 3D test specimen. Fundamental investigations into these structural components with overmolded fiber skeletons demonstrate the potential of continuous fibers in injection molded components made from PP and PPS.

Keywords: Robot-based winding · 3DSW · Commingled yarns · Polyphenylene sulfide

1 Introduction

To meet the demand for structural lightweight components, i.e. for automotive applications, the targeted use of continuous fiber reinforcements in the highly stressed areas is already established in combination with large-scale production processes, such as injection molding. The continuous fiber reinforcements used for this purpose (e.g. pre-impregnated fabrics) must be draped in accordance with the desired component geometry and requirements. In the case of plane continuous fiber reinforcements,

C. Hopmann and R. Dahlmann (Eds.): *Advances in Polymer Processing 2020*,
pp. 150–161, 2020. https://doi.org/10.1007/978-3-662-60809-8_13

such as fabrics, scrims and unidirectional tapes, this can lead to significant restrictions in the design freedom, as it is not always possible to adapt the reinforcements to the component shape in the highly stressed areas. This is particularly the case for complex structural components.

Another possibility to produce continuous-fiber-reinforced components is the filament winding processes, which can be used to manufacture mainly hollow profiles with rotation-symmetric geometries (e.g. pressure vessels, axles, tubes, rollers and shafts). Rotating winding mandrels are usually used for this purpose. Roving guides can be used to automatically place impregnated continuous reinforcing fibers (e.g. glass or carbon fibers) onto these mandrels in a defined pattern and with high accuracy and reproducibility [1, 2].

The 3D skeleton winding technology (3DSW) extends the range of application for continuous fiber reinforcements and enables the use of a filament winding technology for the local continuous fiber reinforcement of structural components. The components made in 3DSW usually consist of defined load application points that are connected via wound fiber structures to transfer external loads. In contrast to the well-known filament winding process, continuous fiber reinforcements in 3DSW are used in a resource-efficient manner, i.e. only locally in the highly stressed areas of a structural component according to the actual component load paths. This corresponds to the preferred load of fibers in general, since fibers can transmit external loads mainly via longitudinal stresses [3].

In comparison to volumetrically short or long fiber reinforced components, the 3DSW components can provide higher stiffness and strength at lower weight. Weight savings can thereby be achieved, since areas of low stress are designed without reinforcement or with low short fiber contents.

2 3DSW Method

Topology optimizations can be used to define the orientation of local continuous fibers according to the external loads. Depending on the component space specifications and the load introduction directions, the resulting fiber skeleton can be a complex truss structure of continuous fibers, which is connected by several nodes and load application points. The resulting fiber skeleton absorbs tension loads, while the component shape and material composition of the overmolding matrix material are selected with reference to compression loads. Metal inserts (e.g. bushings) are used at the nodes and the load application points of the skeleton structure. By deflecting the continuous reinforcing fiber strands at the nodes and winding them around load application elements (eye windings), the reinforcements can be adapted to the actual load path. Compared to other continuous fiber-reinforced components, this offers the advantage that applied forces can be introduced into the reinforcing structure by form-fit manner.

In general, the 3DSW process can be divided into three different process steps: Impregnation, 3D winding and overmolding (see Fig. 1).

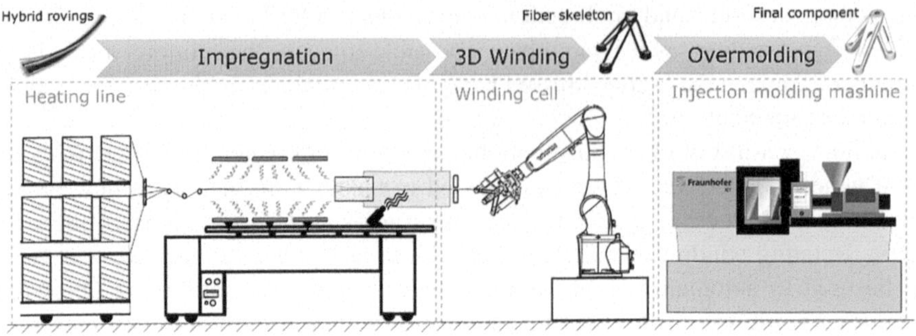

Fig. 1. Process chain of the 3D skeleton winding process

In the first process step (impregnation), the commingled yarns, consisting of continuous thermoplastic and reinforcement filaments, are pulled through a heating line with several heating zones in which the thermoplastic filaments are heated above melting temperature using short-wave infrared heating. The reinforcement filaments are then impregnated with thermoplastic material in a heated die with cross-sectional constriction after the IR heating section. Since the processability of the commingled yarns is mainly dependent on the temperature, the polymer is kept molten using a circulating hot air oven after the impregnation die until the strand reaches the winding area. The necessary pull-off force is achieved by the manipulation of a winding tool attached to a six axis industrial robot (3D winding). The winding tool describes a fixture designed for a specific fiber skeleton structure and contains the load application or node elements (inserts). The skeleton structure is wound around the inserts solely by manipulating the winding tool according to the load paths of the component [4]. Ideally the pull-off speed is kept constant during the winding phase to guarantee constant impregnation quality of the reinforcing fibers. At the same time, fiber pre-tensioning plays an important role in ensuring that the fibers are aligned. The fiber pre-tensioning is mainly defined by the friction in the heated die and the adjustable roll-off resistance on the roving rack (unwinding unit). The heated commingled yarns, which are wound around the inserts, adhere to each other and form a stiff and easy-to-handle skeleton structure (semi-finished product) after cooling. In the third process step (overmolding), the manufactured skeleton structure is placed in an injection mold and overmolded with a thermoplastic matrix to realize the final component shape including the local continuous fiber reinforcements and load introduction elements. The 3D skeleton prototype line developed at Fraunhofer ICT can be seen in Fig. 2.

Fig. 2. 3D skeleton winding prototype line at Fraunhofer ICT

3 Development of a New Generic 3D Test Specimen

A new generic 3D test specimen was developed based on investigations carried out on simple loop structure specimens (see Fig. 4). It essentially connects three simple loop components in one main load application point to examine different load application directions. In order to investigate the influence of obliquely acting forces, the main load application point was displaced in Z-direction. Three mounting points (side inserts) and one main load application point (main insert) introduce external loads, whereby each side insert is connected with the main insert by a closed loop winding structure (see Fig. 3 left). Furthermore, closed windings around each insert (eye windings) increase the dimensional stability of the entire skeleton structure and prevent the fibers from slipping when the fiber arrangement is inclined. The angles of inclination for the middle leg and the side legs were set to 25° and 45° respectively (see Fig. 3 right). As the loop structure provides less rigidity by loading perpendicularly to the fiber direction, the topology of the part was optimized. Consequently, in addition to vertical ribs, which include the continuous fiber structure, horizontal ribs were designed at an angel of 90° to the fiber direction (see Fig. 3 middle). The aim of the horizontal ribs is to increase the component's stiffness perpendicular to the fiber direction, while still allowing high compliance in fiber direction for maximum fiber utilization, i.e. fiber failure before matrix failure. The matrix is injected in the overmolding process using a film gate at the main insert, filling the mold along the continuous fiber reinforcement. Such gate position results in an ideal short fiber orientation in the direction of the legs. Moreover, the flow forces applied to the continuous fiber loops are minimal, since the matrix flow direction is along the fiber orientation, enhancing fiber alignment accuracy. Weld lines emerge behind each insert, representing weak areas which can be eliminated when introducing continuous windings, as explained in Sect. 5.

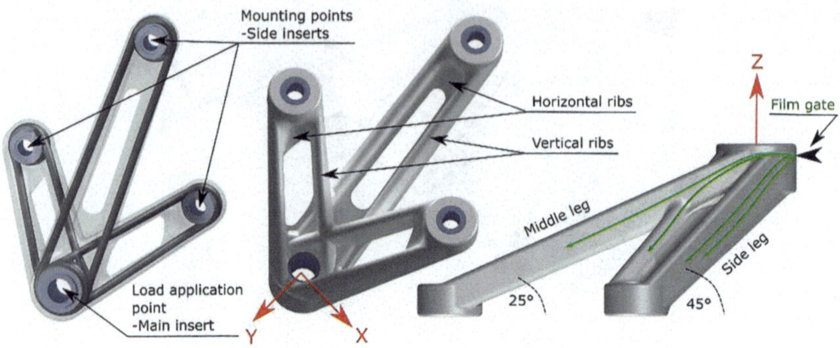

Fig. 3. New generic 3D test specimen

4 Experimental

4.1 Commingled Yarn Processing

The heating line consisted of two heating zones. In the first zone, six IR ceramic heaters with maximum power of 800 W were placed with adjustable distance to the strand. The second heating zone was the impregnation die, tempered with a 600 W heater band and controlled with a temperature control unit (Thermoplay TH-M6).

4.2 Materials

The test specimens manufactured in polypropylene (PP) were made using a PP/glass fiber (GF) commingled yarn of the type Twintex R PP 60 (tex per strand = 2x 1870 tex). The PP/GF skeleton structure was overmolded with Braskem C765-15NA (unreinforced PP) and Hostacom HRG 328T NAT (30 wt.-% GF reinforced). The polyphenylene sulfide specimens (PPS) were made with PPS/GF commingled yarn named COMFIL-G 60G (tex per strand = 8x 476 tex). The overmolding with PPS was carried out with Xytron U3020E (unreinforced PPS) and KOPLA KHG1040 (40 wt.-% GF reinforced).

4.3 Test Specimens

To investigate the potential of wound reinforcement structures, two generic components were used in this study. The first sample was a simple loop structural component, mainly used to investigate the effects of winding numbers and new material combinations. It consisted of two inserts for load application connected through a closed loop winding, which was embedded in the thermoplastic matrix in the injection molding process, see Fig. 4 (left).

The second specimen was a generic 3D component as described in Sect. 3, see Fig. 4 (right). The component weight for unreinforced PP and PPS results in 253 g and 343 g respectively. Reinforcing the structure with short glass fibers (30 wt.-% for PP and 40 wt.-% for PPS) increased the weight by 20% and 27%, while 2 closed loop fiber windings increase the weight by 7% and 2% compared to the unreinforced components.

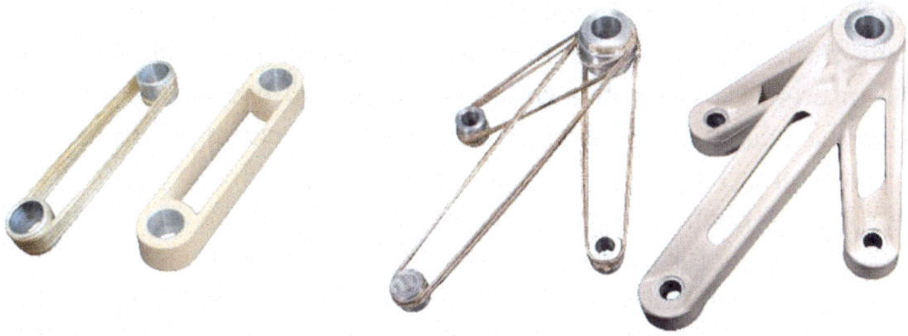

Fig. 4. Simple loop structure before/after overmolding with PPS (left), and 3D generic component before/after overmolding with PPS (right)

4.4 Robot-based 3D Skeleton Winding and Overmolding

The core of the winding cell is an industrial six-axis robot (KUKA KR 10 R1100 sixx) with a modified 6th axis to allow endless rotation. The simple loop structural components were overmolded using an injection molding machine from Arburg (Allrounder 279/320 C), whereas the generic 3D structural components needed to be overmolded with a KraussMaffei 550-2000 GX injection molding machine due to the higher shot volume.

Two different winding tools were developed to produce the test specimens for this study, see Fig. 5. Both tools were equipped with pneumatic grippers to pick up the heated strand from the heating line. The metal inserts, which were used as load application elements in the final component, were attached to the insert holders before the winding process started. Both winding tools had the possibility to pre-tension the wound filaments by displacing one insert holder with an integrated pneumatic cylinder. After cooling, this cylinder slightly released to compensate shrinkage forces allowing a pneumatic ejection of the final wound structure afterwards.

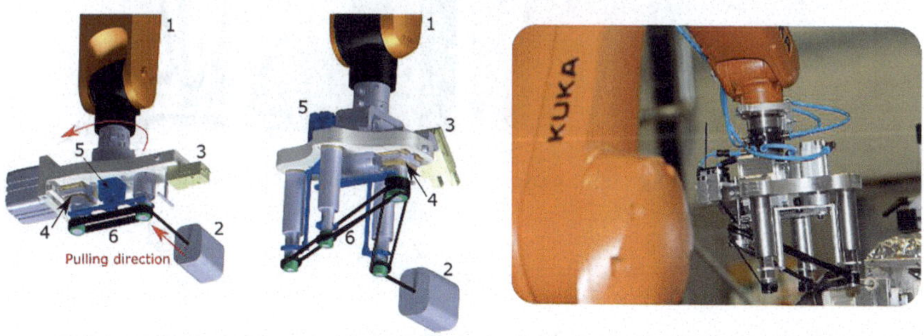

Fig. 5. Winding tools for the simple loop structure and for the 3D component (left): (1) 6-axis robot; (2) Impregnation die; (3) Fiber gripper; (4) Pre-tensioning mechanism; (5) Ejector; (6) Wound fiber skeleton. PP/GF skeleton winding process in progress (right)

The winding tool was manipulated using an industrial robot to pull the impregnated rovings from a fixed strand feed and simultaneously wind them around and between the inserts [4]. The robot was programmed to manipulate the winding tool in a way that the heated strand connected the single load application elements in a closed loop structure. This loop structure also included closed windings around the inserts (eye windings), which were essential for the structure stability during winding and handling, as well as for the load transfer in the final component. Moreover, previous investigations, which tested simple fiber loops with eye windings on both inserts, showed that the breaking load can be increased by 86% compared to loops without windings [5]. During the winding of the fiber skeleton structure, the impregnated yarns needed to be kept straight at the output of the heating line to provide homogeneous fiber distribution over the cross section of the strand. According to Minsch et al., for optimum component quality, the outlet angle at the fiber feed should be kept low during the winding process. Excessive fiber outlet angles increase fiber tension significantly, resulting in fiber damage [6].

Overmolding of the fiber skeleton was carried out in the final production step. Previous experience at Fraunhofer ICT showed that warpage is likely to occur when thermoplastics with high shrinkage are used to overmold components with large span [7]. During cooling, glass fibers with a linear thermal expansion of $4.8 * 10{-}^{6}\,K^{-1}$ (for E-glass) almost maintain their length, while the surrounding thermoplastic matrix shrinks, especially in the case of semi-crystalline thermoplastics [8]. Consequently, for the 3D generic structure components, hydraulic pre-tensioner mechanisms, acting on each loop separately, were integrated in the injection mold. Parts with pre-tensioned fiber skeletons showed high dimensional stability. The overmolding process is illustrated in Fig. 6.

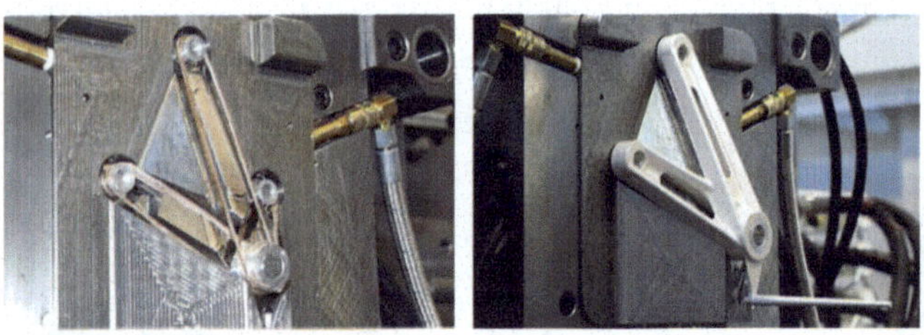

Fig. 6. 3D fiber skeleton before overmolding (left), and after overmolding (right)

4.5 Mechanical Testing

An Inspek Table 50 from Hegewald & Peschke (50 kN) was used to investigate the mechanical performance of the mentioned components. For the simple loop structural component, an U-shaped mounting bracket with a bolt connection was used to apply the load on the component. For the 3D structural component, a test fixture was developed, on which the 3D component is mounted at three fixation points applying the load to the main insert. Figure **7** shows different setups of the testing fixture for the load case FX and FY as well as the corresponding component deformation and stress distribution (blue areas illustrate the compressive stress and red areas illustrate the tensile stress). All tensile tests were carried out based on DIN EN ISO 527 standard with at least five tested specimens at a testing speed of 5 mm/min [9].

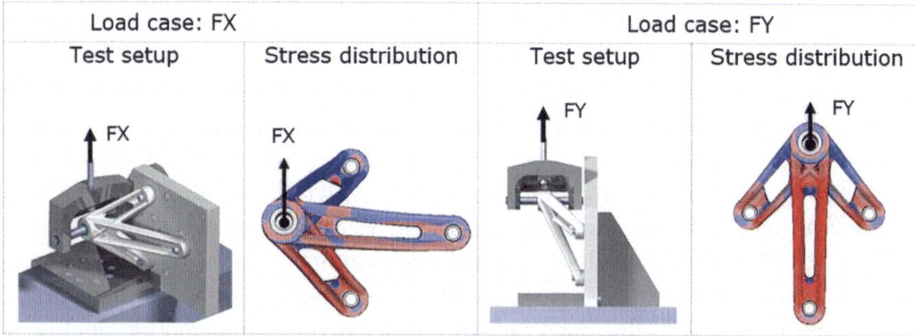

Fig. 7. Testing fixture for FX and FY load cases and stress distributions

5 Results and Discussion

By testing loop structural components with unreinforced PP and 30 wt.-% GF reinforced PP matrix material, it was observed that the higher ductility of the matrix positively increases the reinforcing effect of the wound continuous fibers, as can be seen in Fig. 8. The breaking load of unreinforced PP with 10% elongation at break increased proportionally to the number of continuous windings [10]. While unreinforced PP specimens failed at 3.7 kN, adding two continuous closed loop windings (equivalent to 6 wt.-%), increased the breaking load by 194%. Reinforcing the component with four (12 wt.-%) and six (18 wt.-%) windings, further increased the breaking load compared to the unreinforced reference sample by 425% and 511% respectively.

When conducting the same test campaign with a 30 wt.-% short-fiber-reinforced PP matrix with only 2.6% elongation at break, the increase in breaking force relating to the winding number is significantly lower [11]. Without continuous windings, the short-fiber-reinforced overmolding material showed 82% higher breaking forces than the unreinforced material. When adding six continuous windings, however, the short-fiber-reinforced overmolding material showed 20% lower breaking forces. Consequently, the maximum utilization of wound continuous fibers can be achieved when the overmolded matrix has higher elongation at break than the reinforcing fibers, i.e. approx. 4.8% [2, 12].

A similar investigation was carried out using unreinforced and short-glass-fiber-reinforced PPS materials in injection molding, which further highlighted the disadvantage of brittle matrix materials in combination with continuous windings. By considering a 40 wt.-% short-fiber-reinforced PPS material, no significant improvement in breaking load was determined depending on the winding number. All tested samples showed, as expected, matrix failure instead of fiber failure. The same investigation with impact modified unreinforced PPS material with increased elongation at break (14% [13]) improved the breaking load of the unreinforced PPS reference sample by 37%, 78% and 125% when adding two, four and six windings respectively (see Fig. 8).

Since the short-fiber-reinforced matrix materials have different fiber contents for PP and PPS, the investigations with short-fiber-reinforced matrices are to be regarded independently. Nevertheless, the conclusion can be drawn that the elongation at break of the overmolding material - which can be adjusted, for instance, by the short fiber content - has a decisive influence on the achievable breaking force of the continuous fiber-reinforced components. The fiber weight percentage of the PP/GF and PPS/GF commingled yarns was chosen to be identical at 60 wt.-%, while using similar tex numbers (see Sect. 4.2). However, the density of PPS (1.29 g/cm^3) compared to the density of PP (0.9 g/cm^3) results in higher fiber volume content of the PPS strand compared to the PP strand [10, 13]. As the same impregnation die diameter was used for both materials at the same pull-off speed, higher impregnation quality was achieved with PP. This describes one reason for the higher breaking loads when comparing unreinforced overmolding matrices with four and six continuous windings, while at two windings the better impregnation quality of PP/GF commingled yarns was compensated by the superior mechanical properties of the PPS matrix.

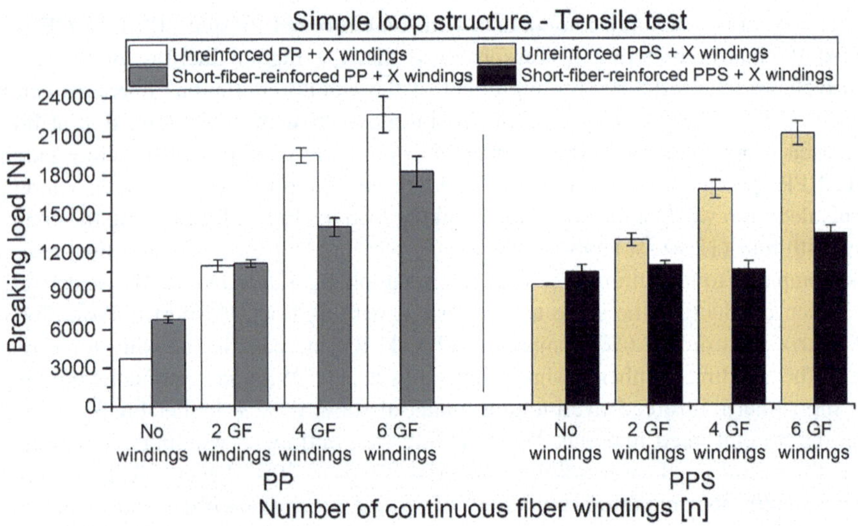

Fig. 8. Simple loop structure tensile test results: PP (left), and PPS (right)

When the reinforced 3D structural component was loaded in the Y-direction, while almost all the continuous fibers were aligned in the load direction, a significant improvement in strength was achieved as well as a remarkable improvement in the failure mode. Where PP specimens were investigated, unreinforced 3D samples failed at 7.2 kN load at weld line A, as shown in Fig. 9 (failure mode A). By reinforcing PP with 30 wt.-% short glass fibers, the breaking force increased by 71% to fail at 12.4 kN load at weld line A (see Fig. 9, failure mode B). Pure PP with wound continuous fibers (equivalent to only 5 wt.-%), increased the breaking load by 147% to fail at 17.8 kN. The fiber skeleton increased the resistance to weld line failure and thereby the fiber breakage indicating full utilization of the continuous fibers (see Fig. 9, failure mode C). Meanwhile, unreinforced PPS components bore up to 27.3 kN load. Due to the complex part geometry, induced bending forces were critical for the load cell of the testing machine. Reinforced PPS components with wound fiber structure reached the machine safety limit by 30 kN and the test was stopped without failure.

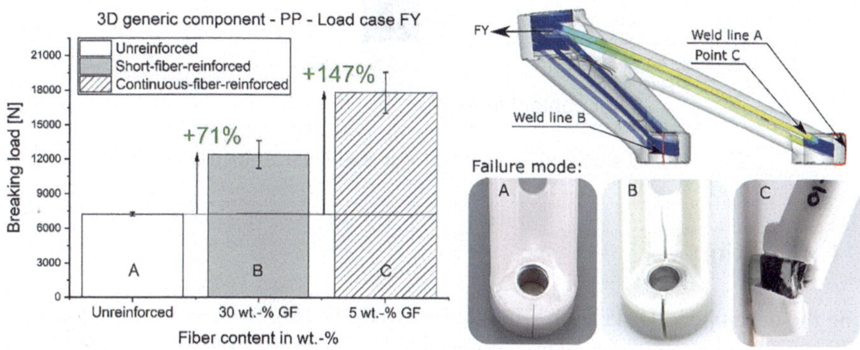

Fig. 9. PP 3D structural component test results under load case FY (left); Stress distribution at weld lines and at fibers (upper right); breaking loads and failure mode (lower right)

Breaking forces, resulting from loading the component in X-direction are generally lower than in Y-direction, as only the fiber loop in the lower leg is under tension, as illustrated in Fig. 7. While the load is applied perpendicularly to the fibers of the middle leg, the side leg in load direction is completely under compression load, whereby the continuous fibers do not significantly contribute to the component's strength. By loading the unreinforced PP 3D structural component in the X-direction, weld line failure of the lower leg occurs at 3.8 kN load. When reinforcing PP with 30 wt.-% short glass fibers, the matrix failed (brittle failure) at 9.3 kN load. When PP is reinforced with two continuous windings (equivalent to 5 wt.-%), the breaking load increased by 191% and the specimen failed at 11.1 kN through fiber failure.

Unreinforced and continuous-fiber-reinforced PPS specimens showed significantly higher breaking force values than PP specimens, as illustrated in Fig. 10. Unreinforced PPS specimens failed (ductile failure) at 16.2 kN load. Adding two continuous windings increased the breaking load by 23% and the specimen failed with fiber fracture at 20 kN. The percentage increase in breaking load by using continuous fiber reinforcements in PPS is low compared to PP due to the relatively high strength of PPS (65 MPa [13]) and thus higher critical fiber content. Consequently, two continuous windings provide lower reinforcing effect to PPS compared to PP. Meanwhile, short fiber reinforced PPS (40 wt.-%) showed remarkable brittle failure behavior.

The matrix failure occurred at the end of the vertical ribs of the side leg at 7 kN load, leading to 57% decrease in breaking load compared to the unreinforced PPS specimens.

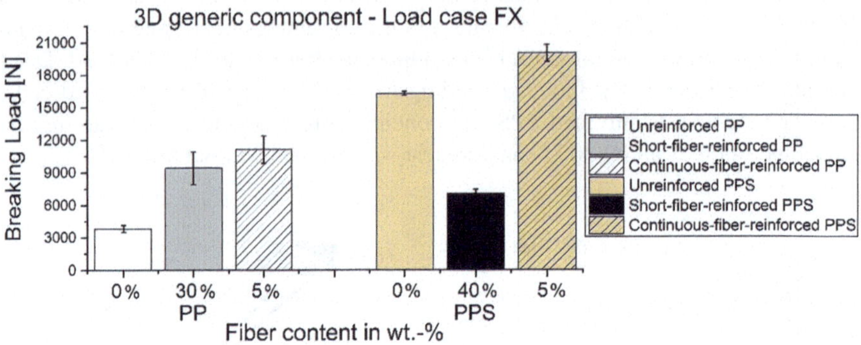

Fig. 10. 3D structural component test results under load case FX

6 Conclusion

Using the developed 3D skeleton winding process (3DSW), two generic structural thermoplastic components were manufactured to investigate the influence of wound continuous fibers in combination with injection molding. It was possible to demonstrate the potential of overmolded fiber skeletons, where wound continuous reinforcements connect load application elements to transfer external loads. By using continuous fibers only locally with the right orientation according to the load path, it was possible to significantly increase the breaking forces of the tested components. However, this reinforcing effect was substantially lower if fewer fibers were oriented in load direction. Additionally, it was shown that high elongation at break of the overmolding matrix played a decisive role in utilizing the material properties of the continuous-fiber-reinforcements to increase the breaking load. Furthermore, it was possible to demonstrate improvements in failure modes when using wound continuous fibers, whereby the resistance to weld line failure increased and fiber fracture occurred instead.

Acknowledgments. This research was supported by the Core Industrial Technology Development Program (Grant No. 10052896) funded by the Ministry of Trade, Industry and Energy (MOTIE) of Korea through Korea Evaluation Institute of Industrial Technology (KEIT).

References

1. Ehrenstein, G.W.: Faserverbund-Kunststoffe – Werkstoffe – Verarbeitung – Eigenschaften. Hanser, München (2006)
2. Witten, E. (Eds.): Handbuch Faserverbundkunststoffe/Composites – Grundlagen, Verarbeitung, Anwendungen. AVK, Industrievereinigung Verstärkte Kunststoffe. Springer Vieweg, Wiesbaden (2014)

3. Flemming, M., Roth, S.: Faserverbundbauweisen Eigenschaften – Mechanische, konstruktive, thermische, elektrische, ökologische, wirtschaftliche Aspekte. Springer, Berlin (2003)
4. Verfahren und Vorrichtung zur Herstellung eines mit Endlosfasern verstärkten Polymer-Formteils. Erfinder: Krause, W., Reif, M., Walch, M. und Henning, F. Anmeldung: 24. November 2004, EP 1 568 473 B1
5. Huber, T.: Einfluss lokaler Endlosfaserverstärkungen auf das Eigenschaftsprofil struktureller Spritzgießbauteile, Wissenschaftliche Schriftenreihe des Fraunhofer ICTBd. 60. Fraunhofer, Stuttgart (2014)
6. Minsch, N., Herrmann, F.H., Gereke, T., et al.: Analysis of filament winding processes and potential equipment technologies. Procedia CIRP **66**, 125–130 (2017)
7. Heinzle, V., Huber, T., Henning, F., et al.: Process development of injection molded parts with wound fiber structures for local reinforcement. In: AIP Conference Proceedings. American Institute of Physics, Nuremberg, Germany, pp. 736–740 (2014)
8. Eyerer, P., Elsner, P., Hirth, T.: Polymer Engineering – Technologien und Praxis. VDI-Buch, Springer, Berlin (2008)
9. DIN EN ISO 527-1: Kunststoffe – Bestimmung der Zugeigenschaften – Teil 1 Allgemeine Grundsätze
10. Braskem Europe GmbH: Polyolefins – product and properties Ausgabe September (2013)
11. Lyondellbasell: Technical Data Sheet – Hostacom HRG 328T NAT Ausgabe Juli (2018)
12. Bonnet, M.: Kunststoffe in der Ingenieuranwendung – Verstehen und zuverlässig auswählen. Vieweg+Teubner Verlag, GWV Fachverlage GmbH Wiesbaden, Wiesbaden (2009)
13. DSM: Technical Data Sheet – Xytron U3020E Ausgabe September (2017)

Towards Process Optimisation of Polyurethane Pultrusion Using 3D Simulation

Benedikt Kilian(✉), Sascha Fröbel, and Dirk Brüning

Covestro Deutschland AG, Leverkusen, Germany
benedikt.kilian@covestro.com

Abstract. To boost high productivity when manufacturing composite profiles with consistent quality in pultrusion, simulation methods are widely used to identify optimum process parameters. Although polyurethane resins are of interest for high speed production due to their low viscosity and high reactivity, no in-depth process simulation of the polyurethane pultrusion process has been published yet. In this paper, a new kinetic model for a polyurethane pultrusion system is presented and applied in a process simulation routine. After comparing simulation results with on-line measurements, the effect of varying process and profile parameters on the resulting degree of cure and temperature distribution is evaluated. It can be confirmed that increasing line speed or profile thickness significantly decreases the average degree of cure at the die exit. This effect can at least partly be compensated by major die temperature adjustments which in turn lead to an overall higher profile temperature. Line speed and die temperatures also determine length and position of the gel zone. The initial material temperature has no significant impact on the resulting degree of cure and temperature distribution.

Keywords: Pultrusion · Polyurethane · Simulation · Kinetic model · Curing · Temperature · Die

1 Introduction

As one of the few processes to produce fibre-reinforced plastic (FRP) parts continuously and thus efficiently, pultrusion has been well established in the market for decades [1]. To enable high quality mass production of complex shapes such as profiles for the automotive and window industry, a thorough understanding of process and profile parameters affecting part quality and process stability is needed. In addition, process productivity has a significant impact on the manufacturing profitability [2, 3]. As a consequence, polyurethane (PU) resin systems have recently gained increasing attention in industry and science [3–11]. In addition to its good mechanical properties such as high tensile and flexural strength and stiffness, the low viscosity and high reactivity of PU offers potential to increase the production speed [4, 5, 7, 12].

C. Hopmann and R. Dahlmann (Eds.): *Advances in Polymer Processing 2020*,
pp. 162–176, 2020. https://doi.org/10.1007/978-3-662-60809-8_14

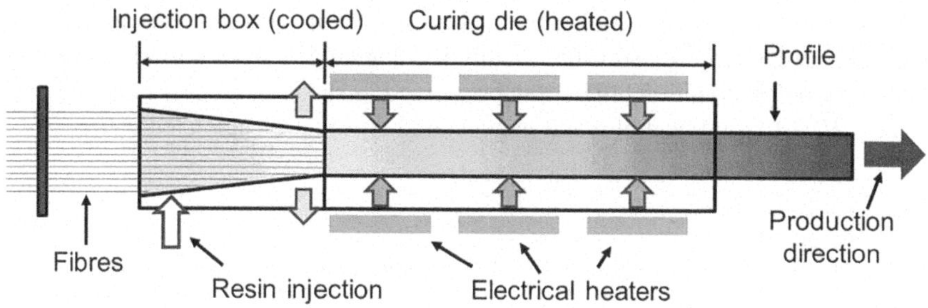

Fig. 1. The resin injection pultrusion process

Nevertheless, the heat transfer and curing taking place inside of the pultrusion die still can be limiting factors for productivity [3, 13, 14]. Modelling and simulation have been shown by various researchers to be powerful tools to reduce the experimental effort needed to identify optimum process conditions for a high line speed, but so far only for unsaturated polyester, vinyl ester and epoxy resin systems [13, 14]. Although PU resins in principle enable a high line speed, no in-depth process simulation of the PU pultrusion process has been published yet.

Figure 1 schematically shows the resin injection pultrusion process which is commonly used for PU pultrusion. After being hauled off of the fibre rack and being aligned, the reinforcing fibres and optionally textiles are guided through the injection box where the already mixed, but still unreacted and low viscous resin impregnates the fibres. The wetted fibre strand is then pulled through the heated curing die. The heat is transferred from electrical heaters to the fibre strand and accelerates the chemical reaction of the resin. After the die, the profile is cooled down and finally cut to length [12].

2 State-of-the-Art

The pultrusion process has been developed in the 1950s [1]. After focusing on improving the process and broadening the range of applications in the first decades, researchers began to model the pultrusion process in the late 1970s [15]. Since then, numerous process models covering different types of analysis (steady-state/transient), dimensionality (1D/2D/3D), numerical models (finite difference/finite element method), resin systems, reinforcements and profile geometries have been published. A comprehensive overview of state-of-the-art process models can be found in [13, 14, 16].

A holistic pultrusion master model consists of various inter-related sub models describing the chemical reaction, the evolution of viscosity, the heat transfer, the resin flow and pressure increase, the pulling force, the mechanical properties evolution and finally stress and strain in the profile at the end of the process [14]. The present paper focuses on modelling the chemical reaction of a PU resin system along with the heat transfer in the pultrusion die.

To describe the exothermal chemical reaction of arbitrary thermoset resins depending on temperature and time, usually the degree of cure *r* is used, representing the extent of polymerization [14]. *r* can be defined as the ratio of heat generated up to the time *t*, *H(t)*, to total heat of reaction generated during a complete cure H_{tr}:

$$r = \frac{H(t)}{H_{tr}} \tag{1}$$

The rate of cure *dr/dt* is assumed to be proportional to the heat flow *dH/dt* during reaction:

$$\frac{dr}{dt} = \frac{1}{H_{tr}} \frac{dH}{dt} \tag{2}$$

Usually, empirical or semi-empirical kinetic models are used to describe the rate of cure by first or second order partial differential equations, combining a conversion function *f(r)* depending on the resin type with an Arrhenius function *K(T)* to account for the reaction's temperature dependence [13, 14, 16]:

$$\frac{dr}{dt} = f(r) \cdot K(T) \tag{3}$$

$$f(r) = (1 - r)^n \quad \text{for n - th order models} \tag{4}$$

$$f(r) = r^m (1 - r)^n \quad \text{for autocatalytic models} \tag{5}$$

$$K(T) = A \cdot exp\left(-\frac{E_a}{RT}\right) \tag{6}$$

In this regard, *m* and *n* are the orders of reaction, *A* is a pre-exponential factor, E_a is the activation energy of the cure reaction and *R* is the universal gas constant. *Kamal & Sourour* developed an autocatalytic model which incorporates an initial rate of cure (at r = 0) differing from zero:

$$\frac{dr}{dt} = \left(K_1(T) + K_2(T) \cdot r^m\right)(1 - r)^n \tag{7}$$

where $K_1(T)$ and $K_2(T)$ are Arrhenius functions [17]. The models described above are widely used in pultrusion modelling with unsaturated polyester and epoxy resins. The model parameters usually are fitted to measurement data of the heat generated during reaction in isothermal and dynamic differential scanning calorimetry tests. As the models do not consider deceleration of the reaction due to the reactants' limited mobility in the later stage, they tend to overestimate the rate of cure to the end of the reaction.

Chen et al. were the first to propose a mathematical process model specifically for PU pultrusion [18–21]. They used a standard autocatalytic model to describe the reaction of a special blocked NCO-terminated PU prepolymer [20]. *Connolly & Heberer* proposed a simplified mechanistic first order model to describe the PU reaction in a pultrusion process:

$$\frac{dr}{dt} = k_{eff}(1-r)\left(\beta^{-1}-r\right) \tag{8}$$

where k_{eff} is an effective curing rate described by an Arrhenius term and β is the initial isocyanate-to-polyol ratio [5]. Their model is able to describe the beginning of the reaction but does not account for diffusion effects to the reaction's end.

To predict the local temperature at any position within the pultrusion profile when it passes through the die, usually a three dimensional energy balance is applied [13]:

$$\rho c_p\left(\frac{\partial T}{\partial t}+v\frac{\partial T}{\partial x}\right)=\frac{\partial}{\partial x}\left(k_x\frac{\partial T}{\partial x}\right)+\frac{\partial}{\partial y}\left(k_y\frac{\partial T}{\partial y}\right)+\frac{\partial}{\partial z}\left(k_z\frac{\partial T}{\partial z}\right)+q \tag{9}$$

where x is the pulling direction and y and z are the coordinates in width and thickness direction, ρ and c_p are density and heat capacity of the composite, k is the thermal conductivity of the composite in the respective direction, v is the pulling speed and q is the source term for heat generated during the exothermal reaction. The bulk properties of the composite can either be measured directly or derived from the components' properties via rules of mixture. The source term can be derived from fibre volume content φ, resin density ρ_r, total heat of reaction H_{tr} and rate of cure dr/dt:

$$q=(1-\varphi)\rho_r H_{tr}\frac{dr}{dt} \tag{10}$$

In the beginning of pultrusion process modelling, Eq. 9 often was simplified to a one dimensional form and the die wall temperature was imposed as a boundary condition [13]. Nowadays, the pultrusion die itself including the electrical heaters can be modelled three dimensionally and the resulting equations for the die and the profile are solved using diverse numerical methods. In addition, sophisticated models also take the temperature dependence of material properties, a variable heat transfer coefficient between die and profile and the thermally and/or chemically induced changes of the profile and die volume into account [13, 14].

3 A New Kinetic Model and Its Parameters

To improve accuracy of prediction, especially in the later stages of reaction, a new kinetic model for polyurethane systems has been developed earlier at Covestro Deutschland AG, Leverkusen, Germany [22, 23]. The semi-empirical model uses the *Rabinowitch* approach to compute the total reaction rate constant $k_{tot}(T)$ by combining a kinetic rate constant $k_{kin}(T)$ with a diffusional rate constant $k_{diff}(T)$. The diffusional rate constant is necessary to describe the reaction hindrance due to diffusion effects at high conversion levels. $k_{kin}(T)$ is expressed as the sum of a traditional and an m-th order autocatalytic Arrhenius term. Another Arrhenius term is used in combination with an empirical sigmoidal term to yield $k_{diff}(T)$:

$$\frac{dr}{dt}=k_{tot}(T)\cdot(1-r)^n \tag{11}$$

$$\frac{1}{k_{tot}(T)} = \frac{1}{k_{kin}(T)} + \frac{1}{k_{diff}(T)} \tag{12}$$

$$k_{kin}(T) = k_{kin,0} \cdot exp\left[-\frac{E_a}{RT}\right] + k_{auto,0} \cdot exp\left[-\frac{E_{auto}}{RT}\right] \cdot r^m \tag{13}$$

$$k_{diff}(T) = k_{diff,0} \cdot exp\left[-\frac{E_{diff}}{RT}\right] \cdot \left(x_{diff} + \frac{1 - x_{diff}}{1 + \left(exp\left[r - r_{infl}\right]\right)^p}\right) \tag{14}$$

$$\frac{dT}{dt} = \Delta T_{ad} \frac{dr}{dt} \tag{15}$$

In this regard $k_{kin,0}$, $k_{auto,0}$ and $k_{diff,0}$ are kinetic, autocatalytic and diffusive pre-factors, E_a, E_{auto} and E_{diff} are kinetic, autocatalytic and diffusive activation energy, R is the universal gas constant, n is the reaction order, m and p are fitting parameters, x_{diff} and r_{infl} are variables to reduce the diffusive rate constant and ΔT_{ad} is the adiabatic temperature increase.

To adjust the general model to a specific PU system, a set of parameters has to be identified which describes best the development of degree of cure and temperature for a broad range of temperature conditions. For this purpose, measurements of temperature and NCO conversion in a defined volume of mixed PU resin under quasi-adiabatic conditions and for multiple initial material temperatures $T_{Material,0}$ are conducted. Pre-heated components of the PU system are mixed in a paper cup, which is then placed in an oven with the same temperature as the initial material temperature. During reaction, the development of the material temperature is measured continuously using temperature sensors placed in the middle of the resin volume. Due to the low thermal conductivity of the resin, the stage of the exothermic temperature increase can be approximated as adiabatic. In addition, material samples of the liquid and solidified polyurethane are extracted discontinuously to measure the NCO conversion with attenuated total reflectance Fourier-transform infrared spectroscopy (ATR-FTIR).

For the experiments and simulations described in this paper, the commercial PU pultrusion system Baydur® PUL from Covestro Deutschland AG is used. For temperature and NCO conversion measurements a total resin mass of 210 g ± 15% is prepared. Temperature measurements are conducted with initial material and oven temperatures of 30, 40, 50, 60, 70 and 80 °C. NCO conversion is measured for the experiments starting at 60, 70 and 80 °C. The higher temperatures lead to reaction times of the resin in the range of common residence times in the pultrusion die for standard line speeds (v ≈ 0.5 m/min). To account for the deviation naturally occurring in experiments, multiple measurements per temperature are averaged. The generalized reduced gradient (GRG) nonlinear solver of Microsoft Excel is used in combination with multiple starting points to identify the optimum set of model parameters for the specific PU system in the described temperature range by minimizing the sum of error squares between measurement and model prediction. The final parameter values are listed in Table 1.

Table 1. Fitted parameter values of the kinetic model

Parameter	Unit	Value
n	–	2
$k_{kin,0}$	1/s	54,523
E_a	J/mol	49,570
R	$(kg * m^2)/(s^2 * mol * K)$	8.31
$k_{auto,0}$	1/s	100
E_{auto}	J/mol	25,239
m	–	1.49
$k_{diff,0}$	1/s	$1.06 * 10^{12}$
E_{diff}	J/mol	88,387
x_{diff}	–	0
r_{infl}	–	0.79
p	–	60
ΔT_{ad}	K	157

To validate the material model, the temperature and conversion in the neat resin during the first 1000 s are exemplarily calculated with the model (T_{calc}/r_{calc}) for $T_{Material,0} = 40$, 60 and 80 °C and compared to the measurements of temperature T_{meas} at $T_{Material,0} = 40$, 60 and 80 °C (Fig. 2) and conversion r_{meas} at $T_{Material,0} = 60$ and 80 °C (Fig. 3). As can be seen in the graphs, the model is able to predict the temperature rise accurately for a broad temperature range with only one set of parameters, although it tends to underestimate the maximum temperature for lower $T_{Material,0}$ and to overestimate the maximum temperature for higher $T_{Material,0}$. After the initial

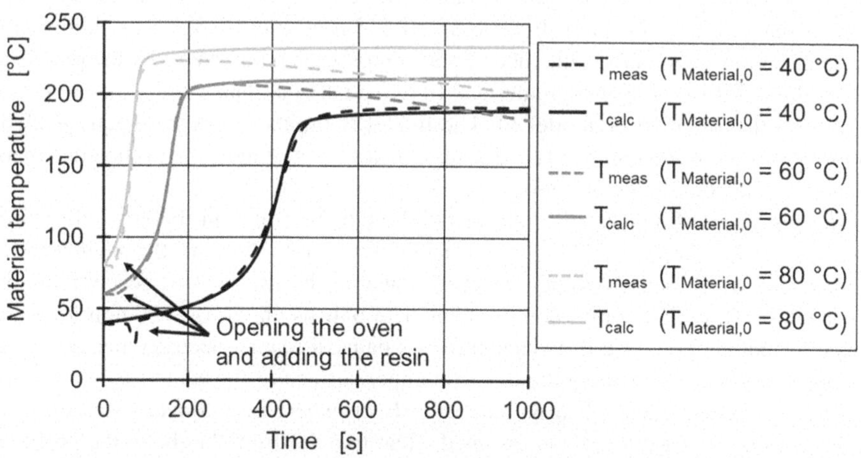

Fig. 2. Comparison of predicted and measured temperature for model validation

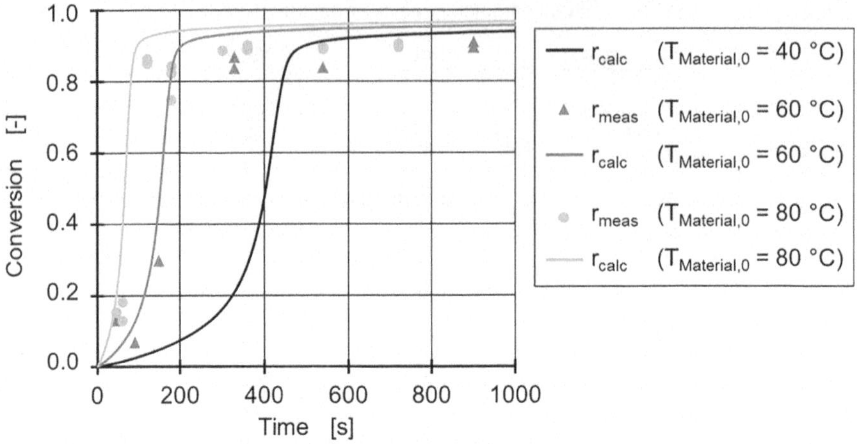

Fig. 3. Comparison of predicted and measured conversion for model validation

temperature rise the measured temperature drops due to cooling of the resin, which is not considered in the model. Regarding conversion, the initial rise is predicted quite well, but the final degree of cure is overestimated (Fig. 3).

4 Simulation Model and Validation

In pultrusion production, besides line speed usually the electrical heater temperatures of the pultrusion die are controlled. For a realistic prediction of temperature and degree of cure evolution in the profile for diverse scenarios, a sophisticated 3D simulation model of the pultrusion die including the electrical heaters is needed. In this paper, a flat profile with 3 mm thickness and 115 mm width is used to validate the simulation and to assess the influence of varying process parameters. As the profile as well as the pultrusion die are symmetrical to two adiabatic planes, only one quarter of die and profile needs to be modelled. The model is implemented in Abaqus R2016x HotFix 9, Dassault Systèmes, France, and shown in Fig. 4 along with the coordinate system and the boundary conditions.

The profile mesh consists in total of 6600 elements (three in thickness direction, eleven in width direction, 200 in pultrusion direction), whereas the element size decreases in width and thickness direction towards the die wall to account for the higher temperature and conversion gradient. The pultrusion die is 1000 mm long and has a constant element size. It is heated by six pairs of planar electrical heaters in six heating zones (T_{H1}–T_{H6}) along the die. The upper surface of the heaters as well as the die front surface are defined to be adiabatic. Everywhere else convective cooling of the die with $\alpha_{d\text{-}a} = 10\ \text{W/(m}^2\text{K)}$ is assumed. Heat transfer from the die to the profile is assumed to be ideal.

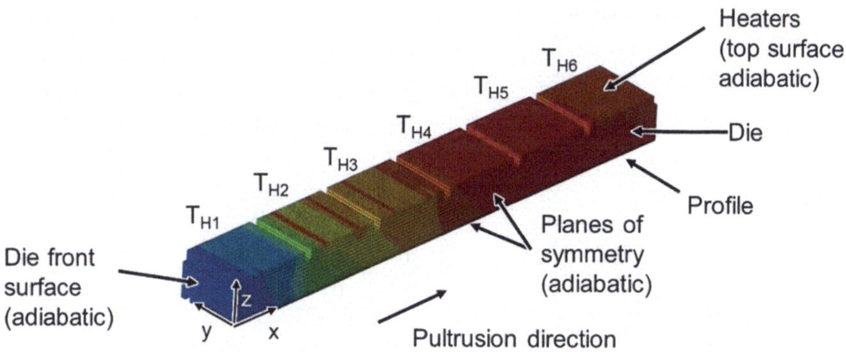

Fig. 4. Simulation model for a flat profile

Table 2. Physical properties of the materials used

Property		Unit	Value
Pultrusion die			
Density	ρ_d	kg/m³	7850
Specific heat capacity	$c_{p,d}$	J/(kgK)	430
Thermal conductivity	k_d	W/(mK)	57
Composite profile (PU+glass fibres)			
Density	ρ_c	kg/m³	2090
Specific heat capacity	$c_{p,c}$	J/(kgK)	855
Thermal conductivities	$k_{c,x}$	W/(mK)	0.678
	$k_{c,y}$	W/(mK)	0.511
	$k_{c,z}$	W/(mK)	0.501
Others			
Heat transfer coefficient die – air	$\alpha_{d\text{-}a}$	W/(m²K)	10
Fibre volume content	φ	%	61

Eq. 9 is solved for the pultrusion die and the profile respectively. The source term equation is slightly modified:

$$q = (1 - \varphi)\Delta T_{ad}\rho_c c_{p,c}\frac{dr}{dt} \tag{16}$$

ρ_c and $c_{p,c}$ are density and heat capacity of the glass fibre reinforced composite. The physical properties for die material (steel) and composite profile are assumed to be constant and listed in Table 2.

Mass transport of the profile through the die is implemented via subroutines of Abaqus. To compare results of different simulation runs, the steady state values of temperature and degree of cure at the integration points (IP) of the elements are used. Due to their slightly inward shifted position within the finite elements, the exact edges

Table 3. Comparison of node and IP coordinates for analysed relevant surfaces and edges

		Node coordinates [mm]			IP coordinates [mm]		
		x	y	z	x	y	z
Profile surfaces	x = max	1000	Var.	Var.	998.9	Var.	Var.
	y = max	Var.	57.5	Var.	Var.	57.43	Var.
	z = max	Var.	Var.	1.5	Var.	Var.	1.433
Edges	y, z = 0	Var.	0	0	Var.	4.1	0.15
	y = 0, z = max	Var.	0	1.5	Var.	4.1	1.433
	y = max, z = 0	Var.	57.5	0	Var.	57.43	0.15
	y, z = max	Var.	57.5	1.5	Var.	57.43	1.433

and outer surfaces of the profile cannot be evaluated (see Table 3). When referring to edges or profile surfaces later, the IP closest to the respective nodes are analysed.

The following boundary conditions are used:

- Profile temperature at x = 0: $T = T_{Material,\ initial}$ (=25 °C for standard conditions)
- Profile degree of cure at x = 0: r = 0
- Heater temperatures T_{H1}–T_{H6}: see Table 4

Table 4. Heater temperatures for different scenarios in °C

Scenario	T_{H1}	T_{H2}	T_{H3}	T_{H4}	T_{H5}	T_{H6}
Standard	52.5	140	160	180	180	170
+20 °C	72.5	160	180	200	200	190
+40 °C	92.5	180	200	220	220	210

To validate the simulation model, the predicted temperature evolution (T_{sim}) for different positions in the profile and for two different line speeds (v = 0.5 m/min and v = 1.5 m/min) is compared to experimental on-line measurements (T_{meas}) with standard temperature conditions (see Table 4). The temperature in the profile is measured by feeding type J thermocouples through the die. Figure 5 exemplarily shows simulated and measured temperatures for two different line speeds and positions. Although the model uses some simplifications, such as constant material properties and no shrinkage of the profile, the simulated temperature evolution qualitatively meets the measured temperature. Nevertheless, an almost constant offset in temperature in pultrusion direction is observed, which is bigger for a higher line speed (~35–50 mm at v = 0.5 m/min; ~105 mm at v = 1.5 m/min). This can at least partly be attributed to inaccuracies in experimental measurement. *Batch & Macosko* estimated the uncertainty of measurement to be ±50 mm in axial direction and ±5 °C in temperature for a line speed of v = 0.6 m/min [24]. Prediction quality of the model can further be improved by using variable material properties and a variable heat transfer coefficient between profile and die [24, 25]. In addition, the adiabatic boundary condition of the die entrance leads to a higher simulated die temperature in the first zones. In total, the model obviously tends to overestimate the temperature rise and, as a consequence, also the resulting degree of cure.

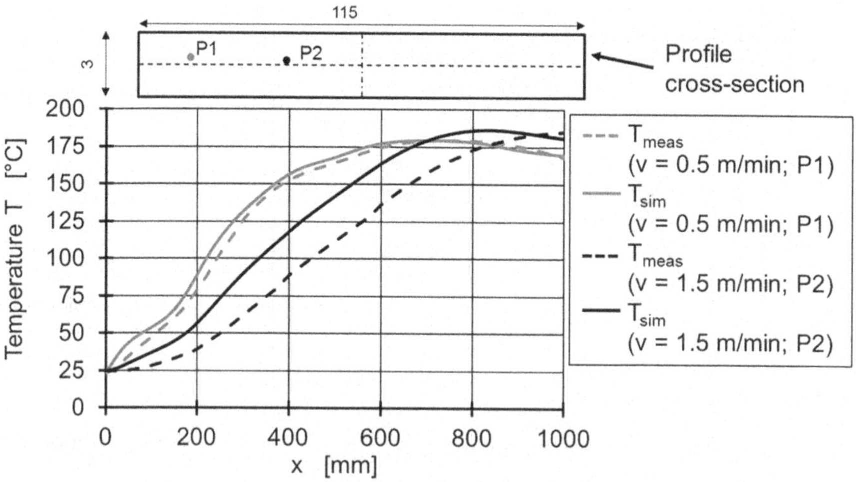

Fig. 5. Comparison of simulated and measured temperature for simulation validation

5 Criteria for Process and Part Quality

To evaluate, compare and optimize the quality of the production process as well as of the parts produced for different simulation scenarios, specific quality criteria have to be defined. Those criteria can be (among others specifically defined by the producer for its process/profile):

- Minimum degree of cure to be reached everywhere in the profile at the die exit (r_{min})
- Average degree of cure to be reached in the profile at the die exit (r_{av})
- Deviation in degree of cure in the profile's cross-section at the die exit (Δr_{av})
- Maximum temperature in the profile when it passes through the die (T_{max})
- Average temperature reached in the profile at the die exit (T_{av})
- Deviation in temperature in the profile's cross-section at the die exit (ΔT_{av})
- Length of the gel zone (L_{gel})
- Position of the gel zone in the die (x_{gel})

The minimum and average degrees of cure at the die exit correlate with the profile's ability to bear the clamping forces of the pullers. The maximum deviation in temperature and degree of cure is important for induced stresses and resulting warpage in more complex profiles [26]. The maximum temperature should not be exceeded to avoid thermal decomposition of the resin. Length and position of the gel zone influence the height of the pull force [13, 14]. The gel zone is here defined as the area in the pultrusion die where at least part of the resin somewhere in the profile currently has the degree of cure at gelation r_{gel}. In a three dimensional visualization, this area is a curved surface of which the extent in x-direction (L_{gel}) as well as the average x-value (x_{gel}) can be analysed. Figure 6 exemplarily shows the evolution of temperature and conversion in the profile middle (y, z = 0) and the profile corner (y, z = max)

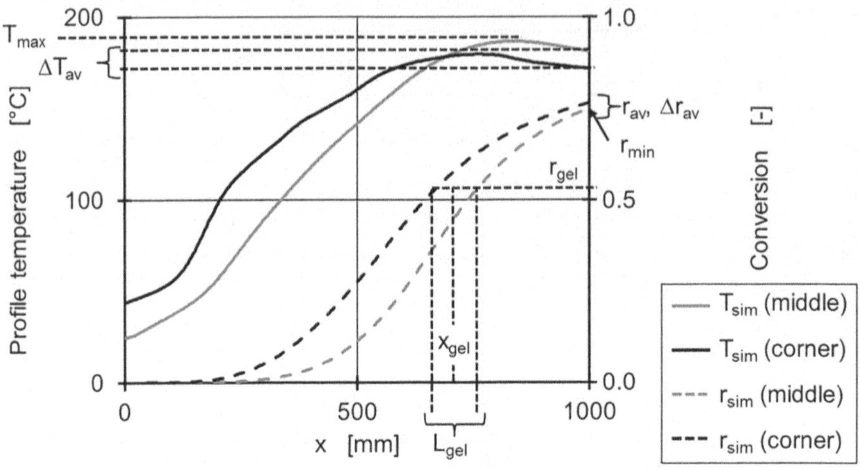

Fig. 6. Exemplary temperature and conversion evolution with quality criteria

together with the quality criteria defined above. The degree of cure at gelation of the PU system is calculated according to the cascade theory of *Miller-Macosko* [27] to be $r_{gel} = 0.56$. Specific target values for the quality criteria need to be specified when the simulation model is used for process optimization.

6 Parameter Study: Influence of Line Speed, Die Temperatures, Profile Thickness and Initial Material Temperature

The simulation model is used to investigate the influence of line speed, die temperatures, profile thickness and initial material temperature on the degree of cure (conversion) and temperature of the profile at the die exit (r_{av}, Δr_{av}, T_{av}, ΔT_{av}), the maximum temperature (T_{max}) and the length and average position of the gel zone (L_{gel}, x_{gel}).

Figure 7 shows r_{av} and its standard deviation Δr_{av} (depicted as error bars) depending on line speed v and for three different die temperature settings (see Table 4). As expected, the average degree of cure decreases with increasing line speed. The deviation in the profile Δr_{av} is increasing with higher line speed, but can generally be considered as low as the profile is relatively thin. With higher die temperatures, r_{av} increases particularly at higher line speeds. For line speeds below $v = 0.75$ m/min, an increase in die temperatures has almost no effect on r_{av} due to the promoting diffusion control in the rate of reaction.

With an increase in line speed, the average temperature in the profile at the die exit T_{av} is increasing due to the exothermal peak being shifted to the die exit (Fig. 8). The same is observed for the maximum temperature in the profile. Due to the exothermic reaction, the profile effectively is cooled by the die in the later heating zones. At a higher line speed, the exothermal peak moves to the die exit and time for cooling is shortened. At very high line speeds, e.g. above $v = 1.5$ m/min at standard temperature

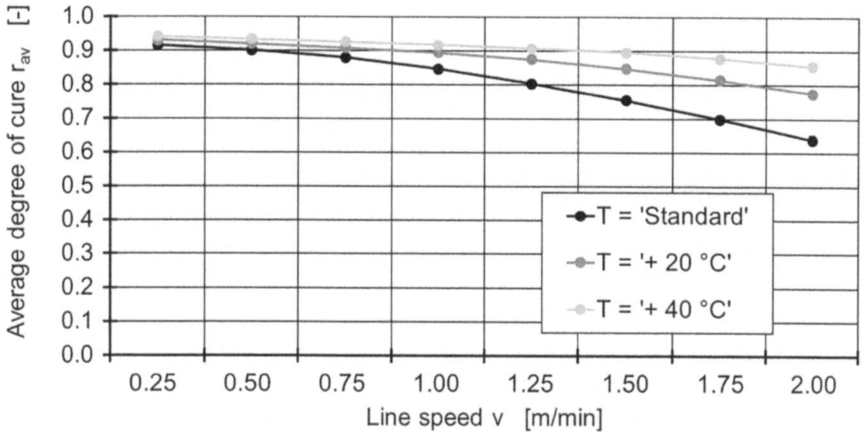

Fig. 7. Conversion at the die exit depending on line speed and die temperatures

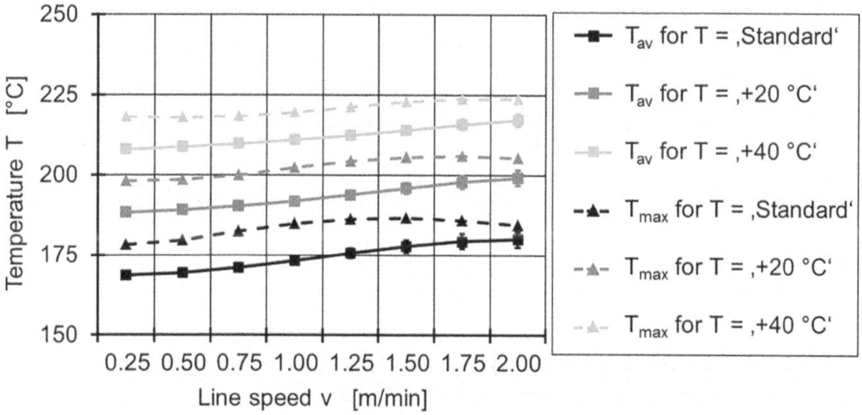

Fig. 8. Temperatures in the profile depending on line speed and die temperatures

conditions, T_{max} slightly decreases again, as the exothermic peak seems to be shifted outwards of the die. An increase of the die temperatures leads to an almost similar increase in T_{av} and T_{max}. As a certain temperature in the resin should not be exceeded to avoid decomposition, this effect will be a limitation for process optimisation. In Fig. 8, ΔT_{av} is depicted as the error bars of T_{av} and increases slightly with increasing line speed, but generally can be considered as low.

Figure 9 shows position and length of the gel zone depending on line speed and die temperatures. The length is depicted as the respective error bars. With an increase in line speed, the gel zone gets longer and is shifted towards the die exit. An increase of the die temperatures moves the gel zone backwards, but does not have a significant effect on the length.

In Fig. 10, T_{av} along with ΔT_{av} (as error bars), T_{max} and r_{av} along with Δr_{av} (as error bars) are shown for three different initial material temperatures (25, 35 and

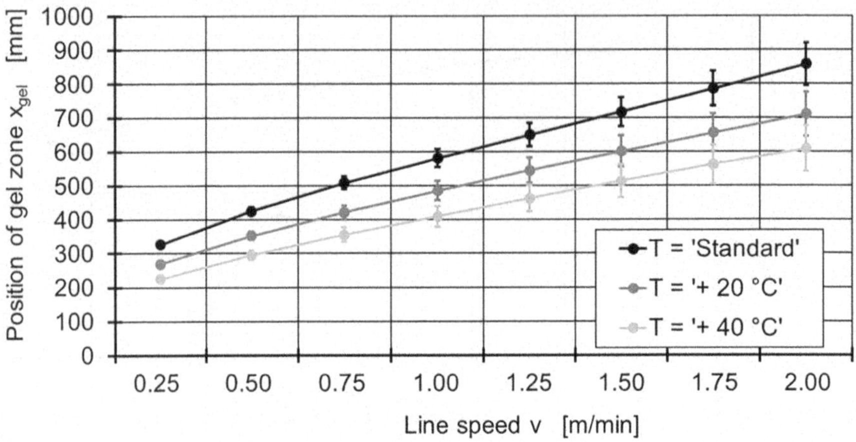

Fig. 9. Average gel point position and gel zone length (error bars) depending on line speed and die temperatures

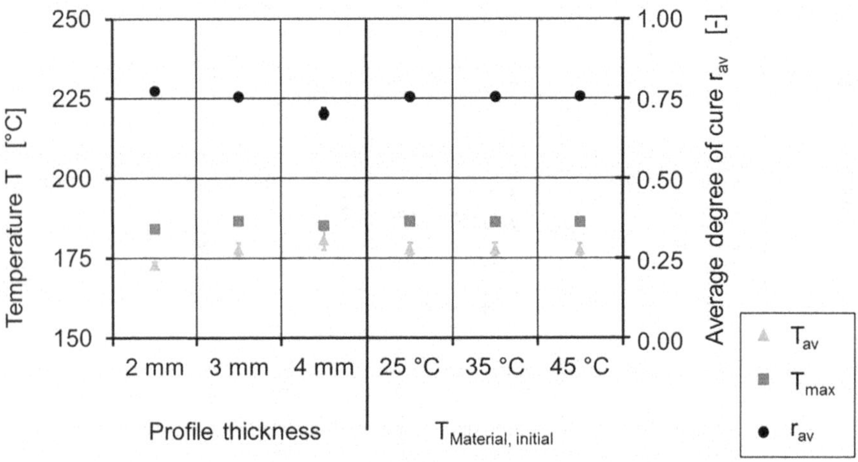

Fig. 10. Influence of the profile thickness and the initial material temperature

45 °C, profile thickness is 3 mm) and three different profile thicknesses (2, 3 and 4 mm, initial material temperature is 25 °C). Line speed is v = 1.5 m/min and the die temperatures are set to standard (see Table 4). With increasing profile thickness, the average conversion at the die exit decreases and at the same time the deviation increases (Fig. 10 left). This can be attributed to the longer time it takes for the heat to be conducted from the die to the profile centre and hence a delay in the chemical reaction. In addition, the exothermal peak moves towards the die exit at a higher profile thickness, which in turn leads to a higher average temperature at the die exit. The initial material temperature at the die inlet does not have a significant effect on T_{av}, ΔT_{av}, T_{max}, r_{av} and Δr_{av} in the simulated range (Fig. 10 right).

7 Summary and Outlook

In this paper, a process simulation tool with a new kinetic model for the exothermic reaction of polyurethane systems has been presented. The model is able to predict the evolution of material temperature and degree of cure for a broad range of initial material temperatures. It also takes the deceleration of the reaction in the later stage due to the reactants' hindered mobility into account. The model parameters were altered to fit to temperature and NCO conversion measurements in neat resin samples with different initial material temperatures. The reaction kinetic model was utilized in a 3D simulation tool to predict heat transfer and cure phenomena in the pultrusion die. Although some simplifications were made, such as constant material properties and no shrinkage of the profile, comparison of simulation runs with on-line temperature measurements during pultrusion trials shows good agreement despite an almost constant offset which at least partly can be attributed to inaccuracies in experimental measurement. The process simulation tool was used to predict the degree of cure and temperature within a flat profile along the pultrusion die depending on line speed, die temperatures, profile thickness and initial material temperature. It was shown that production speed has a significant influence on the average degree of cure at the die exit as well as the length and position of the gel zone, which at least partly can be compensated by die temperature adjustments. In addition, the degree of cure and the temperature in the profile are influenced by the profile thickness, whereas the initial material temperature does not have a significant effect.

To reach better agreement between simulation and measurement, variable material properties and non-ideal heat transfer between die and profile can be incorporated in the model in future. The simulation tool then can be used to optimize the process conditions for a given profile to reach optimum part and process quality at the highest line speed possible.

References

1. Starr, T.: Pultrusion for Engineers. Woodhead Publishing Limited, Abington (2000)
2. Patrawala, T.B.: Decision support tool for costing of the pultrusion process, M.Sc. thesis, College of Engineering and Mineral Resources, West Virginia University, Morgantown, West Virginia, USA (1999)
3. Hopmann, C., Manderscheid, A., Kilian, B., Schneider, D., Fischer, K.: Effiziente Verarbeitungstechnologien für Polyurethan-Leichtbau – Ansätze zur kosteneffizienten Herstellung endlosfaserverstärkter PUR-Profile. In: Umdruck zum 29. Internationalen Kolloquium Kunststofftechnik, ed. Aachen: Institut für Kunststoffverarbeitung (2018)
4. Michaeli, W.: Pultrusion of Composite Profiles – Polyurethane (PU) as Alternative Matrix System. Polym. Polym. Compos. **18**, 537–542 (2010)
5. Connolly, M., Heberer, D.: Advances in polyurethane pultrusion: cure modeling and 'Second Generation' resin systems, presented at the ACMA Composites, Las Vegas, Nevada, USA (2012)
6. Hopmann, C., Schneider, P., Böttcher, A.: Von der Injektionsbox zum komplexen Werkzeug zur Fertigung von Hybridprofilen. VDI Konstruktion **7-8**:IW6–IW9 (2014)

7. Bramante, G., Bertucelli, L., Benvenuti, A., Meyer, K.J.: Polyurethan-Verbundstoffe – Mechanische Analyse von pultrudierten Laminaten. PU Mag. **15**, 136–145 (2015)
8. Hopmann, C., Schneider, P., Neuhaus, B., Goeschel, J., Böttcher, A.: Neue Polyurethansysteme für die Produktion von Composite-Profilen. Lightweight Design **9**, 52–57 (2016)
9. Hinz, W., Kilian, B.: New markets for PU pultrusion. FAPU Eur. Polyure. J. **100**, 32–35 (2017)
10. Heinz, P., Kilian, B., Meisenheimer, R., Mentizi, S.: Performance leap in pultrusion with polyurethane. PU Mag. **15**, 476–479 (2018)
11. Neuhaus, B., Liese, J.: New lightness with PU Pultruded Parts. Presented at the 14th world pultrusion conference. Vienna, Austria (2018)
12. Sumerak, J.E.: Tough choices for pultruders–polyurethane for high performance pultrusions. Presented at the 7th world pultrusion conference. Schiphol, The Netherlands (2004)
13. Baran, I.: Pultrusion: State-of-the-Art Process Models. Smithers Rapra, Shawbury (2015)
14. Safonov, A.A., Carlone, P., Akhatov, I.: Mathematical simulation of pultrusion processes: a review. Compos. Struct. **184**, 153–177 (2018)
15. Price, H.L.: Curing and flow of thermosetting resins for composite material pultrusion. Ph.D. thesis, Old Dominion University, Norfolk, Virginia, USA (1979)
16. Bezerra, R.M.: Modelling and simulation of the closed injection pultrusion process. Ph.D. thesis, Karlsruher Institut für Technologie (KIT), Karlsruhe, Germany (2017)
17. Kamal, M.R., Sourour, S.: Kinetics and thermal characterization of thermoset cure. Polym. Eng. Sci. **13**, 59–64 (1973)
18. Chen, C.-H., Ma, C.-C.M.: Pultruded fibre reinforced polyurethane composites I. process feasibility and morphology. Compos. Sci. Technol. **45**, 335–344 (1992)
19. Chen, C.-H., Ma, C.-C.M.: Pultruded fibre reinforced polyurethane composites II. Effect of processing parameters on mechanical and thermal properties. Compos. Sci. Technol. **45**, 345–352 (1992)
20. Ma, C.C.M., Chen, C.H.: The development of a mathematical model for the pultrusion of blocked polyurethane composites. J. Appl. Polym. Sci. **50**, 759–764 (1993)
21. Chen, C.H., Yen, C.C.: Mathematical model for the pultrusion of blocked PU-UP matrix composites. J. Appl. Polym. Sci. **90**, 1996–2002 (2003)
22. Raffel, B.: Material model for the reaction kinetics of the polyurethane pultrusion system Baydur PUL 20PL07 + desmodur PUL 10PL01 measured in an isolated cup with Adiabatic Temperature Method and ATR-FTIR spectroscopy. Covestro Deutschland AG, Leverkusen, Germany, Internal research report (2010)
23. Raffel, B.: Modeling reaction kinetics of the pultrusion system Baydur PUL 20PL10 as an input for process simulation. Covestro Deutschland AG, Leverkusen, Germany, Internal research report (2017)
24. Batch, G.L., Macosko, C.W.: Heat transfer and cure in pultrusion: model and experimental verification. AIChE J. **39**, 1228–1241 (1993)
25. Baran, I., Tutum, C.C., Hattel, J.H.: The effect of thermal contact resistance on the thermosetting pultrusion process. Compos. Part B Eng. **45**, 995–1000 (2013)
26. Baran, I., Hattel, J.H., Akkerman, R.: Investigation of process induced warpage for pultrusion of a rectangular hollow profile. Compos. Part B Eng. **68**, 365–374 (2015)
27. Macosko, C.W., Miller, D.R.: A new derivation of average molecular weights of nonlinear polymers. Macromolecules **9**, 199–206 (1976)

Assistance Machine Function for BMC Injection Molding

Nicolina Topic[1(✉)], Stefan Kruppa[1], and Dietmar Drummer[2]

[1] KraussMaffei Technologies GmbH, Krauss-Maffei-Str. 2, 80997 Munich, Germany
nicolina.topic@kraussmaffei.com
[2] Institute of Polymer Technology, University of Erlangen-Nürnberg, Am Weichselgarten 9, 91058 Erlangen-Tennenlohe, Germany

Abstract. Bulk molding compounds (BMC) are short fiber reinforced polyester molding compounds and mostly processed by injection molding. They consist essentially of polyester resin, styrene, additives, fillers and glass fibers and are fabricated in a conventional Z-kneader by mixing the resin and the solid fillers. Due to its doughy and moist consistency BMC requires force feeding to the barrel of the injection molding machine by a piston or screw stuffer. Furthermore, the raw material has a limited shelf life and needs special storage conditions. Because of the considerable proportion of volatile styrene, the dwell time of the material during processing is an important parameter. In this study the styrene loss is quantified during the injection molding production with a screw stuffer and evaluated for different loading scenarios. As a result, an assistance machine function is developed which optimizes the load operation in order to reduce the impact of styrene loss.

Keywords: BMC · Injection molding · Styrene evaporation

1 Introduction

Polyester molding compounds form a quarter of Europe's glass fiber reinforced plastics production [1]. Bulk molding compounds (BMC) are short fiber reinforced polyester molding compounds and mostly processed by injection molding. Due to their excellent mechanical and electrical properties as well as very good thermal and dimensional stability, BMC have been developed and used mainly for the production of automotive reflectors, cylinder head covers, tailgates and especially electronic applications. Additionally, the flexibility in formulation and the broad property profile at low costs offer a high potential for BMC to be used in further sectors and particularly for substituting metal parts [1, 2]. However, there are several particularities that make BMC injection molding susceptible to instabilities. Examples are sensitive storage conditions, volatile additives, its limited shelf life, the defined curing conditions and possible batch variations. Furthermore, the development progress and investigations of process to part correlations in the past were mainly project specific and not

C. Hopmann and R. Dahlmann (Eds.): *Advances in Polymer Processing 2020*,
pp. 177–188, 2020. https://doi.org/10.1007/978-3-662-60809-8_15

universally usable [3–5]. In order to enable a reproducible and quality assured process chain a comprehensive knowledge of the correlations between process parameters and the part properties is required [4].

Unsaturated polyester resins are oligomeric linear molecules with reactive double bonds and are used as coatings since the 1920s. Initially described in the invention of Ellis and Foster in 1936 they are solved in the reactive diluent styrene [6], which acts as a solvent for the viscosity modification and as reaction partner for the radicalic copolymerization with the unsaturated polyester resin. The reactivity of the resin is dependent on its styrene content to a certain extent due to an increasing share of double bonds [7], which leads to a higher glass transition temperature [8]. The reaction is started by an initiator, mostly organic peroxides, which are initially consumed by an inhibitor. The reaction starts after an induction time depending on the initiator and inhibitor ratio causing a viscosity and an exothermic temperature increase, which are well investigated and described by mechanistic kinetic models [9–15].

The initial viscosity of the polyester resin is modified by using magnesium oxide as a thickening agent. For a further viscosity increase and cost reduction silanized $CaCO_3$ is added as a filler up to 60 wt%. Glass fibers provide most of the mechanical strength to the compound up to a content of 25 wt% [16, 17]. Polymerisation causes a shrinkage of 7–10% of the resin, which depends on the conversion degree [18–20]. By adding low-profile-additives (LPA), in styrene dispersed thermoplastic additives, the shrinkage of the polyester resin is overcompensated by the formation of a two-phase structure and microvoids [19–21]. Furthermore, the shrinkage is reduced by the volumetric proportion of fillers [22, 23].

Bulk molding compounds are doughy and moist materials, which are fabricated in a conventional Z-kneader by mixing the resin and solid fillers. Afterwards, the compound is portioned and packed into sealed bags in order to reduce styrene evaporation. Due to its high vapor pressure styrene is classified as a volatile organic compound [23, 24]. Additives such as paraffins are used to reduce dynamic emissions by building a tight skin [25]. The bags have to be transported and stored at temperatures of 15–20 °C in order to prevent an early degradation of the inhibitor. Due to the gradual styrene loss and inhibitor degradation their shelf life is limited [16, 17]. After opening the bag styrene is evaporating, which leads to a weight loss of the raw material [4].

Due to its raw material condition BMC requires force feeding to the barrel of the injection molding machine by a piston stuffer or screw stuffer [1, 26]. The piston stuffer pressurizes the material in a cylinder with a constant stuffing pressure, which reaches up to 40 bar dependent on the material and is measured as well as controlled at the feed port to the plastification unit. The wear and stress on the fibers as well as the styrene emission are very low, as the piston is sealed. However, at high material throughputs the piston stuffer reaches its limits since the material charging takes some time and makes the process discontinuous [26]. The screw stuffer operates continuously and is demonstrated in Fig. 1 (KraussMaffei Polyload AZ 100). The bulks are charged through the flap of the storage tank (2) to the electrically driven (5) hopper (3). During the dosing phase the material is transported by the electrically driven (1) rotating feeding screw (6) to the plastification unit. Simultaneously, the counter-rotating movements of the hopper ensure a homogeneity of the material and a self-cleaning of the machine unit [26]. At the feed port a pressure sensor (4) is located, which is used to control the feeding screw speed. This control method ensures a constant flow rate [27, 28]. An additional screw stuffer includes a fixed hopper and two opposite rotating screws [29]. Both screw stuffing systems can be refilled continuously and automatically [30] with virgin material in order to reduce the dwell time and hardening of the material because of styrene loss [1, 26].

In this study the styrene loss is quantified and evaluated for different loading scenarios during the injection molding production with a screw stuffer. Furthermore, an assistance machine function is developed which optimizes the load operation in order to reduce the impact of styrene loss.

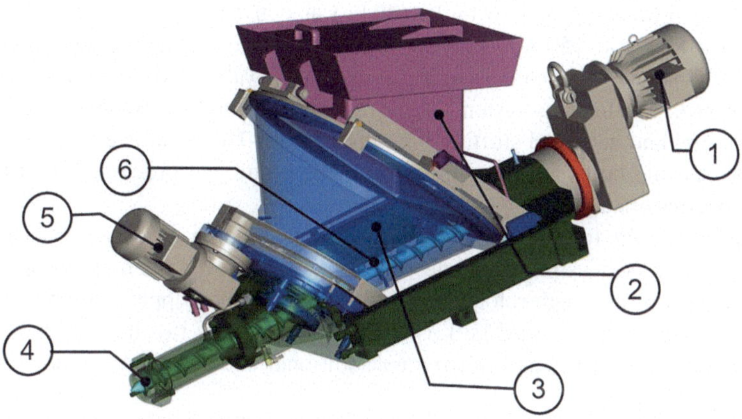

Fig. 1. Screw stuffer AZ 100 Polyload, equipped with a electrically driven (5) hopper (3), a storage tank (2), an electrically driven (1) rotating feeding screw (6) and pressure sensor (4) at the feed port

2 Experimental

The focus is set on the quantification of the styrene loss, occurring from unpacking of the bulk and its loading into the hopper up until the demolding of the part. Hence, the material type, the storage conditions, process parameters and mold are kept constant. Only the dwell time of the rough material in the hopper is changed by varying the time and amount of bulks during reloading phase. The effect of styrene evaporation on the chemical reaction during molding is not part of this study. All trials are carried out with one BMC formulation containing 65 wt% mineral fillers and 10 wt% glass fibers, which is mainly used for the serial automotive light reflector production. The mold geometry forms a test specimen, which consists of two tension bars, bending arms and notched bars, symmetrically placed to the gate (Fig. 2).

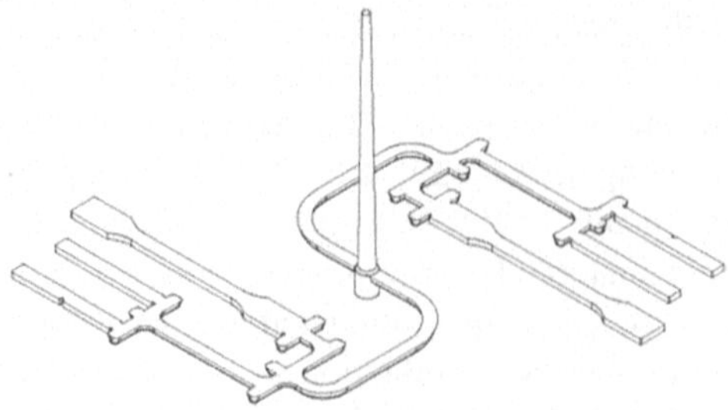

Fig. 2. Test specimen consisting two tension bars, bending arms and notched bars

First, the processing of two bulks (30 kg) at constant process settings is investigated on a KM 200–750 CX PolySet injection molding machine equipped with a special BMC screw (diameter 50 mm), a non-return valve and a watercooled barrel as a reference. The feeding system is a screw stuffer Polyload AZ 50 with a hopper volume of 50 l and activated stuffing pressure control. The mold is electrically heated up to 160 °C. The bulks are thrown in at the beginning and processed until they are completely consumed.

In the next step four bulks with two different loading scenarios (visualized in Fig. 3) are processed using the same process settings. The batch packing of BMC causes the shown saw tooth reloading profile. Initially, the hopper is loaded with two bulks. After 100 cycles a third bulk is added. In the series (a) the hopper is refilled after additional 150 cycles and in (b) after additional 100 cycles. Afterwards the hopper is completely emptied.

The styrene loss is evaluated by gravimetric analysis of the part. The whole specimen and one tension bar of all cycles are measured with a precision scale. The density of 20% of all tension bars is measured based on Archimedes' principle. 20% of the tension bars are incinerated in a pyrolysis furnace at 450 °C and weighted. The precision scale Masterpro LP620S, Sartorius AG, is used for all measurements and has an accuracy of 0.001 g. Additionally, the weight loss of the rough material samples is measured. In addition, three tension bars of test series (a) are characterized by thermogravimetric analysis (TGA) of TA Instruments. All available sensor signals of the injection molding machine and the screw feeding device are recorded.

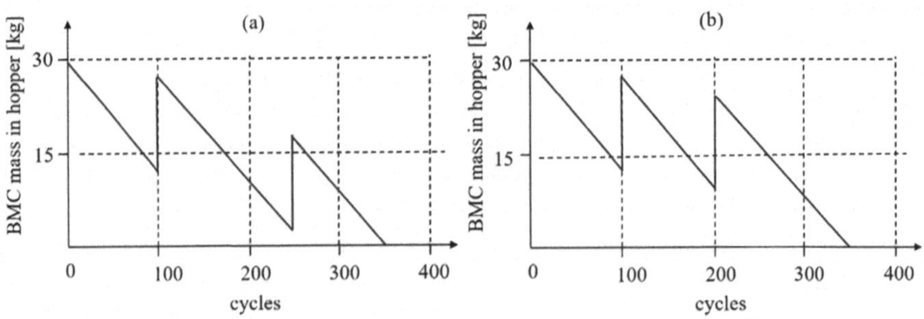

Fig. 3. Loading scenarios for two test series on a KM 200-750 CX PolySet, AZ50 Polyload

3 Results

The resulting part weight of the test specimens and one respective tension bar, resulting from processing two bulks on a KM 200-750 CX PolySet, AZ50 Polyload is displayed in Fig. 4 for 130 consecutive cycles (30 start-up cycles are excluded). Both parameters correlate by 96.9% and are rising by 0.90% with a variation coefficient of 0.05% for the first 10 parts. In the following, only the tension bar will be regarded. The tension bar density increases by 0.87% due to a filler content rise of 0.37 wt% with variation coefficients of 0.04% of both values, which is shown in Fig. 5(a). The part volume is calculated and compared to the part weight in Fig. 5(b). It stays stable with a variation coefficient of 0.10%, which indicates that the weight rise can be traced back to a higher filler content. Figure 5(c) shows the filler weight of the incinerated tension bars and the burnt polymer weight as the difference between the part weight before and after the incineration. The polymer weight is decreasing by 0.85 wt%, whereas the filler weight increases by 1.37 wt%. This indicates, that with increasing dwell time more fillers and less resin is injected. Before the hopper is completely emptied at cycle 100, five raw material samples I were taken from the hopper, whose positions are marked in the right picture of Fig. 6, and compared to five samples II from a packed BMC. The weight loss due to styrene evaporation of the samples I and II is presented in the diagram of Fig. 6. The weight loss of the samples I from the hopper is 0.20% lower compared to the samples II from the package with a total weight loss of 8.85 wt%. The variation coefficient of the first measurements of sample II is higher due to local mixture differences in the hopper. Presuming no styrene residue after 150 h 2.25 wt% of styrene has already evaporated from the raw material in the hopper.

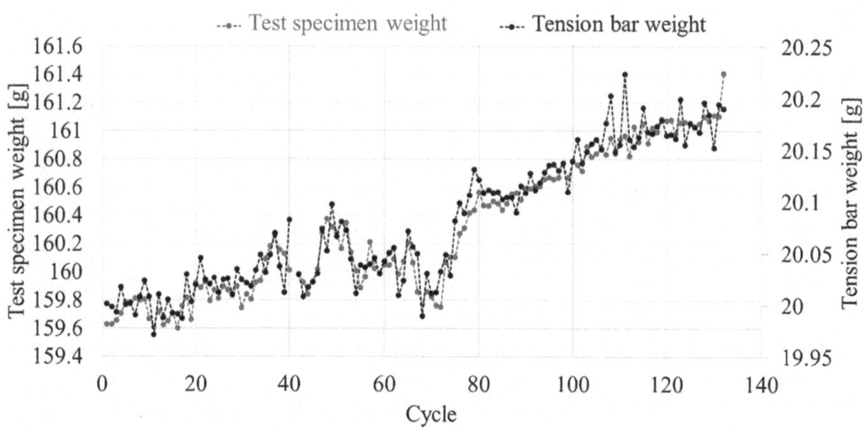

Fig. 4. Weight of the test specimen and one respective tension bar resulting from the processing of two bulks

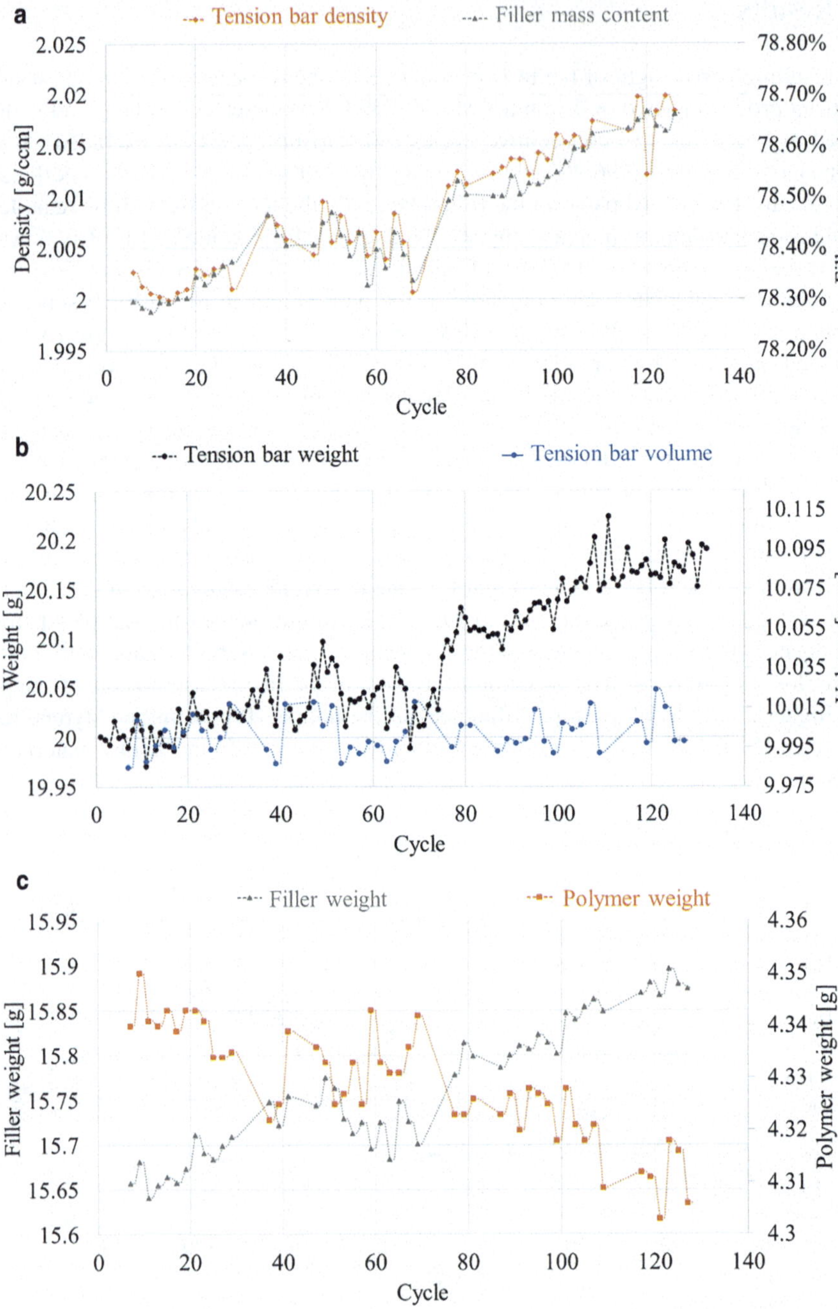

Fig. 5. Density and filler weight content (a), tension bar weight and volume (b), filler and polymer weight of incinerated tension bars (c)

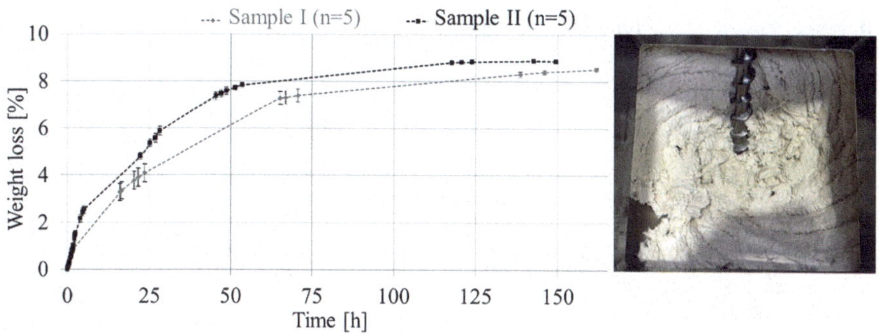

Fig. 6. Weight loss of samples taken from an unpacked bulk (sample I, n = 5) and from hopper at cycle 100 (sample II, n = 5), position marked in the right picture

In the next step the average dwell time of the material in the hopper is changed by varying reloading profiles. In Fig. 7 the tension bar weight is compared to the density measurements and the resulting part volume of the test series (a), presented in Fig. 3. The weight and density are increasing until reloading of the second bulk at cycle 250. As soon as fresh material is processed, weight and density decrease rapidly to their initial value and rise again, when the hopper is completely emptied. The maximal weight and density increase are 0.85% and 0.74% at a variation coefficient of 0.10% for the first 20 parts. The part volume stays constant with a total variation coefficient of 0.10%. Additionally, the weight and its derivative of sample 1 from cycle 50 and sample 2 from cycle 350 in dependence of the temperature are measured by TGA in Fig. 8. The first weight decrease I at approx. 365 °C marks the polymer decomposition, whereas the second drop II at approx. 750 °C is induced by the decarboxylation of $CaCO_3$ to CaO and CO_2. The residue consists of CaO and glass fibers, whose proportion is measured by CO_2 evolution. Sample 1 decomposes generally at lower temperatures and is composed of 20.60 wt% polymer, 68.71 wt% $CaCO_3$ and 10.69 wt% glass fibers. Sample 2 consists of 20.26 wt% polymer, 68.94 wt% of $CaCO_3$ and 10.80 wt% glass fibers, which shows a decrease of 1.68% of polymer and an increase of 1.35% of fillers from the beginning to the end of test series (a). This indicates, that more fillers and less resin are injected due to styrene loss.

In Fig. 9 the tension bar weight is compared to the density measurements and the resulting part volume of the test series (b) with an earlier loading of the second bulk at cycle 200 compared to trial (a). Part weight and density show a slight increase of 0.40% and 0.35% at cycle 200 and a rapid rise by 0.85% and 0.79% up until the end of the test series, respectively at a variation coefficient of 0.10%. Part volume is constant with a variation coefficient of 0.10%. Compared to test series (a) the maximal part weight and density increase have the same relative value but appear later and affect less parts. Concluding, the current proportion of unpacked and virgin material in the hopper influence the effect of styrene evaporation on the part composition in dependence of the reload timing. This becomes visible at a certain filling level of the hopper, which can be compensated by a defined reloading routine.

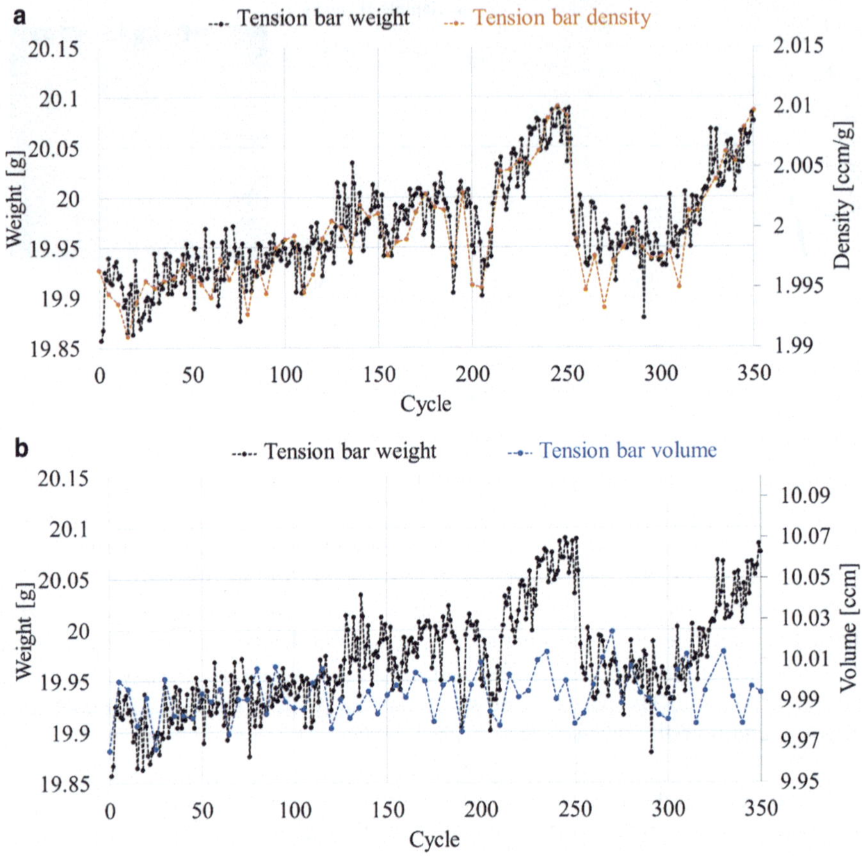

Fig. 7. Tension bar weight, density (a) and part volume (b) for reloading variation (a)

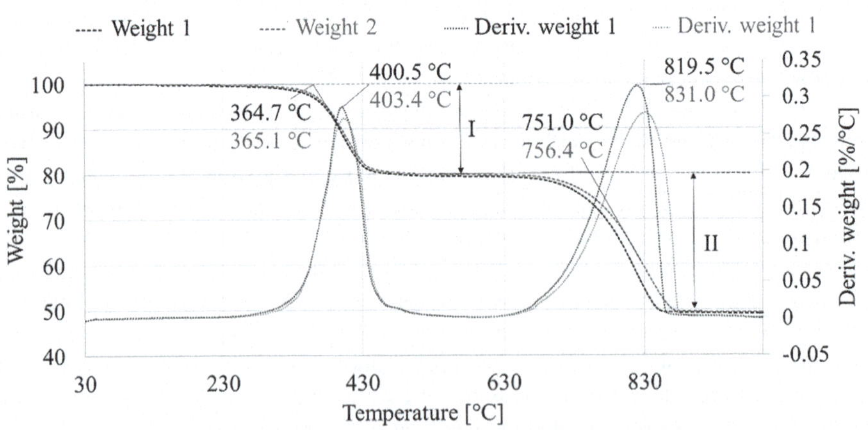

Fig. 8. Weight and its derivative over temperature of sample 1 and sample 2 from test series (a)

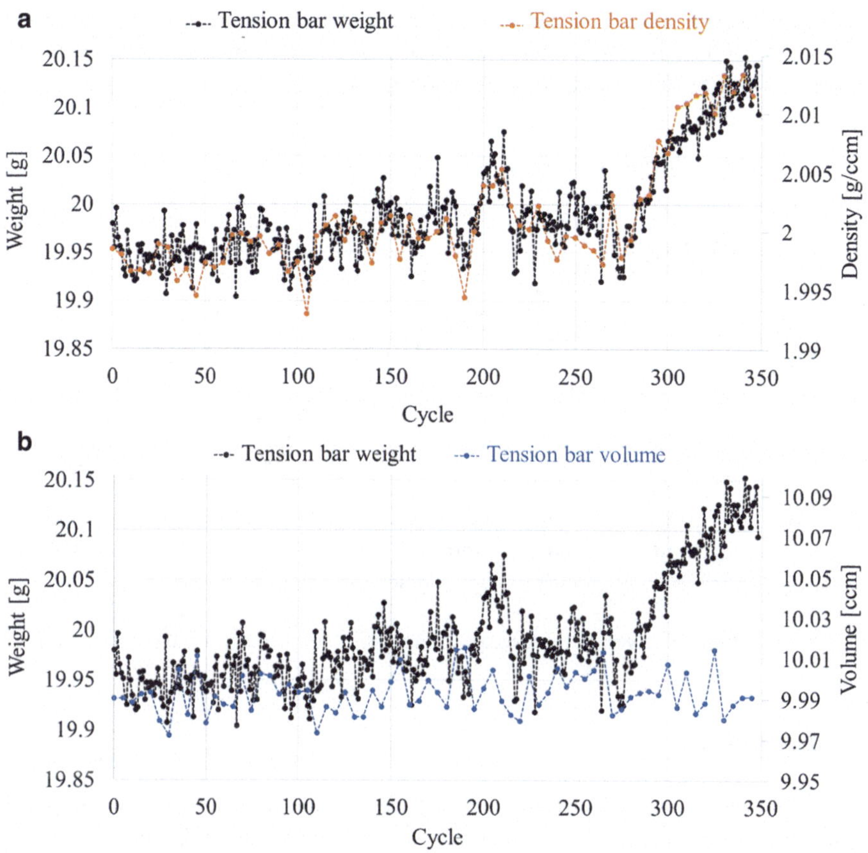

Fig. 9. Tension bar weight, density (a) and part volume (b) for reloading variation (b)

Figure 10 demonstrates the average drive torque of the hopper for test series (a) and (b). According to the Newtonian equation of motion the torque T of a rigid rotating body is the product of its inertia moment I and the angular acceleration α. The torque of the conical rotating hopper T with the mass m_H and the radii r_{min} and r_{max}, loaded with a mass m at the radius r is described by Eq. (1). A decreasing loaded mass m at a resulting lower fill level with smaller radius leads to a decreasing torque T at a constant angular acceleration α by neglecting the starting torque due to friction.

$$T = \left(\frac{1}{2}m_T\left({r_{min}}^2 + {r_{max}}^2\right) + \frac{1}{2}mr^2 + m\,r\,\alpha\right)\alpha \tag{1}$$

Therefore, the drive torque signal of the hopper can be used as a continuous display of the loaded mass. The drive torque in Fig. 10 correlates to the reloading profile in Fig. 3 as the torque decreases during the material consumption and increases as soon as a bulk is reloaded. Below 3 Nm the torque decreases rapidly. From this point on an increase in part weight and density is visible in Figs. 7 and 9. At the end of the test series the torque is constant and the remaining material dwells on the feeding and plastification screw until it is completely used up.

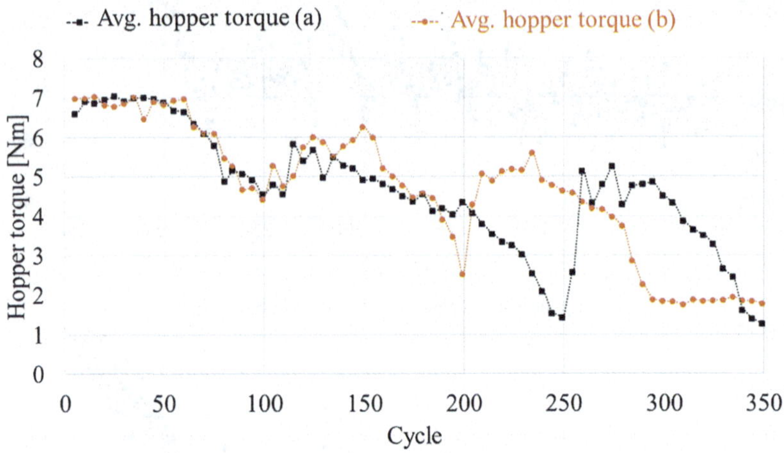

Fig. 10. Average drive torque of hopper of test series (a) and (b)

4 Discussion

The results show a gradual increase of part weight due to a higher filler proportion and lower amount of injected polymer at a constant injected compound volume. The unpacked raw material loses volatile styrene, which is indicated by a gradual weight loss over time. This may lead to the assumption that the effect appears with a longer material dwell in the hopper. But the part weight rise is not as gradual as the dwell time of the material in the hopper. Since it appears at a certain point, the effect of styrene loss is compensated and delayed by reloading the hopper and mixing the unpacked dried material with fresh material. The valuation of the hopper drive torque allows a continuous display of the fill level of the hopper. As soon as a critical fill level is reached, the effect of styrene loss and a different material composition of the part becomes visible. After this point, the torque decreases rapidly. At a high filling mass in the hopper the material is mixed by the hopper rotation. Before the hopper is emptied at a low filling mass, the material sticks to the inner surface of the hopper without relative movement, leading to a large exposed surface of the material. The hopper torque further decreases and the material is pulled-in gradually by the feeding screw. The styrene loss of residual material in the hopper leads to a linear increase of the part weight. An undershot of a hopper torque of 3 Nm signals material sticking to the hopper surface. In order to prevent this effect, raw material should be reloaded to cause blending of the material in the hopper. The hopper torque signal as a display of the filling level can be used as an alarm for refilling in order to prevent the effect of styrene loss. The effect of styrene loss on further part characteristics such as mechanical properties, the degree of conversion and the volume shrinkage are evaluated in prospective analysis.

5 Conclusion

Bulk molding compounds (BMC) essentially consist of polyester resin, styrene, additives, fillers and glass fibers and are fabricated in a conventional Z-kneader by mixing the resin and the solid fillers. Because of the considerable proportion of volatile styrene, the dwell time of the material in the hopper during processing is an important parameter. In this study the styrene loss is quantified during the injection molding production with a screw stuffer and evaluated for different loading scenarios. The results show a gradual increase of the part weight due to a higher proportion of fillers and lower amount of injected polymer at a constant injected compound volume. By a defined reloading routine, the effect of styrene evaporation can be reduced, as fresh and unpacked material is mixed with already dwelled material. Monitoring the hopper drive torque allows a continuous display of the fill level of the hopper. This signal could be used by an assistance machine function, which optimizes the load operation in order to reduce the impact of styrene loss.

Acknowledgement. The presented investigations take place within the research project "Quality-optimized Production of Thermosets", which is supported by the Bavarian Ministry of Economic Affairs and Media, Energy and Technology within the cluster "Neue Werkstoffe (BayNW)". The cooperation partners are University of Applied Sciences Rosenheim, Deckerform Produktionssysteme GmbH and Gubesch Thermoforming GmbH. The authors acknowledge the input and support of the cooperation partners, the promoter as well as the Project Management Jülich.

References

1. AVK – Industrievereinigung Verstärkte Kunststoffe e. V.: Handbuch Faserverbundkunststoffe/ Composites: Grundlagen, Verarbeitung, Anwendung. Springer, Berlin (2014)
2. Palik, M.: Vernetzte Präzision. Kunststoffe **98**(4), 83–85 (2009)
3. Späth, M., Zillmer, M., Karlinger, P., Schemme, M., Gehde, M.: Einfluss des zeitabhängigen Materialzustands beim Spritzgießen von rieselfähigen, duromeren Formmassen auf den Prozess und die Bauteileigenschaften. In: 25. Technomer Conference, Chemnitz (2017)
4. Schemme, M.: Sicherung der Produktqualität bei der Herstellung und Verarbeitung von SMC Werkstoffen. Dissertation, Universität Erlangen-Nürnberg (1996)
5. Lafranche, E., Cilleruelo, L., Ryckebusch, M., Krawczak, P.: A novel adaptive process control for injection moulding of BMC and CIC polyester compounds. Adv. Compos. Lett. **14**(5), 151–161 (2005)
6. Domininghaus, H.: Kunststoffe: Eigenschaften und Anwendungen. Springer, Berlin (1988)
7. Funke, V.W., Knödler, S., Feinauer, R.: Prüfung der Gültigkeit der Copolymerisationsparameter zur Berechnung der Zusammensetzung von Copolymerisaten vernetzter Polyesterharze bei hohen Umsätzen. Makromol. Chem.: Macromol Chem. Phys. **49**(1), 52–61 (1961)

8. Bureau, E., Chebli, K., Cabot, C., Saiter, J.M., Dreux, F., Marais, S., Metayer, M.: Fragility of unsaturated polyester resins cured with styrene: influence of the styrene concentration. Eur. Polym. J. **37**(11), 2169–2176 (2001)
9. González-Romero, V.M., Macosko, C.W.: Viscosity rise during free radical crosslinking polymerization with inhibition. J. Rheol. **29**(3), 259–272 (1985)
10. Yang, Y.S., Suspene, L.: Curing of unsaturated polyester resins: viscosity studies and simulations in pre-gel state. Polym. Eng. Sci. **31**(5), 321–332 (1991)
11. Stevenson, J.K.: Free radical polymerization models for simulating reactive processing. Polym. Eng. Sci. **26**(11), 746–759 (1986)
12. Lee, L.J.: Curing of compression molded sheet molding compound. Polym. Eng. Sci. **21**(8), 483–492 (1981)
13. Blanc, R., Agassant, J.F., Vincent, M.: Injection molding of unsaturated polyester compounds. Polym. Eng. Sci. **32**(19), 1440–1450 (1992)
14. Pusatcioglu, S.Y., Fricke, A.L., Hassler, J.C.: Heats of reaction and kinetics of a thermoset polyester. Polym. Eng. Sci. **24**(4), 937–946 (1979)
15. Yousefi, A., Lafleur, P.G., Gauvin, R.: Kinetic Studies of Thermoset Cure Reactions: A Review. Polym. Compos. **18**(2), 157–168 (1997)
16. Kia, H.G.: Sheet Molding Compounds: Science and Technology. Hanser, München (1993)
17. Burns, R.: Polyester molding compounds. Taylor & Francis, Abingdon (1982)
18. Hill, R.R. Jr., Muzumdar, S.V., Lee, L.J.: Analysis of volumetric changes of unsaturated polyester resins during curing. Polym. Eng. Sci. **35**(10), 852–859 (1995)
19. Bartkus, E.J., Kroekel, C.H.: Low shrink reinforced polyester systems. Appl. Polym. Symp. **15**, 113–135 (1970)
20. Li, W., Lee, L.J.: Low temperature cure of unsaturated polyester resins with thermoplastic additives: I. dilatometry and morphology study. Polymer **41**(2), 685–696 (2000)
21. Bucknall, C.B., Davies, P., Partridge, I.K.: Phase separation in styrenated polyester resin containing a poly (vinyl acetate) low-profile additive. Polymer **26**(1), 109–112 (1985)
22. Boyard, N., Vayer, M., Sinturel, C., Erre, R., Delaunay, D.: Modeling PVTX diagrams: application to various blends based on unsaturated polyester—Influence of thermoplastic additive, fillers, and reinforcements. J. Appl. Polym. Sci. **92**(5), 2976–2988 (2004)
23. Delahaye, N., Marais, S., Saiter, J.M., Metayer, M.: Characterization of unsaturated polyester resin cured with styrene. J. Appl. Polym. Sci. **67**(4), 695–703 (1998)
24. European Community Official Journal, Council Directive 1999/13/EG
25. Zwecker, J., Schreiner, H.: BASF SE: Low-evaporation polyester resins. US Patent 5,132,343 (1992)
26. Hunold, D., Wobbe, H.: Spritzgießen und Vernetzen. Kunststoffe **3**(90), 38–42 (2000)
27. Meixner, F., Brams, P., Schultheis, S.: Vorrichtung zum Beschicken von Spritzgieß- und Extrudiereinheiten, EP0470510 B1 (1995)
28. Drummer, D., Schiffers, R., Topic, N., Kruppa, S.: Stabilization of BMC injection molding by process control measures. SPE Antec, Orlando, 1537–1542 (2018)
29. Jazbinsek, A.: Zuführvorrichtung für eine Extruderschnecke. DE102011122935B3 (2011)
30. Rehmet, P.: Ein Duroplast in der Spritzgießmaschine. K-Zeitung 4 (2017)

The Influence of Hydrothermal Aging on the Material Properties of Continuous Fiber-Reinforced Thermoplastics and its Non-Destructive Characterization

Elmar Moritzer[1(✉)], M. Hüttner[1], B. Henning[2], and M. Webersen[2]

[1] Kunststofftechnik Paderborn (KTP), Paderborn University, Paderborn, Germany
elmar.moritzer@ktp.upb.de
[2] Fachgebiet Elektrische Messtechnik (EMT), Paderborn University, Paderborn, Germany

Abstract. According to Witten [55], the market for fiber-reinforced plastics has been growing steadily in recent years due to advantages such as the combination of lightweight construction, high design freedom and functional integration. The fiber composite parts are exposed to various environmental conditions during their service life, which lead to an aging of the materials and an accompanying change of the material properties up to pre-failure of the components. Therefore, the scope of this article is to investigate the aging behavior and the resulting property changes of selected continuous fiber-reinforced thermoplastics. In addition, an ultrasonic measuring system is presented with which age-related property changes of continuous fiber-reinforced thermoplastics can be determined non-destructively during service. A PP, PA6 and their respective reinforcements with continuous glass fibers in the form of composite sheets were aged in a saline solution at 90 °C up to 3000 h and then examined destructively and non-destructively. The age-related property changes could be determined. Both the conventional destructive and the ultrasonic-based non-destructive measuring methods showed qualitatively consistent results. Only the absolute values were overestimated with the ultrasound method, since this, in contrast to quasi-static destructive testing, takes place in high frequency ranges.

Keywords: Aging · Non-destructive · Ultrasound · Hydrothermal · Fiber-reinforced thermoplastics

1 Introduction

The strong intention of the automotive industry to produce resource-saving and energy-efficient vehicles is one of the main growth drivers for lightweight construction. This is primarily due to the current climate debate and the resulting legal regulations,

C. Hopmann and R. Dahlmann (Eds.): *Advances in Polymer Processing 2020*, pp. 189–204, 2020. https://doi.org/10.1007/978-3-662-60809-8_16

which oblige OEMs to produce vehicles with a maximum CO_2 emission of 95 g per km by the year 2020 [19, 52].

The traffic sector's share in global CO_2 emissions as determined by the International Energy Agency (IEA) is, in total, 24% [2]. Hence, fiber-reinforced plastics are of particular importance as they represent one of the major approaches to reduce CO_2 in automotive construction. According to Vogt et al. [53], it is estimated that by 2035 the proportion of polymers and FRPs in cars will be approximately as high as that of conventional steel and higher than that of aluminum, at around 20% and 15% respectively.

Thermosets or thermoplastics are generally used as matrix materials for fiber composite plastics, whereby thermoset matrix materials still dominate the market. The thermoplastic-based continuous fiber-reinforced plastics – also called composite sheet – are gaining importance because the properties and the processing of this group of materials are particularly favorable. Generally speaking, composite sheet parts are versatile and range from simple cladding elements in the car interior to load-bearing or safety-relevant components in the exterior [26, 51]. Examples are seat shells, door modules, front-end modules, airbag housings, pedals, underbody protection and bumpers [7, 12, 47]. The matrix material is usually a polyamide or polypropylene. Apart from that, parts based on continuous fiber-reinforced plastics are already being utilized in the aviation industry. However, high-priced, high-performance thermoplastics such as PEEK or PPS are used as matrix to meet the extreme thermal and mechanical requirements of the aircraft industry. The wing's front edge on the Airbus A340 represents a corresponding application example [8, 24].

1.1 Influence of the Fiber-Matrix Boundary Layer on the Composite Material

The synergy effects of a fiber composite can only be achieved if the individual components fiber and matrix adhere optimally to each other. Since the chemical affinity between the partners is generally not guaranteed, a chemical modification on the fiber or matrix side is necessary. In practice, the most common solution is to apply a size to the glass fibers. The size has to be adapted to the respective components and the subsequent processes to semi-finished products. Primarily, the fibers need to be protected from abrasion or friction and moisture during further processing. In addition, the individual filaments should be held together to form a more manageable roving and also ensure adhesion to the matrix [15, 30]. The latter is, depending on the composition of the size, not always the case, which means that the textile sizing agent has to be thermally removed after weaving and the fiber prepreg has to be saturated with a finish containing bonding agent. So-called plastic sizes are a compromise between a textile size and a finish [46].

The size is a multi-phase system which is applied as an aqueous emulsion during the spinning process of the filaments. The main constituents are one or more film-forming agents and bonding agents. Furthermore, lubricants, wetting agents and antistatic agents can be incorporated to ensure improved processability of the fiber roving in subsequent processes [44, 50].

The film-forming agent's tasks is to completely cover the glass fiber surface with a thin film so that the fibers stick together and are protected against mechanical stress. In order to ensure optimum compatibility with the later matrix polymer, suitable, chemically similar polymers are used as film-forming agents. Thus, polyurethanes are used for polyamide matrices and polypropylene film-forming agents are used for PP matrices [31, 44].

Organofunctional silanes are typically used as bonding agents. They increase the adhesion between fiber and matrix to create a composite. The silanes are hydrolyzed to silane oils and subsequently condensed to siloxanes. These can be bonded to the silane oils of the glass surface via hydrogen bonds. By splitting off water, they can then be converted into covalent bonds which stabilize the coupling between the adhesion agent and the glass surface [27, 38, 39].

The sizing system consists, to a large extent, of water which accounts for approx. 80–95% by weight. The water is removed by a drying process after application so that the fibers are covered with the remaining multi-phase system. The coating is dispersed inhomogeneously so that the resultant surface roughness and the accompanying increased surface area have a positive effect on the composite's properties [3, 56].

The second possibility to provide the necessary adhesion between the fiber composite components is to modify the matrix. This variant is predominantly used with non-polar plastics such as polypropylene, because the adhesion can only be attained to a limited extent via hydrogen bonds. During polymer synthesis, maleic anhydride groups are grafted onto the PP, which can react with the OH groups on the glass fiber surface. This enables a covalent and a hydrogen bond to be formed between the glass fiber and the maleic anhydride group [15, 31].

1.2 Aging of Fiber-Reinforced Thermoplastics

On condition that fibers and matrix adhere well to one another, a wide variety of components can be developed and produced from the fiber composite plastics. They are used in many different applications and are exposed to all kinds of individual environmental influences during their service life including, for example, temperature change, moisture, UV radiation, salt-containing media and mechanical loads. These result in physical and chemical aging processes in the material and influence the respective material properties [14, 41].

According to standard DIN EN ISO 50035, physical aging is one or several processes that take place by altering the microstructure or the molecular order state, provided that chemical aging processes are not the cause. Aging processes involving a change in the chemical composition or the molecular structure or molecule size of the material are defined as chemical aging. These include, for example, oxidation, chain degradation, hydrolysis, post-condensation and post-polymerization [40].

A comprehensive look at the aging process of fiber-reinforced plastics can only be carried out with all the components involved. Accordingly, the reinforcing fibers and the fiber-matrix adhesion have to be considered, since the failure behavior of the composites is determined by these factors [9].

In the dry state, failure of the fibers themselves often predominates. Aging processes of the matrix do not influence the fiber-matrix adhesion here. However, if moisture or other fluid media are present, the glass fibers can be partially exposed and the fiber-matrix adhesion can be diminished. The size system is of great importance for fiber-matrix adhesion. The better the sizing agent performs at elevated temperatures and high humidity, the more resistant the fiber-reinforced plastic is to hydrothermal aging. Exposed fibers that are in direct contact with the media can be damaged by them [16, 49].

With regard to the application of fiber-reinforced plastics, it can be stated that ageing processes have a proven negative impact on the mechanical material properties of composites. This influence has been investigated and characterized for thermosets and especially for continuous fiber-reinforced thermosets in great detail in various different studies [13, 22, 29, 36, 42].

Numerous studies are also known on the topic of unreinforced and discontinuously fiber-reinforced thermoplastics, all of which show a deterioration of the mechanical properties over aging [10, 17, 18, 20, 21, 32, 40, 48, 54]. The resultant effect is always strongly dependent on the prevailing ambient conditions and the respective material configuration. Bergeret et al. [5, 6] and Sayer [45], for example, examined the influence of hydrothermal aging on the mechanical properties of fiber-reinforced polyamide PA 6.6, polybutylene terephthalate (PBT) and polyethylene terephthalate (PET) and also found a significant decrease in the mechanical properties. The same applies to the studies by Rudzinski [44], Maeder [31] and Lariviere [27], who focused on the resistance of different size systems to various ambient conditions.

Only a few published studies are currently known for continuous fiber-reinforced thermoplastics. Bergamo et al. [4], for example, have studied the influence of material moisture on the mechanical properties of glass fabric-reinforced thermoplastics based on PA 6.6, but did not examine the material-damaging aging over a prolonged period. Arici [1] studied the hydrothermal aging behavior of carbon and glass fiber-reinforced PEI, in which, however, under long-term changing climate conditions, no influence on the mechanical properties was detected. Neft [37] subjected glass fiber laminates with a PP matrix also to changing climate tests and observed a slight decline in the mechanical data. Robert et al. [43] also examined the aging behavior of glass fiber PP laminates under hydrothermal load through immersion in water and was able to show that the mechanical properties can decline by up to 40%. Some studies by the authors [33–35] also show that, for continuous glass fiber-reinforced polyamides, reduced material properties result from a hydrothermal load that can be attributable to reduced fiber-matrix adhesion.

At present, there are only very few known publications that deal with the aging-related long-term behavior of types of continuous fiber-reinforced thermoplastics that could potentially be used in the automotive and aviation industries. Yet, precise information about the long-term behavior of these materials is of elementary importance for their use in safety-related applications. Only in this way can the range of application of these materials be successively extended.

Apart from a knowledge of the effects of aging on continuous fiber-reinforced thermoplastics, it is also important to develop test methods with which material damage can be non-destructively detected [25]. This is the only way to ensure that the safety-related and in some cases cost-intensive parts can be replaced before failure. Several non-destructive test methods for detecting defects or mechanical damage already exist that are also suitable for fiber-reinforced components. However, they cannot detect environment-related material aging before significant macroscopic damage occurs.

A more precise knowledge of the long-term behavior of these materials and the possibility of non-destructive tests on the deterioration of material properties over time would significantly increase their acceptance in industry and thus offer new potential for application.

2 Experimental Investigations

In order to reduce the complexity of the environment-related ageing mechanisms, only the hydrothermal aging process is considered in the present study. In particular, polycondensation-based thermoplastics show a high level of susceptibility to the effects of hot water, as the ensuing hydrolysis reaction splits the polymer chains. For this reason, aging experiments are carried out in which a variety of homogeneous polymers and composite sheets are aged at a constant temperature of 90 °C in a sodium chloride/water mixture for up to 3000 h. The salt concentration is set at a constant 50 g/L and is determined at regular intervals of approx. 3 days using a refractometer. Since the water in the solution evaporates slowly, water must be added if necessary. Table 1 shows the investigated materials and their differences in properties.

In addition to polyamide materials susceptible to hydrolysis, equivalent polypropylene materials are also examined. The aim of this is to effectively eliminate hydrolysis-related chain degradation and the influence of moisture on the mechanical properties, so other effects may be observed. The polyamide and polypropylene composite sheet examined here have identical fabric types of glass roving with a weight per unit area of 600 g/m^2, so that only the fiber size is adapted to the respective matrix type. The fiber volume content, the twill, the fiber undulation and the laminate thicknesses (3 mm) are thus quasi-identical for the materials Tepex 102 RG600 (PA6-based) and Tepex 104 RG600 (PP-based).

Table 1. Investigated materials

Material	Synonym	Fiber content	No. of layers
Moplen HP 500V	PP	-	-
Tepex 104 RG600	PP-CS	47 Vol.-%	6
Durethan B30S	PA6	-	-
Tepex 102 RG600	PA6-CS	47 Vol.-%	6

The composite sheets are aged and characterized in a fabric orientation of 0°/90° to the loading direction. For the chemical, physical, mechanical and acoustic measurements, identical specimens are used. The measuring methods used for material characterization are listed in Table 2.

In addition to the mechanical tensile properties, the polyamide-based materials are tested with regard to their molecular change over time. For this, solution viscometry is used and the viscosity number is determined, which represents a measure for an average molecule chain length. By means of scanning electron microscopy, the fiber-matrix adhesion of the fracture surfaces is also examined.

Table 2. Used test methods

Value	Test method	Norm
Viscosity number	Solution viscosimetry	DIN EN ISO 307
Tensile strength Young's modulus	Tensile test	DIN EN ISO 527
Fiber-matrix-adhesion	Scanning electron microscopy	-
Ultrasound wave velocities	Ultrasound measurement	-

The non-destructive testing is carried out by an ultrasound-based measurement method, with which, among other things, the material's characteristic wave velocities can be determined by applying an inverse procedure. The strong material dependence of Lamb wave modes in plate waveguides is exploited. The measuring system is shown in Fig. 1.

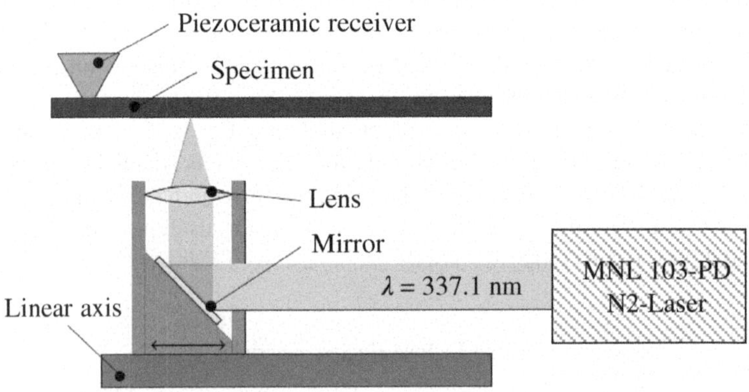

Fig. 1. Ultrasound-based non-destructive measurement system [11]

In the sheet specimens, photo-acoustic ultrasound waves are excited by a pulse laser (225 µJ, 3 ns per pulse). These waves propagate through the material and are detected with a piezo-electric ultrasound transducer. The excitation of the ultrasound waves is carried out at different positions so that spatially resolved measuring data is generated.

The longitudinal and transverse wave velocities can be determined from the data using a two-dimensional Fourier transformation and a suitable material model. With the help of these values it is possible to calculate the Young's modulus [23]. For the composite sheets examined in the present study, a linear-elastic and orthotropic material model was chosen and the material was assumed to be macroscopically homogeneous.

The determined ultrasound wave velocities can, by including the material density, be converted into the elasticity modulus through the following functional relation [28]:

$$E = \frac{\rho \cdot c_T^2 \cdot \left(3c_L^2 - 4c_T^2\right)}{c_L^2 - c_T^2} \tag{1}$$

E: Young's modulus
c_T: transverse wave velocity
c_L: longitudinal wave velocity
ρ: density

In [23] it was shown that since the wave velocities mentioned above depend on the direction of wave propagation with respect to the fiber orientation, a consistent orthotropic material model can be identified if multiple propagation paths are investigated. From this model, the direction-dependent elasticity modulus can be evaluated for arbitrary orientations and displayed, for example, in polar diagrams, as is typical for fiber-reinforced composites.

In this way, the aging-related changes in the properties of the composite sheets is determined in the present study.

3 Results and Discussion

The results of the experiments show that the influence of hydrothermal aging on the chemical and physical properties of the unreinforced polymers and composite sheets with an analog matrix is in principle comparable, but has different effects on the mechanical properties.

For the polyamide 6 materials, a reduction of the average chain length due to the hydrolysis reaction was shown as can be seen by a decreasing viscosity number. In Fig. 2, the viscosity numbers standardized to the starting condition are shown for the unreinforced and fiber-reinforced PA 6 over time.

In addition, the molecular weight loss was validated by additional Gel permeation chromatography (GPC) measurements and a determination of the average molecular weight. Figure 3 shows the comparison between the viscosity numbers (VN) and the average molecular weights (MW) in a normalized representation.

Taking the standard deviation into account, the chemical properties of the non-reinforced polyamide and the relevant composite sheet change approximately to the same extent over the period under investigation.

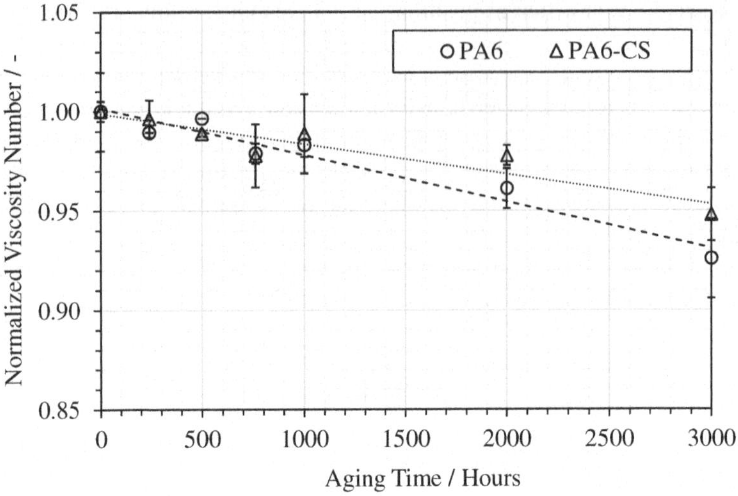

Fig. 2. Normalized viscosity numbers (VN) over the aging time (absolute average VN at 0 h: PA6 – 141.5 cm^3/g and PA6-CS – 134.5 cm^3/g)

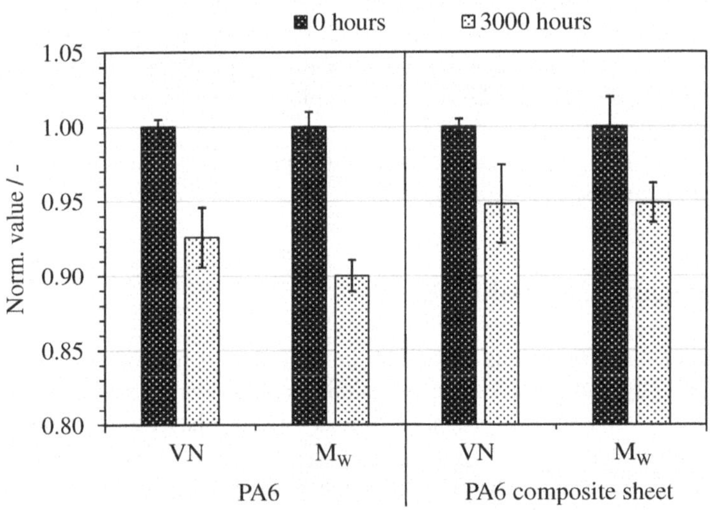

Fig. 3. Normalized viscosity numbers (VN) and average molecular weight (M_w) over the aging time (absolute average MW at 0 h: PA6 – 76188 g/mol and PA6-CS – 75111 g/mol)

The results from the mechanical characterization show different effects. For this, the tensile strengths of the materials in relation to aging time are shown, also in normalized form, in Fig. 4.

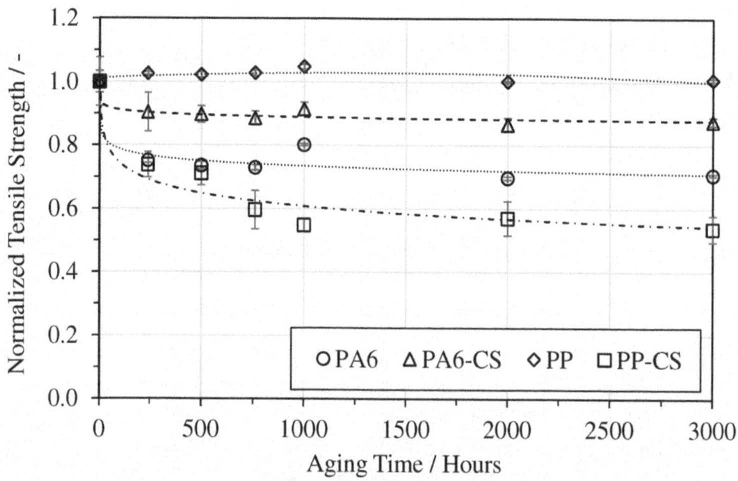

Fig. 4. Normalized tensile strength over the aging time (absolute average Tensile Strength at 0 h: PA6 – 73 MPa, PA6-CS – 417 MPa, PP – 35 MPa and PP-CS – 412 MPa)

Whereas the non-reinforced PP shows virtually no change in mechanical strength, the strength of the PP composite sheet decreases significantly over time. The constant tensile strength of the non-reinforced polypropylene over the aging period confirms the assumption that the hydrothermal aging does not have any significant influence on the properties of this material.

Comparing the non-reinforced polyamide with the respective composite sheet, it can be seen that the strength of the unreinforced material also changes. This is attributable not only to the declining molecular weight but also to the softening effect of water stored in the material, which cannot be avoided even despite pre-drying. Moisture measurements confirm an increased moisture compared with the starting condition, which however is virtually constant after 250 h. The image after 1000 h shows a slightly reduced moisture, which is also reflected in a higher strength value of the non-reinforced PA 6, and can be classified as a statistical outlier. For simplicity, a detailed presentation of the moisture is omitted here.

To explain the changes in the mechanical properties of the composite sheets, the entire composite material has to be investigated. Apart from the polymer matrix, the glass fibers and especially the fiber-matrix adhesion are influenced by aging mechanisms. The fiber-matrix adhesion – which is ensured, for example, by the application after fabric production of a suitable bonding agent containing size – can be visualized, among other things, by scanning electron micrographs. For the examined composite sheets SEM pictures show that the fiber-matrix adhesion decreases significantly over the aging period, so that the reduced mechanical data can also be explained. Figure 5 shows as an example the fracture surfaces of the examined composite sheets and illustrates the declining fiber-matrix adhesion through aging, highlighting selected points in time. The images were taken without any special sample preparation after the mechanical tests.

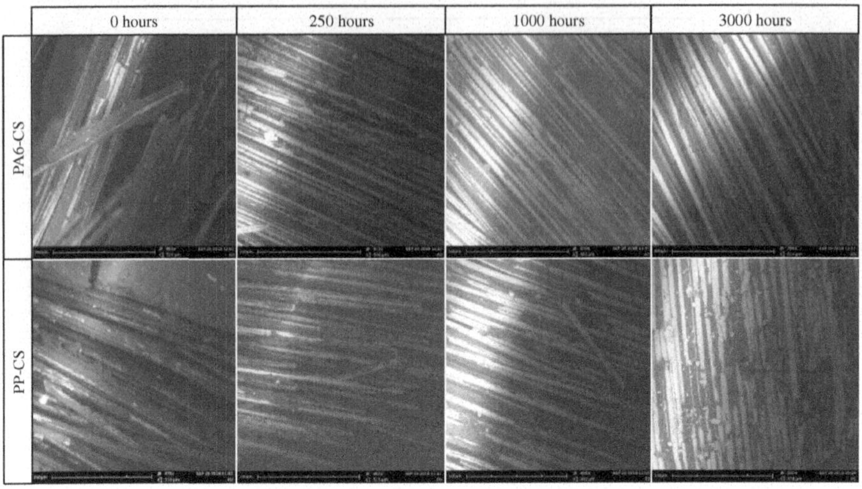

Fig. 5. SEM of the fracture surfaces of the composite sheets over the aging time

Whereas in the unaged condition, the fibers of both types of composite sheet are still covered with polymer matrix (dark areas in Fig. 5), the fibers (bright areas in Fig. 5.) are exposed increasingly during the aging period. Due to the reduced adhesion between fibers and matrix, the occurring stresses can no longer be completely absorbed and transmitted, which leads to a prematurely failure of fiber-plastic composites. The effect of the deterioration in resistance is significantly more pronounced and more critical for the PP-based composite sheets than for the PA6-based composite sheets. This can be attributed to the fact that the sizing system between the polyamide and the glass fibers must be generally resistant to the effect of water and must be adjusted accordingly [44]. The hydrophilic character of the polyamide necessitates a constant presence of water molecules in the material, against which the fiber-matrix adhesion must be protected. Because polypropylene can absorb only very little water, protection against water is only necessary to a limited extent. Furthermore, the fiber-matrix adhesion between PP and glass fibers can only be guaranteed through additional modifications, because of the non-polar character of the polymer.

For this reason, on the matrix side, a PP is used which is specially grafted with maleic anhydride groups (MAH), with the MAH groups ensuring adhesion to the glass [15, 27]. In addition, a PP film-forming agent is typically used in the sizing system in order to increase the compatibility to the PP matrix [31, 44]. If this size is degraded or damaged, for example by aging processes, an adequate fiber-matrix adhesion can no longer be guaranteed.

On the other hand, the effect of the deterioration proves to be less significant for the composite sheet's respective values of Young's modulus. The time dependence is illustrated in Fig. 6, in normalized form.

The Young's modulus is typically determined in the linear-elastic range (up to 0.25% elongation) of a material, so that the material experiences a relatively low stress. The acting forces can still be adequately absorbed by the existing fiber matrix adhesion, so that there is only a small reduction of this value over the aging period.

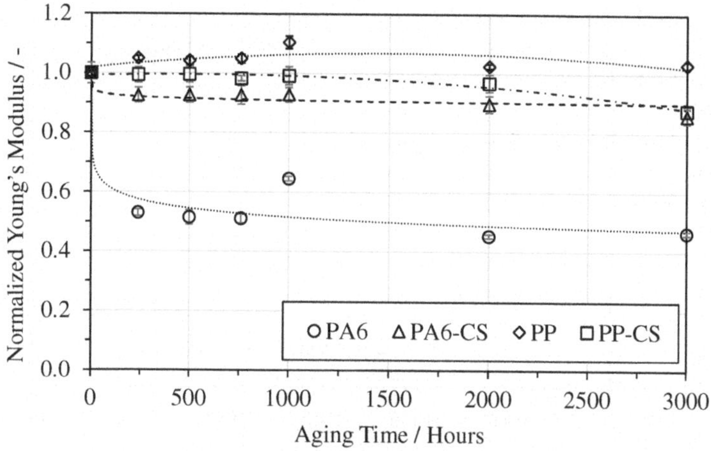

Fig. 6. Normalized Young's modulus over aging time (absolute average Young's modulus at 0 h: PA6 – 2896 MPa, PA6-CS – 22682 MPa, PP – 1522 MPa and PP-CS – 20372 MPa)

However, the determined minor changes in Young's modulus can also be detected with the non-destructive ultrasound measuring system. The graphs of the mechanically (destructively) and the acoustically (non-destructively) determined data are plotted for the two composite sheet types in Fig. 7 for PA 6 and in Fig. 8 for PP.

Considering the respective standard deviations, the results show that the curves for both composite sheet configurations show an adequate level of conformity. Not only the immediate decline in the Young's modulus for the PA6-based composite sheet but also the declining Young's modulus of the PP composite sheet after 1000 h are detected by the ultrasound measurements.

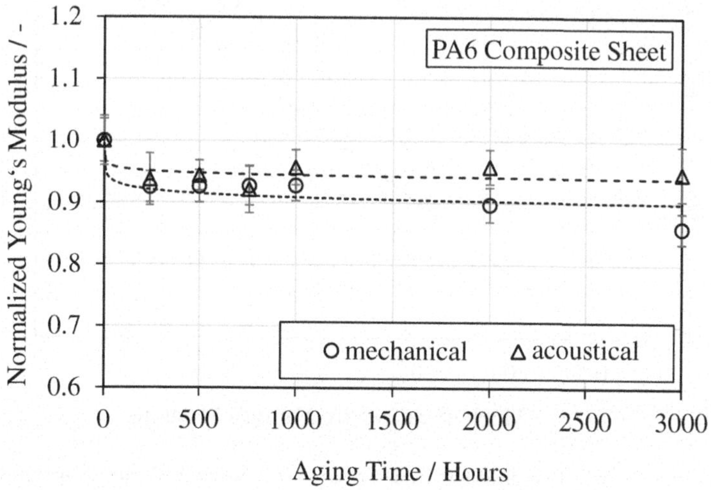

Fig. 7. Destructively and non-destructively determined, normalized Young's modulus of the PA 6 composite sheets over aging time

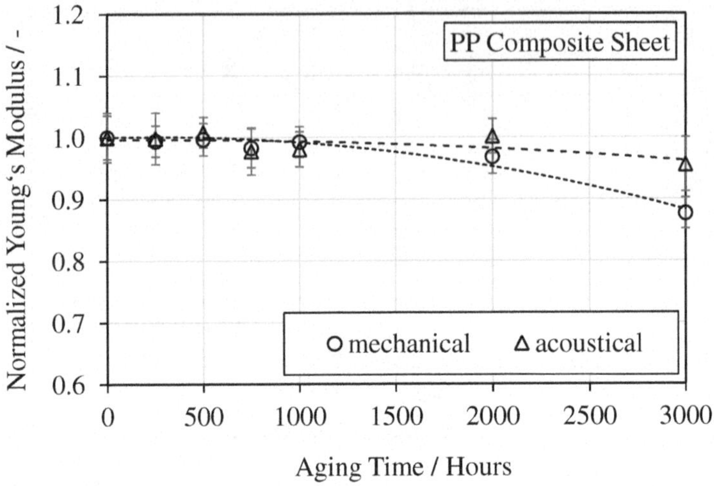

Fig. 8. Destructively and non-destructively determined, normalized Young's modulus of the PP composite sheets over aging time

Overall, it can be said that the acoustically, non-destructively determined values of Young's modulus are higher than the destructively determined values. A comparison of the normalized values determined with the respective measuring process is shown in Fig. 9. The deviations are within 10%.

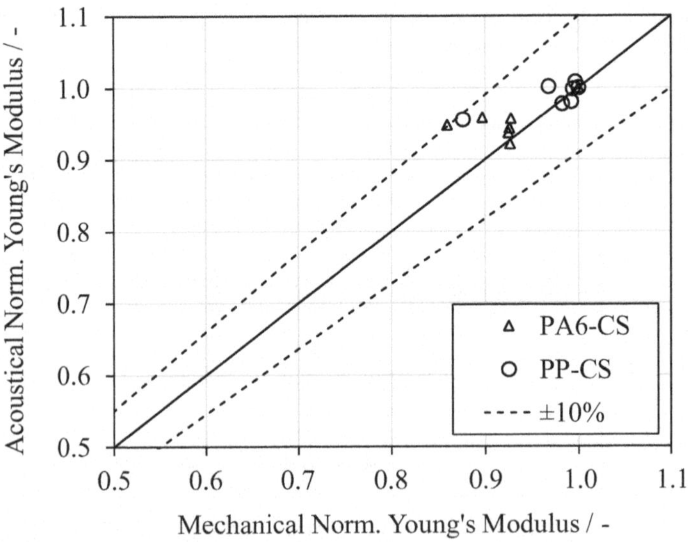

Fig. 9. Destructive and non-destructive determined normalized Young's modulus of both composite sheets

This can be attributed to the fact that the acoustically determined elasticity moduli are generally up to 3 GPa above those of the mechanically determined values. The overestimation is due to the fact that the relevant excitation frequencies are significantly different. Whereas in the mechanical tensile test, a quasi-static excitation occurs, the materials for the ultrasound measurements are excited in the megahertz range. Higher frequencies result in an effective stiffening effect due to the viscoelastic nature of the materials (see also: time-temperature superposition). Hence, a direct comparison between the results of the two measuring processes is only possible to a limited extent, as viscoelastic behavior was neglected in the acoustic material model. For future works, this model will be adapted accordingly. Nevertheless, a qualitative comparison of the normalized curves for Young's modulus is also easily possible in this way, so both procedures yield the same information about the (relative) material degradation process.

4 Conclusion and Outlook

Based on the results it can be summarized that the aging behavior of composite sheet exerts a considerable influence on the mechanical properties. In particular, the fiber-matrix adhesion is significantly weakened in all examined materials, thereby reducing mechanical strength and resistance. The aging effects resulting at microscopic level are, in principle, very similar for non-reinforced polymers and composite sheet of the same matrix type. However, because of the heterogeneity and the additional influences of the other material components, the influence on the resultant mechanical properties is of a different form.

In addition, it was shown that the aging-related changes in Young's modulus can also be measured non-destructively by means of an ultrasound measuring system. This possibility will be developed further in subsequent projects so that even more accurate conformity between the results can be ensured.

Acknowledgements. The results presented here were obtained in the context of the research projects MO 685/13-2 (Project Number 260306237), sponsored by the Deutsche Forschungsgemeinschaft (DFG). We would like to thank the DFG for their support.

References

1. Arici, A.: Effect of hygrothermal aging on polyetherimide composites. J. Reinf. Plast. Compos. **26**(18), 1937–1942 (2007)
2. BDL–Bundesverband der Deutschen Luftverkehrswirtschaft e. V.: Klimaschutz Report 2018 (2018)
3. Berg, J., Jones, F.: The role of sizing resins, coupling agents and their blends on the formation of the interphase in glass fibre composites. Compos. Part A **29A**, 1261–1272 (1998)
4. Bergamo, S., et al.: Effect of relative humidity on mechanical properties of a woven thermoplastic composite for automotive application. Polym. Test. **48**, 160–168 (2015)

5. Bergeret, A., et al.: Influence of the fibre/matrix interface on ageing mechanisms of glass fibre reinforced thermoplastic composites (PA-6,6, PET, PBT) in a hygrothermal environment. Polym. Degrad. Stab. **94**, 1315–1324 (2009)
6. Bergeret, A., Pires, I.: The hygrothermal behavior of glass-fibre-reinforced thermoplastic composites: a prediction of the composite lifetime. Polym. Test. **20**, 753–763 (2001)
7. Bond-Laminates GmbH: Hart im Nehmen. Pressemitteilung vom 29. März 2017. http://bond-laminates.de/de/ueber-bond/pressemitteilungen-medien/aktuelle-pressemitteilungen/2016-00076/. Accessed 04 Oct 2018
8. Bonefeld, D.: Organobleche: Verarbeitung und Anwendungen von endlosfaserverstärkten thermoplastischen FVK, 26. Darmstädter Kunststoff-Kolloquium (15.–16. September 2011)
9. Campeotto, A., Ferguson, G. M., Jefferson, A.: Weathering of engineering thermoplastics. Materials Forum, 267–275 (1992)
10. Chaupart, N., et al.: Molecular weight distribution and mass changes during polyamide hydrolysis. Polymer **39**(6–7), 1375–1380 (1998)
11. Claes, L., et al.: Determination of the material properties of polymers using laser-generated broadband ultrasound. J. Sens. Sens. Syst. **5**, 187–196 (2016)
12. Dallner, C., et al.: Werkstoffe für die Zukunft des Leichtbaus. Kunststoffe 3. Hanser, München (2012)
13. Davies, P., Pomies, F., Carlsson, L.: Influence of water and accelerated aging on the shear fracture properties of glass/epoxy composite. Appl. Compos. Mater. 3 (1996).
14. Deutsches Institut für Normung: DIN EN ISO 50035 – Begriffe auf dem Gebiet der Alterung von Materialien – Polymere Werkstoffe (2011)
15. Ehrenstein, G.: Faserverbund-Kunststoffe. Hanser, München (2006)
16. Ehrenstein, G., Pongratz, S.: Beständigkeit von Kunststoffen, Bd. 1. Hanser, München (2007)
17. Eibeck, P.: Schaden durch Feuchtigkeit vermeiden. Kunststoffe **4**, 83–86 (2012)
18. Eipper, A., et al.: Die Tests werden immer härter. Kunststoffe **5**, 98–101 (2008)
19. Europa-Parlament: CO_2-Emissionen von PKW ab 1. Januar 2015 auf 125 g/km begrenzen. Pressemitteilung des Europa-Parlaments. http://www.europarl.europa.eu/sides/getDoc.do?language=de&type=IMPRESS&reference=20071023IPR12110. Accessed 31 Okt 2018
20. Gardner, R., Martin, J.: Humid aging of plastics: effect of molecular weight on mechanical properties and fracture morphology of polycarbonate. J. Appl. Polym. Sci. **24** (1979)
21. Illing, T.: Bewertung von mechanischen und thermischen Eigenschaften glasfaserverstärkter Polyamid-Werkstoffe unter besonderer Berücksichtigung des Alterungsverhaltens von Bauteilen in der Automobilindustrie. Dissertation, Martin-Luther-Universität Halle-Wittenberg (2015)
22. Jana, R.N., Bhunia, H.: Hygrothermal Degradation of the composite laminates from woven carbon/SC-15 epoxy resin and woven glass/SC-15 epoxy resin. Polym. Compos. **29**(6), 664–669 (2008)
23. Johannesmann, S., et al.: An acoustic waveguide-based approach to the complete characterisation of linear elastic, orthotropic material behaviour. Ein Wellenleiter-basierter Ansatz zur akustischen Charakterisierung des kompletten linear-elastischen, orthotropen Materialverhaltens. Technisches Messen (2018)
24. Keck, R.: Fest und nachgiebig zugleich - Vorflügel in Kohlefaserverbund-Bauweise und mit integrierter Enteisung. DLR Nachrichten 124 (2009)
25. Kreutzbruck, M.: Herausforderungen an die zerstörungsfreie Prüfung im Trend moderner Leichtbaustrategien. 24. Stuttgarter Kunststoffkolloquium, Stuttgart (2015)
26. Laessig, R., et al.: Serienproduktion von hochfesten Faserverbundbauteilen – Perspektiven für den deutschen Maschinen und Anlagenbau. VDMA (2012)

27. Larivière, D., et al.: Hydrothermal ageing of GF/PP composites: when glass/polymer adhesion favours water entrapment. Polym. Polym. Compos. **13**(1) (2005)
28. Lerch, R., et al. 2009: Technische Akustik. Springer (2009)
29. Liao, K.: In-situ strength degradation of glass fibers in a pultruded composite by environmental aging. J. Mat. Sci. Lett. **18** (1999)
30. Loewenstein, K.L.: The Manufacturing Technology of Continuous Glass Fibres, 3rd edn. Elsevier, Amsterdam (1993)
31. Maeder, E., et al.: Role of film formers in glass fibre reinforced polypropylene – new insights and relation to mechanical properties. Compos. Part A **32** (2001)
32. Meister, S.: Alterung spritzgegossener thermoplastischer Mikrobauteile. Dissertation, Technische Universität Erlangen-Nürnberg (2016)
33. Moritzer, E., et al: An approach to non-destructive testing of aged polymers. In: 32nd International Conference of the Polymer Processing Society (PPS), Lyon, Frankreich (2016)
34. Moritzer, E., et al.: The influence of environmental aging on the material properties of back-molded composite sheets. In: 34th International Conference of the Polymer Processing Society (PPS), Taipei, Taiwan (2018)
35. Moritzer, E., Huettner, M.: The influence of hygrothermal aging on the material properties of continuous fiber-reinforced thermoplastics. In: 76th Annual Technical Conference of the Society of Plastics Engineers (ANTEC), Orlando (USA) (2018)
36. Nagae, S., Otsuka, Y.: Effect of sizing agent on corrosion of glass fibre reinforced plastics in water. J. Mat. Sci. Lett. **15** (1996)
37. Neft, J.F.: Einflüsse durch Alterung und Konditionierung auf die mechanischen Eigenschaften endlos glasfaserverstärkter Thermoplaste. Verbundwerkstoffe und Werkstoffverbunde, pp. 18–23. Wiley-VCH, New York (1999)
38. Olmos, D., et al.: Hydrolytic damage study of the silane coupling region in coated silica microfibres: pH and coating type effects. J. Mat. Process. Technol. **143–144**, 82–86 (2003)
39. Plueddemann, E.P.: Silane Coupling Agents. 3rd. Plenum Press, New York (1991)
40. Pongratz, S.: Alterung von Kunststoffen während der Verarbeitung und im Gebrauch. Dissertation, Universität Erlangen Nürnberg (2000)
41. Potthoff, P.: Alterung und Recycling von HDPE für Flaschentransportkästen. Kunststoffe **75**(8), 481–486 (1985)
42. Selzer, R., Friedrich, K.: Mechanical properties and failure behavior of carbon fibre-reinforced polymer composites under the influence of moisture. Compos. Part A **28**(6), 595–604 (1996)
43. Robert, M., et al.: Environmental effects on glass fiber reinforced polypropylene thermoplastic composite laminate for structural applications. Polym. Compos. **31**(4), 604–611 (2010)
44. Rudzinski, S.: Untersuchung des thermischen und hydrothermischen Alterungsverhaltens von Glasfaser-Polyamid66-Verbundwerkstoffen. Dissertation, TU Dresden (2011)
45. Sayer, S.: Mechanical performance of polyamide 66 and influence of glass fiber content on moisture absorption. Mech. Test. **56**(4), 325–330 (2014)
46. Schuermann, H.: Konstruieren mit Faser-Kunststoff-Verbunden. Springer, Berlin (2007)
47. Selig, J., Radeck, A.: Mehr Potenzial für Leichtbau. Kunststoffe 3. Hanser, München (2012)
48. Shu, Y., L, X., Ye, Lin: Study on the long-term acid rain aging behaviour of polyamide 6. J. Macromol. Sci., Part b **48**(3), 526–536 (2009)
49. Spaude, R.: Korrosion und Alterung von Glasfasern und Glasfaserverstärkten Duroplasten. Dissertation, Universität Gesamthochschule Kassel (1984)
50. Teschner, R.: Glasfasern. Springer Vieweg, Berlin (2013)

51. Thomas, D., et al.: Beitrag zum Fortschritt im Automobilleichtbau durch belastungsgerechte Gestaltung und innovative Lösungen für lokale Verstärkungen von Fahrzeugstrukturen in Mischbauweise. FAT Schriftenreihe 244, Forschungsvereinigung Automobiltechnik e. V. (FAT) (2012)
52. Umweltbundesamt: Pkw und leichte Nutzfahrzeuge. Pressemitteilung des Umweltbundesamtes vom 17.08.2016. https://www.umweltbundesamt.de/themen/verkehr-laerm/emissionsstandards/pkw-leichte-nutzfahrzeuge. Accessed 20 Sept 2018
53. Vogt, M., et al.: Bestandsaufnahme Leichtbau in Deutschland. Studie im Auftrag des Bundesministeriums für Wirtschaft und Energie. VDI Verlag, Berlin (2015)
54. Wang, Z., et al.: Dissolution, hydrolysis and crystallization behavior of polyamide 6 in superheated water. Chin. J. Polym. Sci. **33**(9), 1334–1343 (2015)
55. Witten, E.: Der GFK-Markt Europa 2017, Composites-Marktbericht 2017 – Marktentwicklungen, Trends, Ausblicke und Herausforderungen. http://www.avk-tv.de/infothek.php. Accessed: 05 Feb 2018
56. Wu, H.F., et al.: Effects of silane coupling agents on the interphase and performance of glass fiber reinforced polymer composites. Compos. Sci. Technol. **57**, 975–983 (1997)

Integration Concept of Injection, Forming and Foaming: A Practical Approach to Manufacture Hybrid Structures

Saeed Farahani[1,2(✉)], Sai Aditya Pradeep[1,2], and Srikanth Pilla[1,2,3,4]

[1] Department of Automotive Engineering, Clemson University, Greenville, SC 29607, USA
{sfaraha,spradee,spilla}@clemson.edu
[2] Clemson Composites Center, Clemson University, Greenville, SC 29607, USA
[3] Department of Material Science and Engineering, Clemson University, 29634 Greenville, SC, USA
[4] Department of Mechanical Engineering, Clemson University, Clemson, SC 29634, USA

Abstract. Motivated by the concept of the integrative production systems, the hybrid process of polymer injection molding and sheet metal forming, known as polymer injection forming (PIF), has been introduced to manufacture sheet metal-polymer components in a single operation. Despite the wide potential application of this technology, its implementation in actual industrial production has been hindered due to several challenges; a thick layer of polymer where there is deep deformation, non-uniform deformation due to pressure loss and the opposite phenomena of shrinkage and springback. To mitigate these practical issues, the novel idea of integrating supercritical fluid (Sc.F.) technology with the PIF process is introduced in this work. As the proposed technology is a manufacturing innovation, with no available information in the literature correlating to this concept, two sets of experiments are designed to investigate the feasibility of this integration. In the first set, the effect of blank material and shot volume as design variables were investigated over a range of Sc.F. weight percentage. To improve the cell morphology in experiments with the low-strength sheet material, several other processing scenarios are explored in the second set of experiments. The results of this study clearly demonstrate the capabilities of this concept manufacturing process in terms of initiating the foaming process within the simultaneous injection/forming process, ensuring weight reduction (of up to 16%) and complete elimination of issues related to shrinkage.

Keywords: Polymer injection forming (PIF) · Supercritical fluid (Sc.F.) technology · Hybrid structures

C. Hopmann and R. Dahlmann (Eds.): *Advances in Polymer Processing 2020*, pp. 205–218, 2020. https://doi.org/10.1007/978-3-662-60809-8_17

1 Introduction

1.1 Polymer Injection Forming (PIF)

Hybrid production system (HPS) involves the combination of multiple, diverse material systems (i.e. plastics and metals), which is of great interest for the complementary characteristics they offer to a single hybrid component [1]. Although several processes are used to produce metal-polymer hybrid components, all of them involve issues such as a large number of processing steps and limitations in terms of both productivity and complexity for the component produced [2]. Motivated by the aforementioned aspects, a new platform technology has been recently developed to manufacture sheet metal-polymer macro composites in a single operation – known as Polymer Injection Forming (PIF) [3]. PIF is a hybrid production system that integrates the best-in-class manufacturing technologies in polymers and metals, viz., injection-molding and sheet metal forming. However, injection molding and sheet metal forming have differing physical principles (materials and process levels) and working mechanisms. Hence, while combining these methods can lead to increased output – which meets the objectives of using HPS – it is also a challenging initiative [4]. A schematic of the PIF process is shown in Fig. 1.

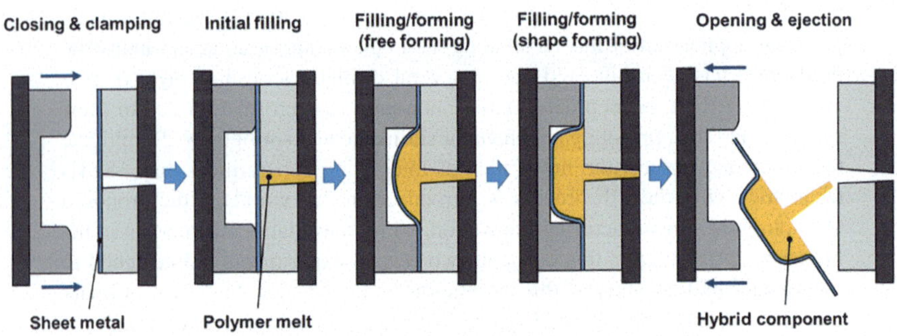

Fig. 1. Schematic of the PIF process

During this process, the sheet metal blank is first placed into the mold cavity, followed by the injection of polymer melt. This melt serves as a pressure medium to deform/shape the blank during the filling stage of the injection molding process. After the injection stage, the polymer-metal joint is also achieved after solidification of the system by either thermal bonding, mechanical interlock or adhesive coating [5]. The solidified polymer remains as a reinforcing or functional element, depending upon the targeted design. PIF improves the production process/cycle by reducing the number of production steps while facilitating easy assembly via embedding several functionality features into a single product. Only a single tool is needed in PIF, thus greatly reducing tool costs [6].

1.2 Supercritical Fluid (Sc.F.) Assisted Injection Molding

Supercritical fluid (Sc.F.) assisted injection molding is a unique technology in which CO_2 or N_2 in a supercritical state is used to form structural foamed products with superior strength-to-weight and cost-to-performance ratios to that of conventional injection-molded products [7]. In this process, after solid polymer reaches its molten state at the midpoint of the injection molding barrel, Sc.F. is introduced into the barrel via an auxiliary metering system. Both Sc.F. and the polymer melt continue through the barrel, undergoing shear mixing in which the polymer melt is super-saturated with the Sc.F. fluid. This high-pressure single-phase solution is then injected into the mold cavity at atmospheric pressure (below the gas saturation pressure). This pressure drop below the saturation point triggers thermodynamic instability, inducing cell nucleation. Cell growth is controlled by gas diffusion rate and the stiffness of polymer-gas solution which directly influence the morphology of the part after solidification [8]. The benefits of Sc.F. assisted injection molding (SFAIM) technology are listed in Table 1 along with a brief reasoning for each benefit [9].

Table 1. A list of benefits achievable by implementing Sc.F. assisted injection molding technology

Benefits	Reasoning
Part weight reduction	Due to creating the foamed core of the injected part
Faster cycles	Due to the elimination of packing phase and reduction in cooling time
Lower injection pressure	Due to lower viscosity of polymer melts with dissolved Sc.F.
Lower clamping force	Due to lower injection pressure
Reduced energy consumption	Due to the reduction in required injection pressure, clamping force and cooling time
Less shrinkage	Polymer melts with dissolved Sc.F. tends to expand rather than shrink
Less warpage and sink marks	Due to the less shrinkage and more uniform cooling condition

1.3 Supercritical Fluid-Assisted Polymer Injection Forming (SFPIF)

PIF is an HPS with the potential for expanding manufacturing research through the creation of multi-material constructs. However, several practical issues hinder its use in industrial applications. Comparing these hindrances with the benefits of Sc.F. technology (Table 1) shows the synergy of integrating these two processes. For example:

- A thick polymeric part resides (after solidification) at the location where the sheet is considered to deform deeply by means of melt pressure. This thick layer of polymer is not desired in conventional injection molding process for multiple reasons: (i) it increases the weight of the part; and (ii) It causes several issues such as increasing cooling time, possibility of warpage and local excessive shrinkage or sink marks. Such issues were observed in some of our initial experiments as shown in Fig. 2(a). Therefore, integration of PIF with Sc.F. technology can maintain the lightweight condition owing to the nature of foaming process and eliminate the issues such as warpage and sink marks due to the significantly lesser shrinkage in the SFAIM process.

- The flow of polymer melt through a thin channel increases pressure loss along the flow path due to the viscous nature of melt. This excessive pressure loss causes a non-uniform pressure distribution and consequently non-uniform deformation as reported in prior works [10] and schematically illustrated in Fig. 2(b). Therefore, combining PIF with Sc.F. injection molding can help overcome this drawback as the dissolution of Sc.F. into the polymer melt significantly reduces its viscosity and thereby ensures uniformity in sheet metal deformation.
- Shrinkage and springback are common problems in injection molding and sheet metal forming, both of which play opposite roles in the hybrid PIF process, inducing significant residual stresses on the contact area that in turn reduces the bonding strength and leads to the delamination of sheet metal from the polymeric part (see Fig. 2(c)). Although reverse geometrical modification can compensate springback in conventional sheet metal forming processes, it is an imperfect solution for the PIF process as the injected polymer takes the deformed sheet shape and shrinks from that point. Therefore, reducing the residual stress and geometrical instability is possible by integrating PIF with Sc.F. injection molding as the polymer melt with dissolved Sc.F. tends to expand more (and shrink less), and such less shrinkage is one of the main benefits of Sc.F. assisted technology.

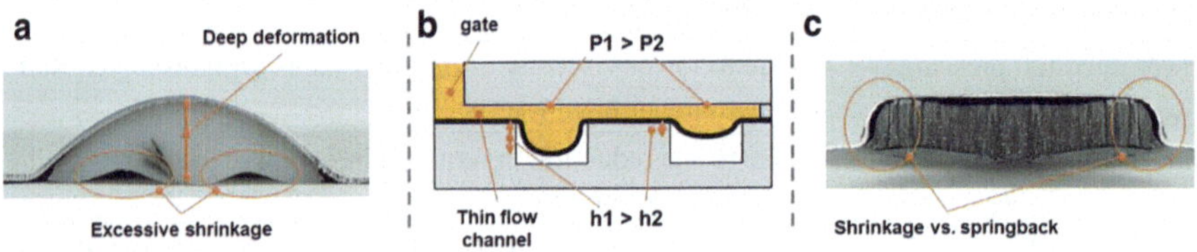

Fig. 2. Major practical issues associated with the application of the PIF process (Fig. 2(b) was reproduced from [10])

Given the aforementioned benefits of conducting PIF process with Sc.F. technology, SFPIF – the integration of both processes – will yield lightweight, hybrid polymer-metal components for use in automotive, aerospace, and home appliance applications. As the proposed technology is a transformative manufacturing innovation, with nothing in literature relating to this concept, two sets of experiments based on different processing scenarios are designed to investigate the feasibility of this integration.

2 Design of Experiments and Process Settings

In the first set of experiments, it was assumed that the optimum process parameters of the SFPIF process is identical to that of the regular SFAIM process. Hence, all the parameters except the Sc.F. dosing time and shot volume were set based on the optimum condition previously obtained in our experiments related to regular SFAIM process as listed in Table 2. Unless otherwise mentioned, all parameters listed in this table remain identical for all the experiments.

Table 2. A summary of fixed process parameters in this study

Clamp force	Inj. rate	Melt temp.	Mold temp.	switch over	Cool. time	Back press.	Sc.F. press.	Sc.F. Rate
100 kN	30 cm3/s	240 °C	35 °C	99% volume	120 s	150 bar	200 bar	0.07 kg/h

The weight percentage of the Sc.F. (defined by dosing time) was considered as one of the plasticizing parameters in order to examine its influence on the deformation of the sheet metal and the morphology of the final foamed part. Regarding the molding parameters, the shot volume was chosen as another variable parameter given its direct effect on the depth of deformation and consequently the thickness of the polymeric region. Two aluminium alloys (AA1100 and AA6061) exhibiting a significant difference in the strength and formability were considered for the sheet metal blank to investigate the expected effect of the sheet strength on the melt pressure and consequently on the cell morphology of the foamed part. The polymer selected as the injected material is ADX-2075 from Advanced Composites, Inc. It is an impact-resistant thermoplastic composed of polypropylene, rubber, and talc as fillers. The high melt flow rate (MFR) of 29 g/10 min and the fillers make this polymer a suitable option for the Sc.F. foaming process. Given all, the first set of the experiments were designed as listed in Table 3.

Table 3. Variable parameters in the first design of experiments (E1)

Effect of the Sc.F. dosing time					
Blank: AA1100				Blank: AA6061	
Shot volume: 30 cm^3		Shot volume: 20 cm^3		Shot volume: 20 cm^3	
Exp. #	Dosing time (s)	Exp. #	Dosing time (s)	Exp. #	Dosing time (s)
E1-1	0	E1-5	0	E1-8	0
E1-2	2	E1-6	4	E1-9	4
E1-3	4	E1-7	6	E1-10	6
E1-4	6				
Effect of shot volume at constant dosing time of 4 s					
Blank: AA1100			Blank: AA6061		
Exp. #	Shot volume (cm^3)		Exp. #	Shot volume (cm^3)	
E1-11	20		E1-14	20	
E1-12	26		E1-15	26	
E1-13	30		E1-16	30	

The results of the first set, as will be later shown and discussed in Sect. 4.4 (Fig. 8), revealed that a microcellular morphology cannot be achieved by only variation of Sc.F. percentage, especially in the experiments with low strength sheet materials (AA1100). Hence, in the second set of the experiments, several other processing scenarios were examined to see whether it is possible to improve the cell morphology in low cavity pressure condition. In this regard, influence of injection speed, eliminating packing phase and adding decompression action (before plasticizing phase) were examined in the second set of experiments as listed in Table 4.

Table 4. Variable parameters in the second design of experiments (E2)

Blank: AA1100			
Shot volume: 20 cm^3			
Injection rate: 30 cm3/s		Injection rate: 300 cm3/s	
Exp. #	Variable parameters	Exp. #	Variable parameters
E2-1	5 s packing	E2-4	5 s packing
E2-2	No packing	E2-5	No packing
E2-3	No pack + decompress.	E2-6	No pack + decompress.

3 Measurements and Sample Preparation

Both the sheet metal blank (before the experiment) and final hybrid part (after completion of each experiment) were weighed to record the weight of the injected part. Then, the height and volume of the deformation were determined by measuring the deform sheet metal. To assess the light-weighting potential of the proposed SCPIF process, the density of the injected parts was calculated and compared with the density of the injected parts with zero percent Sc.F. (regular PIF). Using these values, the light-weighting percentage of each experiment was calculated using Eq. (1) where ρ_0 is the density of the solid part and ρ is the density of the foam part.

$$Lightweighting = \left(\frac{\rho_0 - \rho}{\rho_0}\right) * 10 \qquad (1)$$

The supercritically foamed samples were imaged at the cross-section using the following procedure, as demonstrated in Fig. 3:

1. A blade is used to notch the flat face of the samples along their diameters.
2. The samples are immersed in a dewar of liquid nitrogen for 45 min.
3. The samples are removed, secured in a vise and cryogenically fractured via rubber mallet.
4. Double-sided carbon tape is used to secure the fractured samples on an SEM sample mount.
5. The samples are inside a Hummer 6.2 sputtering system for 3 mins to deposit a thin layer of platinum on the non-conducting polymeric part of the hybrid samples.

6. The sputter-coated samples are imaged in a Hitachi 3400S scanning electron microscope at a maximum working distance to maximize the field of view. They are then subjected to an accelerating voltage of 5 kV at different magnifications to characterize cell size and density.

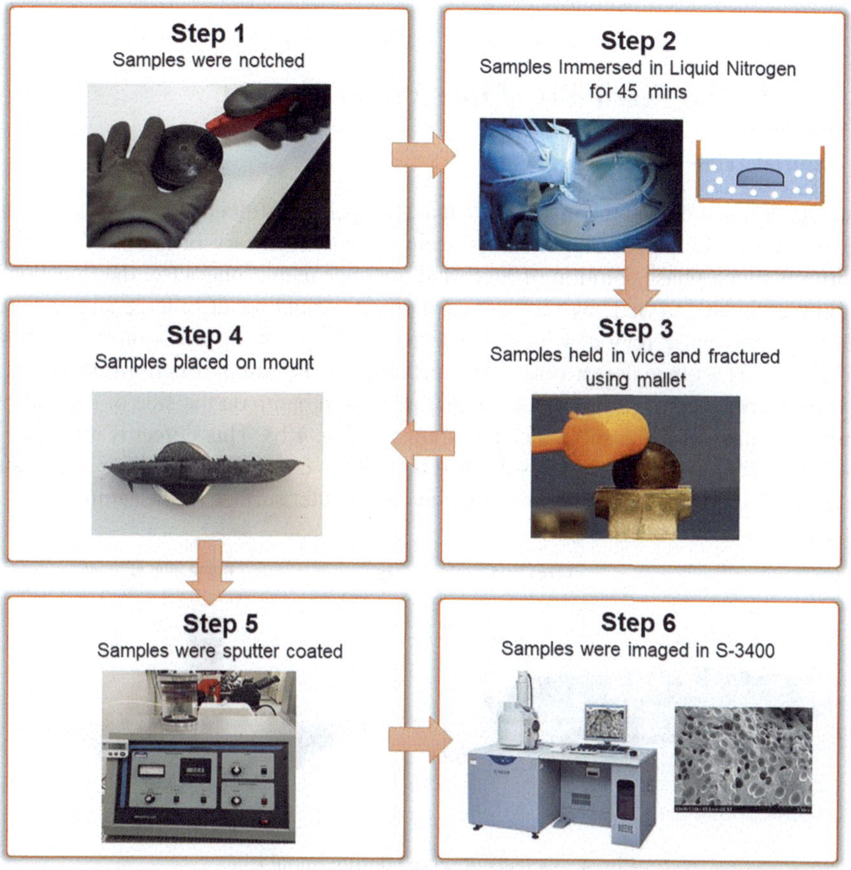

Fig. 3. Procedure followed to study foam morphology

The imaged samples were processed using Image J analysis tool to calculate the average cell size and cell density. Cell density, in particular, was calculated using Eq. (2), where N is the number of cells, L is the linear length of the area, and M is a unit conversion, resulting in cell density being expressed as the number of cells per cubic centimeter [11]. In order to avoid skewing of data, a few abnormally large voids observed in some specimens were excluded from the calculation of average cell size and cell density.

$$Cell\,density = \left(\frac{N}{L^2}\right)^{\frac{3}{2}} M \qquad (2)$$

As the main objective of this work is to explore the feasibility of this integration (the SFPIF process) and its capabilities in eliminating the PIF issues, the replication of the experiments has been limited to the plasticizing stage in order to make sure about processing a uniform solution of polymer melt and Sc.F. before injection stage. But the rest of the experimental procedure and measurements have been restricted to a single experiment with no replications.

4 Results and Feasibility of Integration Concept

4.1 Initial Trials and Observation

It was determined during the initial trials that a proper adjustment of the process parameters related to the plasticizing stage is deemed most necessary in terms of achieving a uniform single-phase solution of Sc.F. and polymer melt. Otherwise, the injection of the polymer melt and Sc.F. as two separated phases would result in large empty spaces within the polymeric region, which is detected only by cutting the samples as shown in Fig. 4(a). The other defect observed especially on the sample with a thick layer of polymer (experiments with shot volume 30 cm^3) was a bump on the side of the part that is not in contact with the sheet metal as shown in Fig. 4(b). This defect is attributed to insufficient cooling time as the unsolidified melt at the core of the sample expands after the ejection with a manifestation of the bulge on the outer surface of the sample.

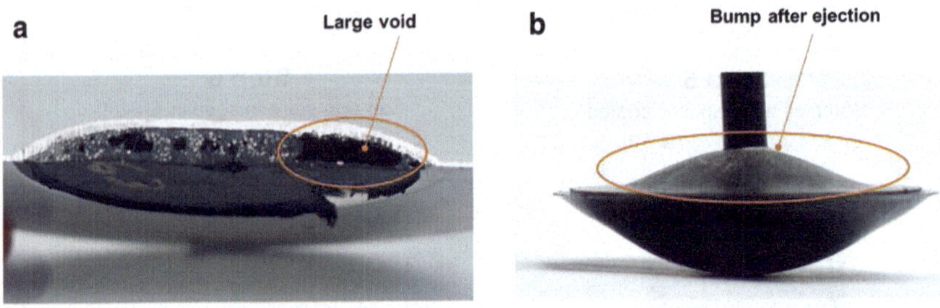

Fig. 4. Defects observed during initial trials

4.2 Dimensional Properties and Shrinkage

The height and volume of the deformation shown in Fig. 8 is an important result of this study as it is demonstrated that the application of the Sc.F. yields a notable increase in both the height and volume of the deformation. In the first set of experiments (E1) with AA1100 sheets, an increase of height up to 32% and deformation volume up to 47% was observed. The deformation of the AA6061 blank was less than the AA1100, however, falling less than 26% and 22% for the additional height of the dome shape and increase of the deformation volume respectively. Although the additional deformation from the application of the Sc.F. is dependent on the blank material, no consistent trend is observed in the deformation of the sheet metal by increasing the weight percentage of the Sc.F.

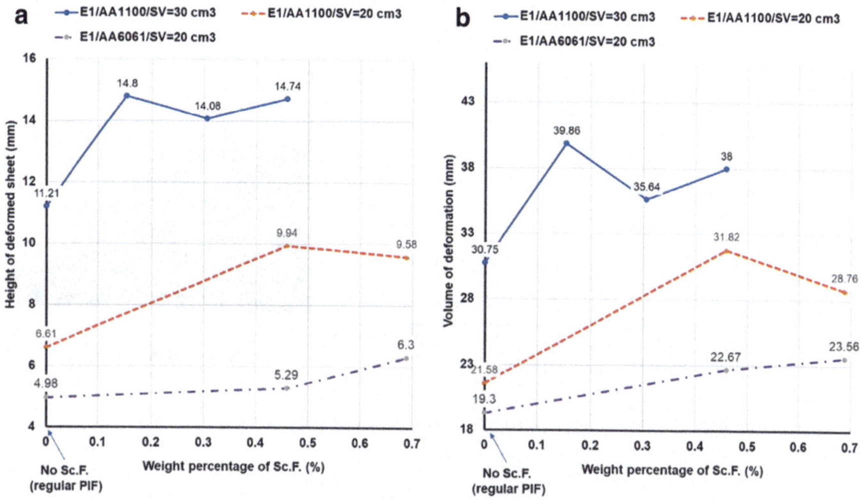

Fig. 5. (a) Height and (b) volume of the deformation

Shrinkage is usually a prominent challenge in the injection molding process. To quantify shrinkage in this study, diameter of the injected parts was measured one day after the experiment. The results of this measurement, as reported in Fig. 6(a), clearly demonstrate that using the Sc.F. technology resulted in a larger part due to less shrinkage of samples. However, no consistent trend was observed upon increasing the weight percentage of Sc.F. Shrinkage is more critical in the PIF process as the springback phenomena in sheet metal deformation would act in the reverse direction, which can result in delamination or a gap between the injected part and deformed metal. In order to investigate this phenomenon, a layer of adhesive coating was added on the surface of several blanks. These blanks were later processed in the same condition as main experiments. After one day, the hybrid samples were cut from the center. Investigating the cut section of these sheet metal-polymer samples provided further proof that the polymer melt with dissolved Sc.F. completely filled the deformed area and there was no gap or delamination observed between the sheet metal and the polymeric part (see Fig. 6(b)). The layer of adhesive has been applied to make bonding between the injected part and the deformed sheet metal and keep them together to investigate the opposing effect of shrinkage vs. springback. Hence, the effects of this adhesive layer on the other aspects of the process were out of the scope of this work.

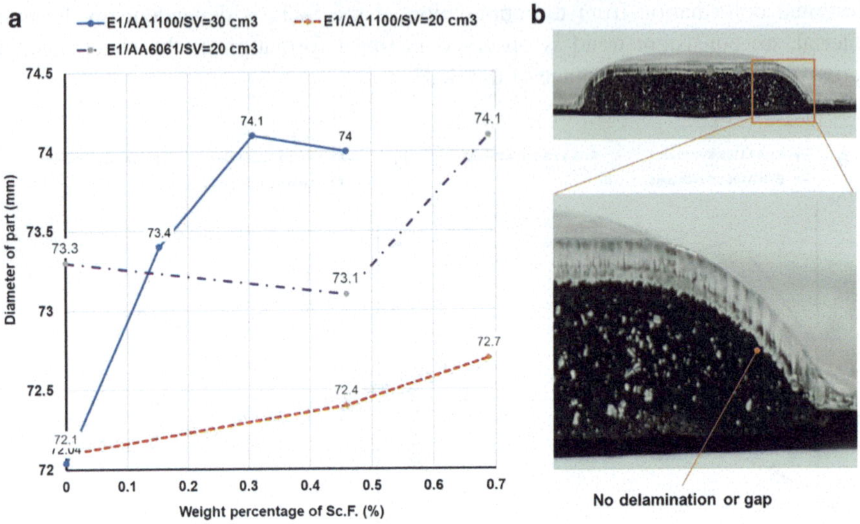

Fig. 6. (a) Diameter of the injected samples. (b) Cut section view of a hybrid sample produced by SCPIF

4.3 Weight and Weight Reduction

In previous studies, it was determined that the weight of the injected part in the PIF process is not only dependent on the shot volume but also on the formability of the sheet metal blank owing to the coupled filling/forming condition during the injection phase [12]. Hence, it is important to investigate the weight of the injected part to determine the consistency of the experiments and light-weighting capability of the SFPIF process. As seen in Fig. 7(a), the injected samples with Sc.F. assisted technology exhibit a higher weight for an identical shot volume. This phenomenon occurs as the use of Sc.F. increase the sheet metal deformation (see additional deformation in Fig. 5) and thus expanding the region in which the melt can flow into the cavity. In other words, the lessening of resistance enhances the flow of polymer melt into the cavity and reduces that which would otherwise escape through the gap between the barrel and the fights to the other side of the injection screw.

It was observed that although the weight of the injected part showed an increase upon the introduction of Sc.F., the total density of the hybrid part showed a reduction of up to 16%. Moreover, this result shows that light-weighting increased with increase in weight percentage of Sc.F. until an optimum point, but then showed a decline. This result can be further explained by correlating the results of deformation (see Fig. 5(b)) and the weight of injected parts (Fig. 7(a)). This correlation indicates a decrease in deformation volume and an increase in sample weights, which in turn reduces light-weighting after the inflection point.

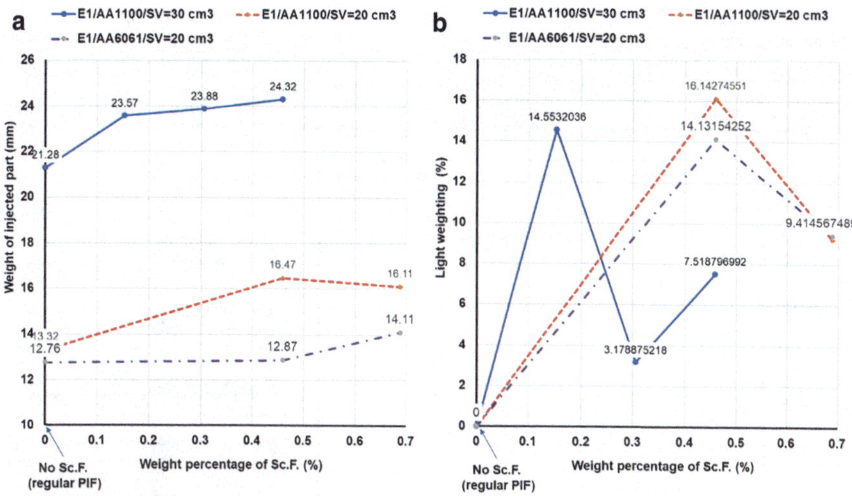

Fig. 7. (a) Weight of injected parts and (b) Light-weighting achieved with SCPIF

4.4 Micro-Structure: Cell Size and Density

As the aim of the first set of experiments (E1) was to broadly understand the effects of integrating PIF and Sc.F. technology, the experiments were designed to elucidate changes in foam morphology with variances in both Sc.F. wt % and shot volume upon two aluminum alloys. A map of the cell densities vs cell sizes from E1 experiments plotted in Fig. 8 shows the holistic effect of process and material variables on that foam morphology.

Out of the 9 samples selected for SEM imaging, three samples exhibited a micro-cellular foam morphology (i.e. average cell diameters <100 μm and cell densities $>1 \times 10^6$ cells/cm^3). These samples were produced in experiments (E1-10, 14 & 15) in which the AA6061 alloy was used as the sheet metal blank. But, all the samples were produced using the AA1100 sheets (E1-4, 7, 11, 12, 13 & 16) showed inferior cell morphology (large cells with poor density) and blowholes underscoring the challenges associated with the integration concept of PIF and Sc.F. processes.

Two major reasons are hypothesized as to why the samples foamed in experiments with AA6061 sheet metal exhibited a consistently superior cellular morphology:

1. The use of the stiffer AA6061 sheet metal creates a higher pressure within the cavity and consequently a higher pressure drop at the end of the injection from suction induced by the plasticizing phase and/or the solidification of the gate. As a result, the foaming stage as the supercritical fluid in the polymer melt falls below the critical pressure resulting in either a diffusion into nucleated cells or nucleation of new cells [13].
2. The coupled filling/forming phase of the PIF process exhibits a similar set of conditions as the counter pressure method, caused by the resistance of the blank against the melt flow. This condition prevents the Sc.F. from escaping from the

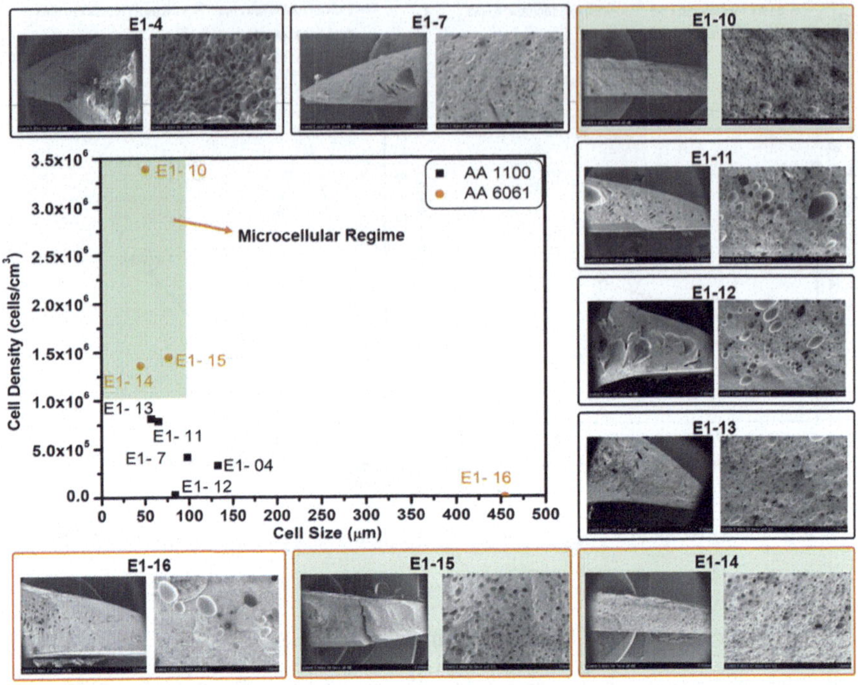

Fig. 8. Compilation of foam morphologies obtained in E1

melt flow front and keeps the melt in a pressure higher than the supercritical until the end of the injection [14]. Clearly, it is now known that this condition will improve with the use of AA6061 given its far greater yield stress over AA1100 which results in the application of the higher pressure inside the cavity from the beginning of the filling/forming phase.

As it was not possible to achieve the microcellular structure in the experiments conducted with AA1100 sheets, the second set of experiments was designed and performed to understand the effect of injection speed in different processing scenarios: (i) with 5 s packing pressure, (ii) no packing phase, and (iii) no packing plus decompression before plasticizing phase. A map of the cell density vs. cell size for E2 experiments is shown in Fig. 9.

It is clearly evident from Fig. 9 that samples produced with no packing and/or decompression performed significantly better than the samples subjected to packing pressure vis-a-viz their smaller cell sizes and larger cell densities. Additionally, these experiments proofed that it is possible to achieve truly microcellular morphology with low strength sheet metals only by adjusting the process sequences and parameters. Such behavior is attributable to the acceptable performance of the second set of the experiments to control the pressure drop rate assuming the amount of the pressure drop would be the same as the first set. Other studies undertaken in microcellular injection molding also reported the importance of the drop rate, especially under low-pressure drop conditions [9].

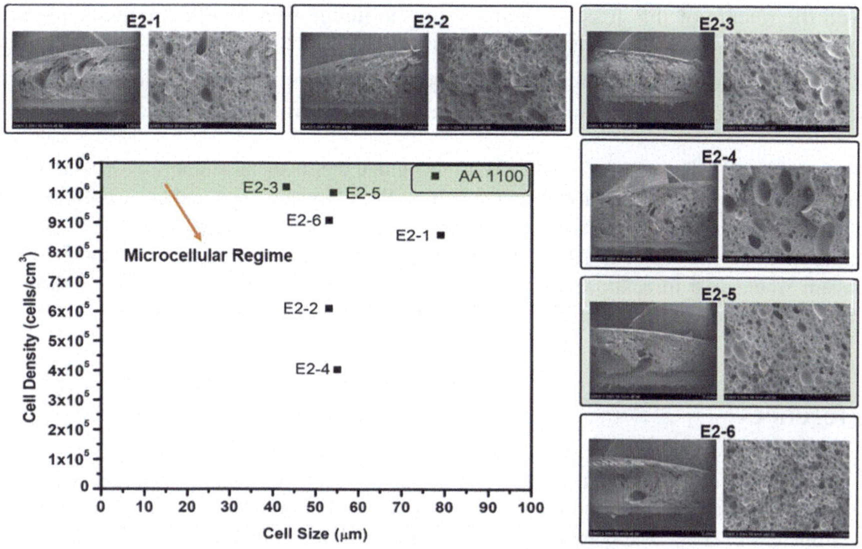

Fig. 9. Compilation of foam morphologies obtained in E2

5 Summary

The integration concept of PIF process with Sc.F. technology was successfully realized using the first set of the experiments. In the second set of the experiments, three processing scenarios were proposed to improve the cell nucleation and control the foaming process in SFPIF process. The effect of blank material and shot volume as design variables were investigated over a range of Sc.F. weight percentage. The following findings were derived from this study.

- Additional deformation due to the application of the Sc.F. was observed. Despite the dependency of deformation on blank material, no consistent trend was detected in the deformation of sheet metal to increase in weight percentage of Sc.F.
- Investigating the diameter of injected parts and cut section of hybrid samples further proved that this integration concept could completely eliminate the shrinkage issue, as no gap or delamination was observed on the hybrid parts due to the opposite reaction of shrinkage and springback.
- Despite the increase in the weight of the injected parts, density results demonstrated a good capability of this integration for lightweighting as up to 16% weight reduction was achieved.
- The microstructure of Sc.F. foamed samples investigated by SEM showed that truly microcellular cell morphology was obtained as a result of conditions similar to counter pressure process created by use of the stiffer AA6061 sheet metal and/or higher pressure drop rates exhibited by eliminating the packing phase and setting decompression action before plasticizing.

Given the results of this feasibility study on the integration of PIF process with Sc.F. technology, future work should focus on improving this integration by further process optimization and introduction of more effective manufacturing procedures in order to control the nucleation process and attain the more-uniform microcellular morphology.

Acknowledgment. The authors would like to recognize the financial support by Robert Patrick Jenkins Professorship, and Dean's Faculty Fellow Professorship. The authors would also like to acknowledge HASCO America Inc. for providing the metric mold base and standard mold components and Mr. Vinayak Khade for his help in the measurement of the samples. Saeed Farahani would like to acknowledge the financial support provided by Firestone Fellowship Award in Advanced Materials and Sonoco Fellowship in Automotive Engineering.

References

1. Kopp, G., Beeh, E.: New multi-material design concepts and high integration light metal applications for lightweight body structures. Mat. Sci. Forum **638–642**, 437–442 (2010)
2. Grujicic, M., et al.: The potential of a clinch-lock polymer metal hybrid technology for use in load-bearing automotive components. J. Mater. Eng. Perform. **18**(7), 893–902 (2009)
3. Hussain, M.M., Rauscher, B., Trompeter, M., Tekkaya, A.E.: Potential of melted polymer as pressure medium in sheet metal forming. Key Eng. Mater. **410–411**, 493–501 (2009)
4. Brecher, C., et al.: Integrative production technology for high-wage countries. In: Brecher, C. (ed.) Integrative Production Technology for High-Wage Countries, pp. 17–76. Springer, Berlin (2012)
5. Farahani, S., Arezoodar, A.F., Dariani, B.M., Pilla, S.: An analytical model for nonhydrostatic sheet metal bulging process by means of polymer melt pressure. J. Manuf. Sci. Eng. **140**(9), 091010 (2018)
6. Michaeli, W., Maesing, R.: Injection moulding and metal forming in one process step. Prog. Rubber, Plast. Recycl. Technol. **26**(4), 155–166 (2010)
7. Kazarian, S.: Polymer Processing with Supercritical Fluids (2000)
8. Aditya Pradeep, S.: An Investigation of Bimodal Cellular Distributions via Supercritical Fluid Assisted (SCF) Foam Injection Molding (2016)
9. Xu, J.: Microcellular Injection Molding (2011)
10. Tekkaya, A.E., Hussain, M.M., Witulski, J.: The non-hydrostatic response of polymer melts as a pressure medium in sheet metal forming. Prod. Eng. **6**(4–5), 385–394 (2012)
11. Pilla, S., Kramschuster, A., Yang, L., Lee, J., Gong, S., Turng, L.S.: Microcellular injection-molding of polylactide with chain-extender. Mater. Sci. Eng. C **29**(4), 1258–1265 (2009)
12. Farahani, S., Zheng, T., Pilla, S.: Hybrid process of forming-injection molding – investigation of polymer melt behaviour and final injected part. In: SPE-ANTEC (2019)
13. Guo, Q., Wang, J., Park, C.B., Ohshima, M.: A microcellular foaming simulation system with a high pressure-drop rate. Ind. Eng. Chem. Res. **45**(18), 6153–6161 (2006)
14. Chen, S.C., Hsu, P.S., Hwang, S.S.: The effects of gas counter pressure and mold temperature variation on the surface quality and morphology of the microcellular polystyrene foams. J. Appl. Polym. Sci. **127**(6), 4769–4776 (2013)

Potential of Mesoscale Structural Elements in the Interface of Hybrid CFRP-Metal-Parts on the Load Transfer

Fabian Günther(✉), Jan Ewens, and Markus Stommel

Chair of Plastics Technology, TU Dortmund University,
Leonhard-Euler-Str. 5, 44227 Dortmund, Germany
fabian.guenther@tu-dortmund.de

Abstract. The applicability and the influence of pin structures in dependence on the adhesive properties in hybrid joints are investigated in the course of this paper. First, single lap joint experiments are performed and a Finite Element model is used to obtain force-displacement curves, and thus the specific simulation parameters. The comparison of different models with a pin a semi-spherical pin, and without a pin show that the pin has a positive influence, especially on low adhesive strengths. In contrast, the pin has a lesser effect for adhesive joints with high strength. In the following study, different pin geometries, pin sizes and adhesive bonding parameters are tested. It is shown that the negative influence of a pin for high adhesive strength can be compensated by the specific design of a pin. Furthermore, small pins ensure the best force transmission performance. Finally, the alignment of elliptical pins depends on the bonding properties, as for high strength parameters, the pin should be oriented in load direction, and for low strength parameters, it is ideal to orient the pin transversely to the load direction.

Keywords: CFRP hybrid joint · Pin structure · FEM simulation · Parameter variation · Cohesive zone modelling

1 Introduction

A multi-material design approach is used in many lightweight concepts to combine the specific properties of different materials favorably. The type of connection of the different materials is of particular importance and, in addition to established connection techniques for conventional materials (screws, rivets, gluing, etc.), novel techniques and processes are developed for new material combinations [1]. A distinction can be made between extrinsically and intrinsically manufactured components, depending on whether a downstream assembly for the component is necessary or the joining is created during the consolidation of individual components.

A carbon fiber reinforced plastic (CFRP)-metal hybrid composite in which a thermoplastic component between CFRP and metal functionally supports the composite at various levels serves as an application example [2]. In addition to the macroscale

C. Hopmann and R. Dahlmann (Eds.): *Advances in Polymer Processing 2020*,
pp. 219–232, 2020. https://doi.org/10.1007/978-3-662-60809-8_18

form-fit (cm) and the microscale roughness interlocking (μm) as well as the adhesive bonding of the connections, a form-fit in a mesoscale plane by structural elements of millimeter size is possible. They are also known as integrated microscopic mechanical interlockings [3] or Z-pins [4–9]. In the following, these mesoscale form-fit elements will be investigated and called pin structures.

1.1 Pin Structures in CFRP-Connections

Pin structures are investigated in different applications and material combinations mostly in CFRP-metal joints [4–9]. All methods have the intrinsic production in common, where the connection partners are joined during the production process e.g. in Resin Transfer Molding (RTM). Therefore, a combination of adhesive bonding and form-fit is achieved without damaging the reinforcing fibers.

The pin-based CFRP-metal hybrid joint is a concept that has already been conventionally applied [10, 11]. However, various complex welding processes that are necessary for the production of the metallic pin structures are disadvantageous. In extensive experimental tests the pin size is examined in Double Cantilever Beam (DCB) tests [4–6], in Single Lap Joint (SLJ) tests [9] and in Double Lap Joint tests [7, 8]. All show rapid increases of the bonding strength by means of pins. In [12] another metal pin structure for hybrid joints is examined, where the pins are created by a combined stamping forming process. The strength increase of different arrangements in SLJ is investigated experimentally and simulative.

In [13] SLJ tests show the effect of different surface structures on CFRP-thermoplastic joints. Beside other structures, thermoplastic pins have shown a highly reinforcing effect to the hybrid connection. Furthermore, the influence of single pin geometry on the joint properties is investigated in initial simulative approaches.

1.2 Cohesive Zone Models and Parameters

In [2, 14] different methods are presented to model the interface of hybrid parts. Cohesive Zone Models (CZM) are used in Finite Element Method (FEM) simulations to describe the bonding behavior or crack propagation between two adhesively bonded or intrinsically produced parts [4, 15–26]. The CZM use cohesive elements to describe stiffness, strength and damage behavior. Therefore, different cohesive parameters are necessary, that are highly dependent on the FEM model design and simulation parameters such as mesh size and time step size. These parameters cannot be determined directly. In experiments, e.g. DCB tests or End Notched Flexure tests, the energy release rate for different stress modes can be determined [27, 28] and subsequently adjusted in simulative models [2, 4–6].

Another method to determine the cohesive parameters is a direct fitting to the specific application by testing SLJ and fitting the parameters with simulations [2].

An overview of different cohesive parameters is given in Fig. 1. As it can be seen in Fig. 1, the CZM is used in various engineering disciplines and for a multitude of different material combinations. Depending on the materials and the joining processes, the adhesive strength and stiffness vary widely. Hence, the corresponding cohesive parameters are application-specific.

adherents	test type	K_n [MPa/mm]	K_s [MPa/mm]	σ_{n_max} [MPa]	σ_{s_max} [MPa]	G_{n_c} [N/mm]	G_{s_c} [N/mm]	ref.
cfrp	DCB	35700	35700	42,2	81	0,19	1,1	[4]
cfrp-steel	single lap	3700	1300	30	14,25	4	4,7	[15]
cfrp-steel	single lap	2176	765	25,6	15	4	4,8	[15]
cfrp-steel	double lap	18667	6763	33	25	0,6	0,975	[15]
Al	single lap	4890	1560	39,45	30,2	0,2	0,38	[16]
Al	single lap	1850	560	21,63	17,9	0,43	4,7	[16]
Al	single lap	1890	710	28,6	20	1,18	8,72	[16]
not spec.	DCB	10000	-	30	-	0,26	-	[17]
cfrp	DCB	100000	100000	45	75	0,26	1,002	[18]
cfrp-steel	ENF	7,6	54	230	300	0,28	0,14	[19]
glass fibre-epoxy	microdroplet test	-	-	50	-	0,1	-	[20]
cfrp	single lap	1850	560	21,6	17,9	0,43	4,7	[21]
cfrp	DCB	10000	10000	30	30	0,266	1,002	[22]
cfrp-metal	complex panel	1540	1540	25,1	50	0,3	0,8	[23]
cfrp-Al	single lap	1850	560	21,6	17,9	0,43	4,7	[24]
concrete	DCB like	7500000	-	3	-	0,08	-	[25]
concrete	DCB like	60000	-	3	-	0,1	-	[25]
concrete	DCB like	100000	-	3	-	0,152	-	[25]
cfrp-epoxy	DCB	1000000	1000000	45	45	0,268	0,632	[26]
cfrp-plastic	single lap	1203	405	9,5	10,5	0,005	0,043	Sec. 2
cfrp-plastic	single lap	1250	400	9	7,5	0,05	0,05	Sec. 3

Fig. 1. Cohesive parameters of various authors [4, 15–26] (K = stiffness, σ = strength, G = energy release rate; indices: n = normal, s = shear, c = critical)

1.3 Content of the Contributions

As shown in Sect. 1.1, different concepts of pin structures already exist depending on the application and material connection. The aim of this contribution is to determine the geometric design of a pin as a function of the adhesive properties. As shown in Sect. 1.2, CZM are proven for simulating combined bonds of form-fit and adhesion. Thus, a FEM model is created to simulate the hybrid connection. The parameters and the simulation model are fitted and validated with experimental SLJ tests. The influence of a cylindrical and a semi-spherical pin in comparison to a model without a pin is investigated. Finally, the potential of cylindrical pin structures of different geometric shapes is shown in a parametric study. The (maximum) load transferred by the joint is evaluated to make conclusive statements about the dimensioning and use of pins.

2 Experimental Tests

In order to prepare the simulation study, experimental SLJ tests and corresponding FEM simulations are performed to determine the adhesive parameters. The specimens consist of a thermoplastic component and CFRP with a single pin in the joint surface (see Fig. 2, right). The thermoplastic component is made by injection molding and the specimen consolidation takes place intrinsically in the RTM process. The specimens geometry is based on the geometry used in [13]. The thermoplastic consists of PPAGF30 (Vestamid M1033 from Evonik) and the CFRP consists of two 0°–90° woven fabrics (FT300B from Torayca) with epoxy resin (Biresin CR170 and hardener CH150-3 from Sika).

The simulation model represents the SLJ specimen (see Fig. 2, right). The simulation set up is similar as the one described in Sect. 3.

Figure 2 (left) shows the force-displacement course of three characteristic SLJ tests. The curves show an increase of about 500 N up to a displacement of about 0.1 mm and then to a drop to lower values and a plateau, which is determined by the pin. These curves are characteristic for the series of tests, which show a wide variation due to different failure behaviors. Therefore, the joints are simulated using numerical optimized parameters for the cohesive elements (see Fig. 1).

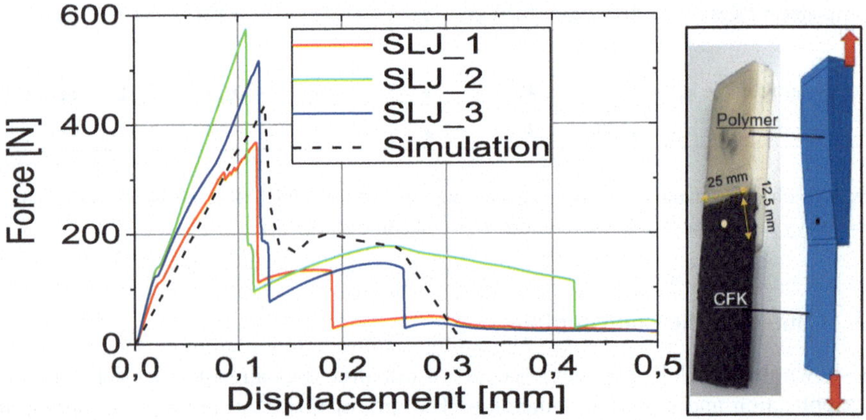

Fig. 2. left) Force-displacement curves of SLJ experiments and simulation, right) SLJ specimen and FEM model

3 FEM Analysis Study

A FEM analysis is used to investigate the effect of pins and their geometric dimensions on the connection strength considering the parameters of the adhesive bonding by using CZM. To determine the potential on load transmission in CFRP-thermoplastic connection with pin structures, two different simulation studies are carried out with *Abaqus (Simulia)* [29]. A reduced representative model of the connection is used as described in Sect. 3.1. First, models with and without a pin are compared (see Sect. 3.2). Further, a parameter variation for determining the pin geometry is investigated (see Sect. 3.3).

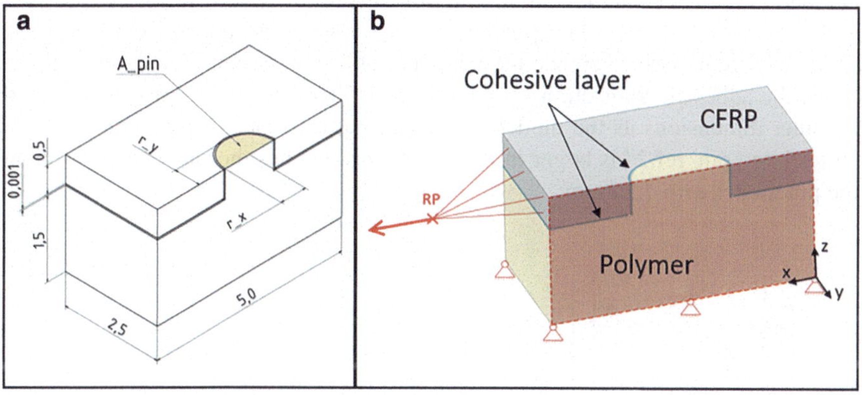

Fig. 3. FEM model – a) dimensions, b) set up and boundaries

3.1 FEM Model

The FEM model consists of a polymer base part with a pin, two CFRP layers stacked in z-direction and a cohesive layer in between. The dimensions are shown in Fig. 3(a). In Fig. 3(b), the boundary conditions are represented. The model is fixed on the bottom surface while load is applied through a projected displacement from a reference point to the side surface of the CFRP layers in x-direction. This allows to reproduce the typical load case and failure behavior. By considering the symmetric load case, only half of the model is created with symmetric boundary conditions. The mesh element size is 0,07 mm and C3D8R and COH3D8 elements are used. To model the materials, the polymer part is considered as linear elastic and the composite material behavior is implemented as multi continuum material considering fiber and matrix properties as well as damage behavior with *Helius PFA* (Autodesk) [30]. Two layers oriented in 0°–90° constitue the CFRP. The material parameters are shown in Table 1.

Table 1. Material parameter

Materials	Young's modulus [GPa] $E_{11}/E_{22}/E_{33}$	Poisson's ratio [–] $\nu_{12}/\nu_{13}/\nu_{23}$	Shear modulus [MPa] $G_{12}/G_{13}/G_{23}$	Strength [MPa] $R_{11+}/R_{11-}/R_{12}$
Polymer	9.1	0.3	–	–
CFRP	22.97/22.97/2.92	0.04/0.63/0.63	732/637/637	230/–110/24

3.2 Comparative Simulation Models with and Without a Pin

To see the influence of a cylindrical pin, models with and without a pin as well as a model with a semi-spherical pin are created and compared. The side view of all three quadratic models and the additional dimensions are shown in Fig. 4. The depth in y-direction is 1.85 mm. The pin and the semi-spherical pin are circular with a diameter of 1 mm. Two reference models without a pin are considered, since the pin enlarges the adhesive connection area additionally. The model "no pin_ref.1" has the same outer dimensions as the model with a pin but a surface of 6.84 mm^2. Whereas the model "no pin_ref.2" is larger in its dimensions and has the same connection area as the pin model with 7.24 mm^2.

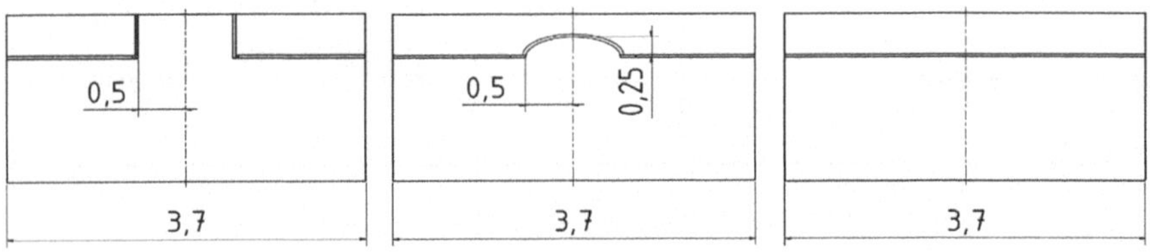

Fig. 4. Dimension of cylindrical pin, semi-spherical pin and no pin FEM models (further dimension in Fig. 3)

3.3 Parametric Study

A full factorial parametric study is performed to investigate the interaction of the pin geometry, pin size and cohesive parameters (see Table 3).

The geometry of the pins is elliptical and is determined by the pin surface A_{pin} (Eq. 1) (see Fig. 3(a)) and the pin radius ratio $p_{x,y}$ (Eq. 2). This parameter selection allows a simple differentiation of the influence by the size and geometry of the pin. The surfaces are selected so that pins with a circular base and radii of 0.5, 0.75, 1.00, 1.25 and 1.50 mm are considered. The radii ratios are considered between 0.25 and 4. The ratio 4 corresponds to a pin oriented in the load direction (x), 0.25 to a pin transverse to the load direction (y). The corresponding radii can be calculated afterwards by Eqs. 3 and 4. Due to the size of the FEM model, no radii above 2.0 mm are considered. For reasons of manufacturing feasibility, no radii below 0.3 mm are considered.

$$A_{pin} = \pi * r_x * r_y \tag{1}$$

$$p_{x,y} = \frac{r_x}{r_y} \tag{2}$$

$$r_x = \sqrt{\frac{A_{pin} * p_{x,y}}{\pi}} \tag{3}$$

$$r_y = \sqrt{\frac{A_{pin}}{\pi * p_{x,y}}} \tag{4}$$

For the parameter variation, a parameter set corresponding to the fitting is selected for the cohesive layer, as shown in Table 2. Because of the wide variation of cohesive parameters as shown in Sect. 1.2, all cohesive parameters are varied equally. K_n is given as the reference (see Table 3).

Table 2. Cohesive parameter

K_n	K_s	σ_{n_max}	σ_{s_max}	G_{n_c}	G_{s_c}
[MPa/mm]	[MPa/mm]	[MPa]	[MPa]	[N/mm]	[N/mm]
1250	400	9	7,5	0,05	0,05

Table 3. Variated parameter

K_n [MPa/mm]	A_{pin} [mm^2]	$p_{x,y}$
156.25	0.79	0.25
312.5	1.77	0.4
625	3.14	0.63
1250	4.91	1
2500	7.07	1.59
5000		2.52
10000		4

4 Results

The evaluation of the first study of cylindrical pin, semi-spherical pin and flat models (without pin) shows that the transmitted force of all models increases with increasing stiffness (K_n) (see Fig. 5). From a threshold of about 5000 MPa/mm the transmitted force decreases due to the failure of the CFRP laminate. A closer look at the curves shows that for large K_n the connections can transmit larger forces without a pin. Here the influence of the connection area size is obvious, since the model without a pin with a larger total connection area (no pin_Ref2) transmits more force than the normal reference model without a pin (no pin_Ref1). For low K_n values, the influence of the form-fit of the pin is clearly evident. At very low K_n values, the pin carries the structure almost completely. Between 1250 and 2500 MPa/mm a transition of the best performing models from models with pins to models without pins takes place. Moreover, comparing the models with semi-spherical pin and cylindrical pin shows the potential of the cylindrical pin shape. At low K_n, the reinforcing influence of semi-spherical pins is recognizable due to an area enlargement and a low form fit compared to the flat models. However, the significant form-fit of the cylindrical pin is much more effective. This changes with large K_n, where however the flat connection is more effective. These results motivate the further simulative research in which the effect of the pin size and geometry on the connection behavior is investigated.

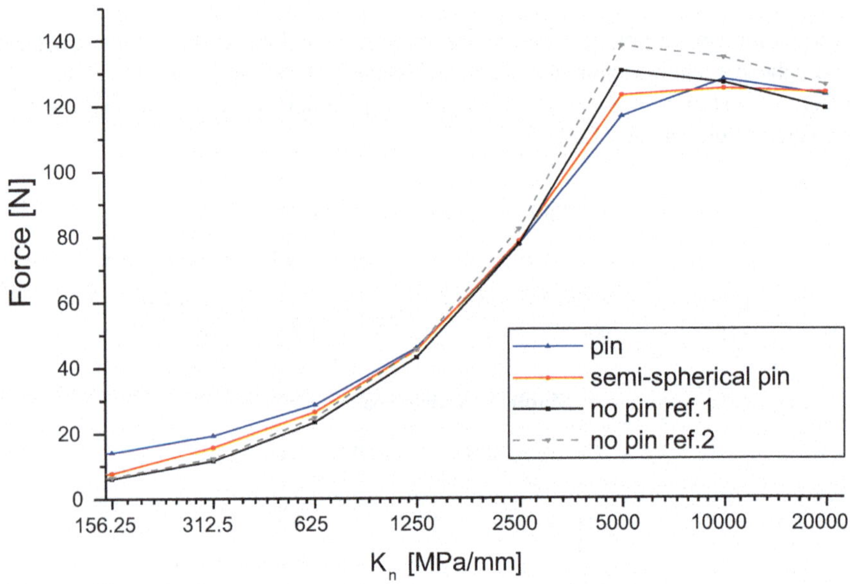

Fig. 5. Force plotted over cohesive layer stiffness parameter for cylindrical pin, semi-spherical pin and no pin

The results of the parametric study shown in Fig. 6 present a similar course as in the previous investigations. For example, in the case of low K_n the connection is very weak in the flat model, while it shows good performance in the case of high K_n. However, it can be seen that the different pin geometries have a significant effect on the force transfer, seen through the wide range of results, shown as black circles in Fig. 6. In addition, it can be seen that the pin simulations are not worse than the flat models even in the very high areas of K_n 5000 MPa/mm. Thus, the best pin model transfers a load of 191.57 N, while the model without pin transfers 191.93 N. The radius ratio of the pin is 2.52 and the pin surface is 0.79 mm^2.

For 2500 MPa, the model with a radius ratio of 2.52 and a pin surface of 0.79 mm^2 transfers 125.82 N, while the flat model achieves 124.37 N. The model with a radius ratio of 4 and a surface area of 1.77 mm^2 also exceeds the flat model with 125.18 N. These interrelations can be seen even better in Figs. 7 and 8.

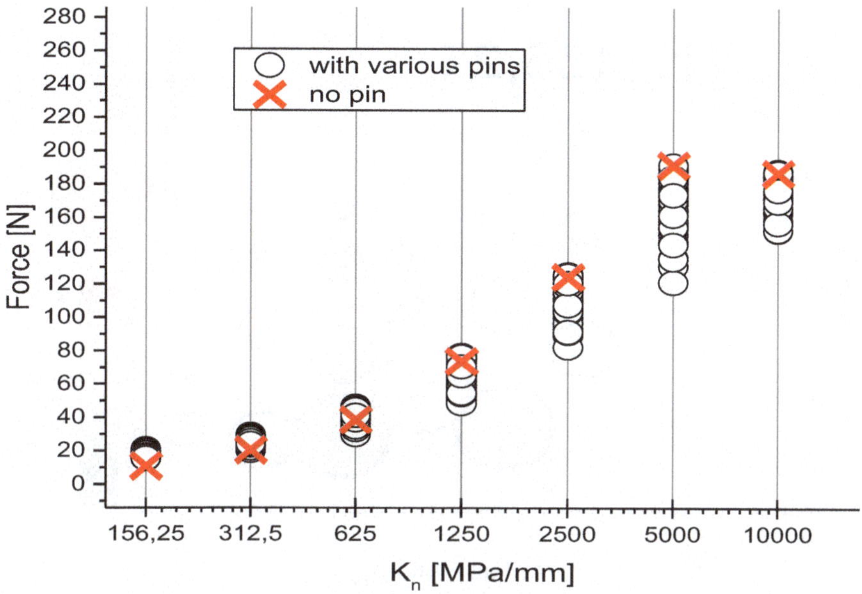

Fig. 6. Transmitted force over cohesive layer stiffness parameter for different pin geometries

The simulations with K_n 10,000 MPa/mm show that no higher forces can be transmitted due to the global failure of the laminate. In addition, the maximum transferable load of the base material must also be taken into account when considering the influence of the pin geometry. Therefore, the area of high K_n is not considered more closely.

Exemplarily, in Figs. 7 and 8 the transmitted force is plotted over the pin radius ratio for K_n 156.25 and 2500 MPa/mm. The size of the data points is selected depending on the pin surface size and the coloration represents the maximum principal stress state of the polymer on a scale of 0–100 MPa.

Regardless of K_n, both simulations show that the size of the pin surface (bubble size) has a negative influence on the transferable force. The smallest pins transmit the highest forces. The influence of the radius ratio is dependent on K_n. This shows that for large K_n a radius ratio as large as possible should be chosen. The radius ratio 4 shows that a pin with a distinct elliptical shape in the load direction is favorable for load transmission with large K_n. For small K_n the radius ratio is particularly good at 0.4. For the radius ratio of 0.25, a negative trend is already apparent when looking at the models with pin surface 1.77 mm^2. The small radius ratio corresponds to an orientation of the pin transverse to the load direction under small connection forces.

The consideration of the maximum principal stress occurring in the polymer pin, shown in color on a scale from 0 to 100 MPa, shows that the small pins in particular are highly loaded. With a small K_n, the pins with a radius ratio of 0.63 are the most

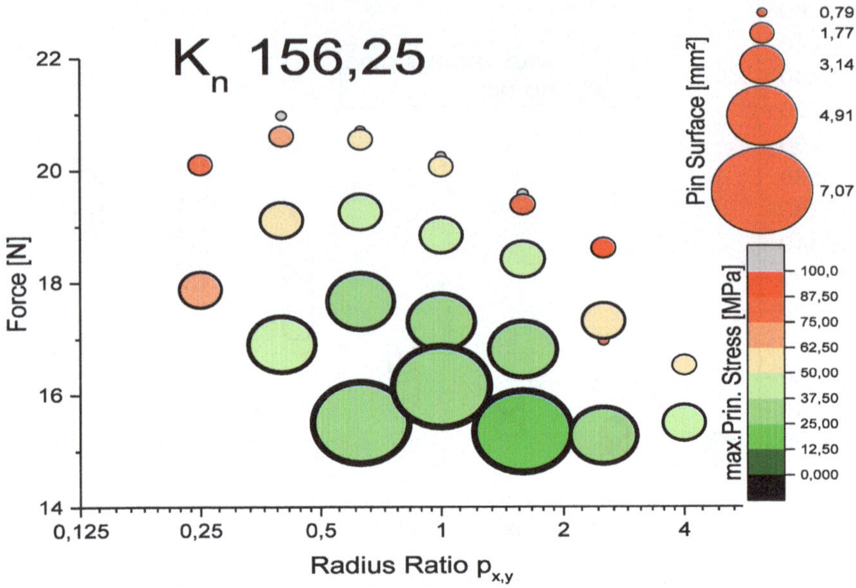

Fig. 7. Force-radius ratio plot for K_n 156.25

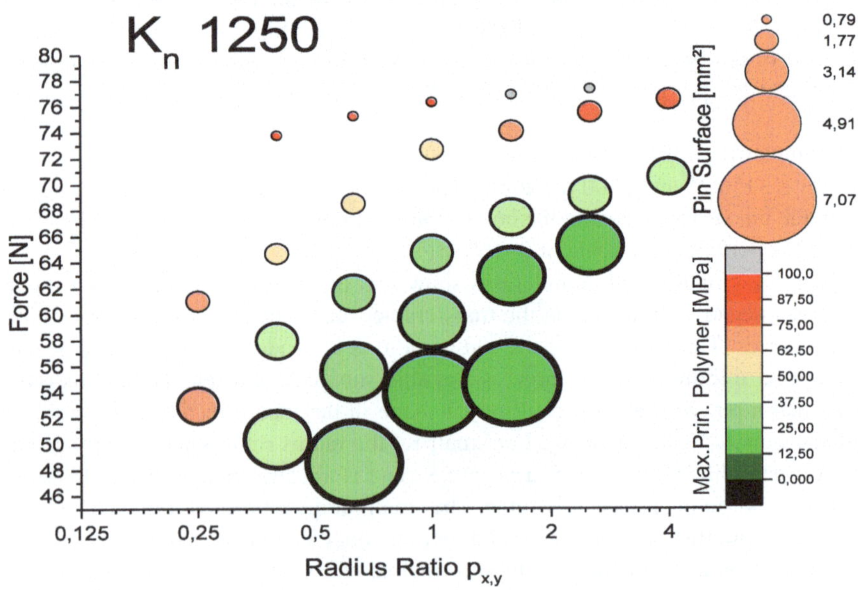

Fig. 8. Force-radius ratio plot for K_n 2500

favorable variant. For large K_n, a positive trend can be seen with respect to the occurring stress for the circular pins with a ratio of 1. With larger and smaller radius ratios, the maximum principal stress increases.

5 Conclusion

In this contribution, the influence of a pin in a CFRP-polymer connection is investigated by means of simulation, considering the adhesive properties.

The typical force-displacement behavior of the connection is first determined in experimental SLJ tests with a single pin. By means of a parameter fitting in a corresponding FEM model, the characteristic cohesive parameters for the investigated CFRP-polymer connection are determined. Furthermore, the general influence of a pin in comparison to a model with a semi-spherical pin and a model without pin in dependence of the cohesive parameter is considered in a simulation study. The load-increasing effect at low bonding strengths and the potential for medium-high value ranges can be clearly shown. This also includes the parameter set determined by the simulative fitting from the SLJ tests for the investigated CFRP-polymer connection. However, there are some limitations in this simulation study: the adhesive parameter of stiffness and strength are varied proportionally and only a simple failure behavior of the adhesive bonding is used. Nevertheless, not all influences and dependencies can be considered in the study and the failure criterion is sufficient to achieve appropriate results.

In addition, a parameter variation is performed for the simulative determination of the influence of the pin geometry as a function of the adhesive connection properties. Here, the findings of the large influence of the pin with low bonding parameters (K_n) are confirmed. It is also shown that, depending on the geometric shape of the pins, the load transfer at high K_n up to 5000 MPa/mm is as good as the simulations without pin. This shows that pins at small K_n have a strengthening effect, which can also be used at larger K_n without weakening the connection in case of quality deviations in production. The further evaluation shows that, independent of K_n, the pin in particular is dimensioned as small as possible in order to achieve high load transfer rates. The dimensioning of the pin is limited by the properties of the base material, which must be taken into account. The consideration of the geometrical shape of the pins for high K_n shows a distinct orientation in the load direction (see Fig. 9). For small K_n, the radius ratio 0.4 is particularly favorable for load transfer and 0.63 to avoid stress in the polymer pin. This shows that the pin should be designed for small K_n perpendicular to the load direction.

In future works, not only a single pin but also the interaction of several pins in a connecting area need to be investigated to gain an understanding of the pin arrangement effects.

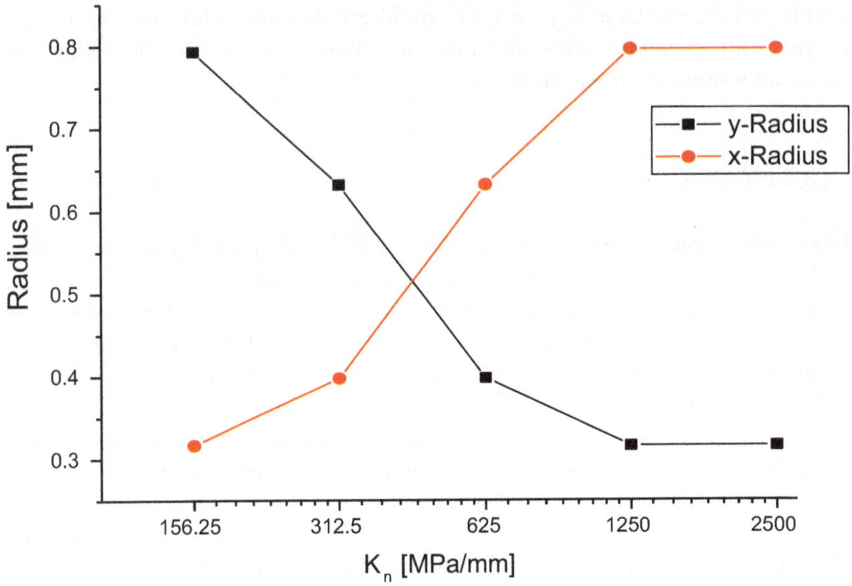

Fig. 9. Pin radii over K_n for the best simulations

Acknowledgment. The funding by the German Research Foundation (DFG) for the project SPP1712 "Intrinsic Hybrid Composites for Lightweight Structures – Fundamentals of Manufacturing, Characterization and Design" is gratefully acknowledged.

References

1. Fleischer, J., Nieschlag, J.: Introduction to CFRP-metal hybrids for lightweight structures. Prod. Eng. **12**(2), 109–111 (2018). https://doi.org/10.1007/s11740-018-0825-0
2. Kießling, R., Ihlemann, J., Pohl, M., Stommel, M., Dammann, C., Mahnken, R., Bobbert, M., Meschut, G., Hirsch, F., Kästner, M.: On the design, characterization and simulation of hybrid metal-composite interfaces. Appl. Compos. Mater. **24**(1), 251–269 (2017). https://doi.org/10.1007/s10443-016-9526-z
3. Messler, R.W.: Joining composite materials and structures: some thought-provoking possibilities. J. Thermoplast. Compos. Mater. **17**(1), 51–75 (2004). https://doi.org/10.1177/0892705704033336
4. Ngyuen, A., Brandt, M., Feih, S., Orifici, A.: Hierarchical surface features for improved bonding and fracturetoughness of metal–metal and metal–composite bonded joints. Int. J. Adhes. Adhes. **66**, 81–92 (2016)
5. Ngyuen, A., Brandt, M., Feih, S., Orifici, A., Amarasinghe, C.: Pin pull-out behaviour for hybrid metal-composite joints with integrated reinforcements. Compos. Struct. **155**, 160–172 (2016)
6. Ngyuen, A., Brandt, M., Feih, S., Orifici, A., Amarasinghe, C.: Loading, support and geometry effects for pin-reinforced hybrid metal-composite joints. Compos.: Part A **98**, 192–206 (2017)

7. Ucsnik, S., Scheerer, M., Zaremba, S., Pahr, D.H.: Experimental investigation of a novel hybrid metal-composite joining technology. Compos.: Part A **41**, 369–374 (2010)
8. Ucsnik, S.: Pin-based hybrid joining. 5th Anniversary of the Institute of Carbon Composite, München (2014)
9. Huelsbusch, D., Haack, M., Solbach, A., Emmelmann, C., Walther, F.: Influence of pin size on tensile and fatigue behavior of TI-CFRP hybrid structures produced by laser additive manufacturing. In: ICCM International Conferences on Composite Materials, July 2015
10. Teufelberger Composite GmbH: T-IGEL® Anbindung. https://www.teufelberger.com/ (2019)
11. Fronius International GmbH: CMT-Pin® Anbindung. https://www.fronius.com/de/schweisstechnik/kompetenzen/schweissprozesse/cmt-pin (2019)
12. Dröder, K., Brand, M., Kühn, M.: Numerical and experimental analyses on the influence of array patterns in hybrid metal-FRP materials interlocked by mechanical undercuts. Procedia CIRP **62**, 51–55 (2017)
13. Günther, F., Pohl, M., Stommel, M., Kretzschmar, V., Scheuermann, G.: Optimising mechanical interlocking interface of CFRP-hybrids. Hybrid Materials and Structures, Bremen (2018)
14. Dassault Systemes Simulia: Course – modeling fracture and failure with abaqus. Lecture Notes. https://www.3ds.com/products-services/simulia/services/training-courses/course-descriptions/modeling-fracture-and-failure-with-abaqus/ (2019). Zugegriffen: 15. Juli 2019
15. Turon, A., Costa, J., Camanho, P.P., Dávila, C.G.: Simulation of delamination in composites under high-cycle fatigue. Compos. Part A: Appl. Sci. Manuf. **38**(11), 2270–2282 (2007). https://doi.org/10.1016/j.compositesa.2006.11.009
16. Su, X., Yang, Z., Liu, G.: Finite element modelling of complex 3D static and dynamic crack propagation by embedding cohesive elements in abaqus. Acta Mech. Solida Sin. **23**, 271–282 (2010)
17. Stuparu, F.A., Apostol, D.A., Constantinescu, D.M., Picu, C.R., Sandu, M., Sorohan, S.: Cohesive and XFEM evaluation of adhesive failure for dissimilar single-lap joints. Procedia Struct. Int. **2**, 316–325 (2016). https://doi.org/10.1016/j.prostr.2016.06.041
18. Reinoso, J., Blázquez, A., París, F.: Damage simulations in composite structures in the presence of stress gradients. In: Modeling Damage, Fatigue and Failure of Composite Materials, pp. 391–424 (2015)
19. Pirondi, A., Giuliese, G., Moroni, F., Bernasconi, A., Jamil, A.: Simulating the mixed-mode fatigue delamination/debonding in adhesively-bonded composite joints. In: Fatigue and Fracture of Adhesively-Bonded Composite Joints, pp. 369–400 (2014)
20. Moya-Sanz, E.M., Ivañez, I., Garcia-Castillo, S.K.: Effect of the geometry in the strength of single-lap adhesive joints of composite laminates under uniaxial tensile load. Int. J. Adhes. Adhes. **72**, 23–29 (2017). https://doi.org/10.1016/j.ijadhadh.2016.10.009
21. Minnicino, M.A., Santare, M.H.: Modeling the progressive damage of the microdroplet test using contact surfaces with cohesive behavior. Compos. Sci. Technol. **72**(16), 2024–2031 (2012). https://doi.org/10.1016/j.compscitech.2012.09.009
22. Jung Lee, M., Min Cho, T., Seock Kim, W., Chai Lee, B., Ju Lee, J.: Determination of cohesive parameters for a mixed-mode cohesive zone model. Int. J. Adhes. Adhes. **30**(5), 322–328 (2010). https://doi.org/10.1016/j.ijadhadh.2009.10.005
23. Harper, P.W., Sun, L., Hallett, S.R.: A study on the influence of cohesive zone interface element strength parameters on mixed mode behaviour. Compos. Part A: Appl. Sci. Manuf. **43**(4), 722–734 (2012). https://doi.org/10.1016/j.compositesa.2011.12.016
24. Giuliese, G., Pirondi, A., Moroni, F.: A cohesive zone model for three-dimensional fatigue debonding/delamination. Procedia Mater. Sci. **3**, 1473–1478 (2014)

25. Campilho, R.D., Fernandes, T.A.: Comparative evaluation of single-lap joints bonded with different adhesives by cohesive zone modelling. Procedia Eng. **114**, 102–109 (2015). https://doi.org/10.1016/j.proeng.2015.08.047
26. Anyfantis, K.N.: Finite element predictions of composite-to-metal bonded joints with ductile adhesive materials. Compos. Struct. **94**(8), 2632–2639 (2012). https://doi.org/10.1016/j.compstruct.2012.03.002
27. DIN EN 6034: Bestimmung der interlaminaren Energiefreisetzungsrate, Mode II (2015)
28. DIN EN 6033: Bestimmung der interlaminaren Energiefreisetzungsrate, Mode I (2015)
29. Dassault Systems: Simulia Abaqus Unified FEA (2017)
30. Autodesk: Helius PFA 2019. https://www.autodesk.com/products/helius-pfa/overview (2019). Accessed 17 July 2019

Interaction Between Foam Injection Molding and Welding Process

Analysis of the Process - Material - Structure - Property Relations

Karoline Hofmann[1(✉)], Christian Brütting[2], Michael Gehde[1], and Volker Altstädt[2]

[1] Professorship of Plastics Engineering, Chemnitz University of Technology, Chemnitz, Germany
karoline.hofmann@mb.tu-chemnitz.de
[2] Polymer Engineering Bayreuth, University of Bayreuth, Bayreuth, Germany

Abstract. Due to increasing demands on component integration, functionalization, saving weight or material, the density and weight of thermoplastic parts could be influenced significantly by using the thermoplastic foam injection molding process.

The characteristic three-layer structure offers numerous advantages for applications, such as weight reduction, increasing the specific bending stiffness with a simultaneous low tendency to warp and optimizing thermal and acoustic properties. For a subsequent joining process, however, difficulties arise due to the thin solid skin layer. Minimum joining distances during welding can no longer be met geometrically and the mechanical properties of the components are reduced.

The present study is intended to analyze the interaction between the microcellular structure of the injection molded parts, the influence on the joining process and the resulting mechanical properties. Therefore, a cooperation of Chemnitz University of Technology and University of Bayreuth was founded to investigate the correlation between injection molding and vibration welding for microcellular polypropylene and polyamide materials.

In addition to various materials with and without fiber reinforcement, the influence of storage time and different joint types were investigated in this study.

The aim of this study was to gain a deeper understanding of the relationships between process, material and its structure as well as the development of processing guidelines which can be transferred to industrial applications.

The results show that the characteristic three-layer structure has a considerable inhomogeneity, depending on the used material, the foaming process and the process parameters.

C. Hopmann and R. Dahlmann (Eds.): *Advances in Polymer Processing 2020*, pp. 233–246, 2020. https://doi.org/10.1007/978-3-662-60809-8_19

However, the weldability of foamed thermoplastics strongly depends on the internal structure due to the injection molding parameters, the storage time between foam injection molding and welding process, the joint type and the welding process itself.

Keywords: Foam injection molding · Physical foaming · Chemical foaming · Vibration welding · Parameter variation

1 Introduction

Due to increasing demands of component complexity, functional integration and assembly requirements, plastic components offer decisive advantages over conventional materials. In addition, there is a trend towards lightweight construction and the optimization of thermal, acoustic and mechanical properties, so that solid components reach their limits. Advantages were offered by the Foam Injection Molding process (FIM). Parts produced by foam injection molding fulfill the requirements of reduced density and thus associated lower weight and less material needed. Until a critical degree of foaming they are also distinguished by a high specific flexural stiffness. Moreover, there is reduced warpage and the probability of sink marks is lower.

For this reason, especially in sectors like Automotive Engineering, Medicine, Housing, Packaging and Transportation, thermoplastic parts are increasingly replaced by their microcellular foamed counterparts.

The basis is the characteristic three-layer structure which is originated during the foam injection molding process: a microcellular foamed core surrounded by a non-cellular solid skin layer. The achievable lightweight effect results from the displacement of the matrix material as from the core into the solid skin layers. The result is an increase in the specific bending stiffness with low tendency to warp.

Depending on the boundary conditions during the injection molding process the appearance of the three-layer structure can differ significantly. Characteristics to be considered for the welding process, among other things, are cell size and their distribution as well as thickness of the solid skin layer. Numerous investigations are dealing with the correlation between injection molding parameters and the resulting part structure. Especially the works of Altstädt and Mantey [1] as well as Okamoto [2] provides extensive information about the structure-property relations. According to the literature, the process temperatures (melt, mold) as well as the injection speed have a high impact on the resulting part structure during injection molding. Furthermore, the material and the metering of the physical and chemical foaming agent are particularly important [2].

Due to this variety of possibilities on influencing the structure, the welding of microcellular foamed parts is a challenge. The welding process has to be precisely adapted to these special circumstances in order to ensure a high weld strength. Due to decades of intense research and development, the welding of solid thermoplastics is quite common. Both infrared and vibration welding are examples for welding processes which are successfully used in a wide range of industrial applications. In contrast to this, there is only a small amount of experience in the welding of their microcellular foamed counterparts. Only a couple of investigations are dealing with this [3–5].

Thus, the aim of this study was to gain a deeper understanding of the relationships between process, material and its structure of foam injection molded parts in comparison to their solid counterparts.

2 Experimental

To evaluate the reference material and the following welding behavior of microcellular foamed parts in dependence on storage time, joint type and processing parameters, several types of polypropylene manufactured by LyondellBasell and a polyamide manufactured by Lanxess, were used:

- PP-GF20: Polypropylene + 20 wt.-% Glass Fiber (Hostacom ERG 393F black)
- PP-Talc12: Polypropylene + 12 wt.-% Talcum (Hostacom TYC 469F)
- PA6-GF35: Polyamide 6 + 35 wt.-% Glass Fiber (Durethan BKV35H2.0)

All materials were processed by using the Cellmould® technology (Wittman Battenfeld GmbH, Austria) for the physical foaming process and using Hydrocerol ITP825 (Clariant) as a chemical foaming agent for chemical foaming process. Therefore a standard injection molding machine (Battenfeld HM 110/525, Wittman Battenfeld GmbH, Austria) were used to manufacture foamed parts (specimens $150 \times 75 \times 4$ mm^3) at the Polymer Engineering department in Bayreuth. Within the injection molding process, specific parameters like injection speed, mold and melt temperature, weight reduction and type of blowing agent were varied for all materials within the reachable processing limits. The used physical foaming agent was nitrogen with a proportion of 0.2% by weight to reach a 10% weight reduction of the part. For chemical foaming, a chemical blowing agent (CBA, 2.0% by weight) was needed to reach the same weight reduction. The maximum achievable foaming degrees, using the same process parameters, were also manufactured in order to determine the influence of the foaming degree on mechanical and optical properties. Non-foamed pendants were manufactured as reference.

The aim was to obtain a combination of good mechanical strength and the best possible three-layer structure for welding: A thick solid skin layer combined with homogenously distributed microcellular foam as core.

For Determination of puncture impact behavior the instrumented puncture test (DIN EN ISO 6603-2:2000 [6]; device: CEAST Fractovis Plus) and the bending test (DIN EN ISO 178 [7], device: ZMART.PRO Z1464) were carried out in order to validate the mechanical properties during the parameter variation in the foam injection molding process.

In order to investigate the characteristic three-layer structure as a result of various foam injection molding parameters and the homogeneity of the cell structure within the parts via the flow path, 15 specimens were extracted at several points of the parts along the flow path. These were examined in a light microscope (type: BX51, Olympus). Afterwards the microscopy pictures were investigated by means of the image evaluation software ImageJ. With this software it was possible to determine foam structures such as the solid layer thickness along the flow path, cell size and density (Fig. 1).

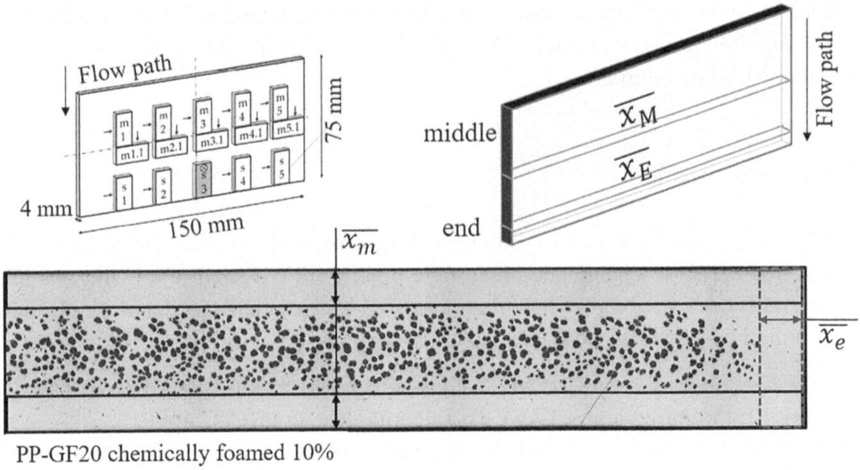

Fig. 1. Sample preparation out of the foamed parts for the microscopic investigations (top, left), measurement of the solid layer thickness representative for PP-GF20 (10%) (bottom), and the resulting mean value in the middle ($\overline{x}_m$) and at the end of the flow path ($\overline{x}_e$) (top, right)

The knowledge of the solid layer thickness in the middle of the plate and at the end of the flow path will help to improve the understanding for the subsequent investigations of welding foamed thermoplastic materials.

After manufacturing and evaluating the parts, investigations by vibration (VIB) welding were accomplished. Therefore, the vibration welding system of Branson (type: M624-HRSi, fabricated by Branson AG) was used. The chosen process parameters are shown in Table 1 and are divided into standard parameters as well as a specific pressure and path variation.

Table 1. Variable process parameters for vibration welding

	Vibration welding
Welding pressure p_F	Standard: 0.5 MPa Variation: 2.0 MPa (pressure ++)
Welding path s_F	Standard: 1.0 mm Variation: 0.5/ 1.4 mm (path −−/++)
Amplitude	0.9 mm
Frequency	240 Hz
Welding direction	Parallel (180°)
Joint type	Butt-joint/T-joint

For all combinations mentioned, the joint type was varied as well. In Fig. 2 the two types are shown. In case of the butt joint arrangement the two flow path ends (dashed) are welded. For the T-joint arrangement the flow path end of the first mating partner is welded to the middle of the long side of the second one.

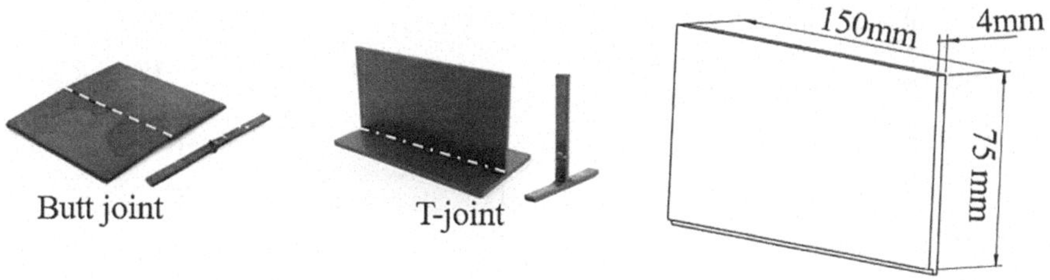

Fig. 2. Joint types in relation to flow path end (dashed, left and middle) and plate specimen dimensions (right).

The mechanical properties were analyzed by using a universal testing machine of Zwick/Roell (device: ZMART.PRO Z1464). The basic material strength was determined according to DIN EN ISO 527 [8] for tensile strength and DIN EN ISO 178 [7] for bending strength. In order to determine the weld strength, the VIB-welded plates were sawn into samples with a width of 10 mm. Each mean value results from tensile test according to the Standard DVS 2203-2 [9].

In addition, the outermost strips were discarded because they have a greater proportion of solid skin layer.

3 Results and Discussion

3.1 Mechanical and Morphological Examinations of Injection Molded Parts

Preliminary investigations in the foam injection molding process showed that the morphology of the foamed parts can be significantly influenced by variation of process parameters. Especially the weight reduction due to the increased foaming degree influences the mechanical and optical properties. By increasing the degree of foaming, the mechanical strength and the solid layer thickness decrease. An increase in the blowing agent content also leads to the reduction of the solid layer thickness. The mould temperature also has an influence on the formation of the foamed core, but does not significantly affect the mechanical properties.

Due to the complex property relationships between process and structure, a targeted adjustment of the component structure by simultaneous manipulation of several parameters is difficult to implement. After the evaluation of the results from parameter variation in the foam injection molding process, a parameter setting has to be chosen for the following welding process. Regarding the purpose: A thick solid skin layer combined with homogeneously distributed microcellular foam as the core layer and good mechanical strength. The selected parameter settings are shown in Table 2 and 3.

Table 2. Optimal process parameter settings, physical foaming process

Material	PP – GF20		PP – Talc12		PA6-GF35	
Foaming degree [%]	10	40	10	36	10	36
Tmelt [°C]	280	260	260	240	290	290
p [bar]	110		110		110	
Tmould [°C]	30		30		70	
SCF Level [%]	0,2		0,2		0,2	
Vinj [cm³/s]	100		100		100	

Table 3. Optimal process parameter settings, chemical foaming process

Material	PP – GF20		PP – Talc12		PA6-GF35	
Foaming degree [%]	10	25	10	27	10	21
Tmelt [°C]	260	240	240	240	270	270
p [bar]	110		110		110	
Tmould [°C]	30		30		70	
CBA content [%]	2.0		2.0		2.0	
Vinj [cm³/s]	100		30	100	30	100

Figure 3 shows the resulting cell morphology in the middle of the plate depending on the material, the foaming degree and the foaming process. The foamed material has a decisive influence on the resulting structure. PA6-GF35 has much smaller and finer distributed cells at the same degree of foaming than the PP materials. Also the degree of foaming has a major influence on the resulting cell morphology. Independent of the material and foaming process, the solid layer thickness reduces with increasing foaming and the number of cells increases.

Comparison of chemical and physical foaming shows significant differences regarding the cell distribution and size in PP materials. While the PA6 material has similar cell structures independent of the foaming process (chemical or physical). In general, chemically foamed parts have a homogeneous foam distribution with many small cells, a pronounced 3-layer structures and high solid layer thicknesses regardless the material. The foam structure of the physically foamed parts is highly inhomogeneous throughout the plate, very coarse cellular and the surface layer thicknesses are low. The maximum-foamed parts were partially foamed to the edge. This inhomogeneity within the plate illustrates the challenge of welding such foam injection molded parts, in particular in view of different joint types.

For the selection of process parameters in vibration welding, it is of great importance to study the thickness of the solid skin layer over the entire plate geometry. Figure 2 shows the two positions which are important for the welding process (Butt joint and T-joint) and Fig. 1 the measuring method to determine cell morphology and layer thicknesses.

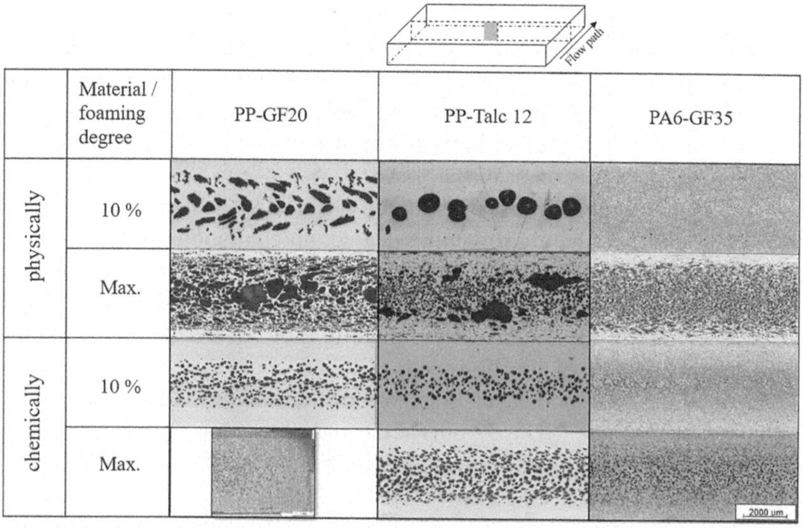

Fig. 3. Cell structure in the center of the plate depending on the foaming process, material and foaming degree

To get an impression of the solid layer thickness of the materials as a function of the foaming process and the degree of foaming, Fig. 4 shows the results of the measurements. The displayed value is the mean value and standard deviation at the specific position (middle, end of flow path) for the material and 10% foaming degree, determined by several microscopy images along the plate (Fig. 1, top, left).

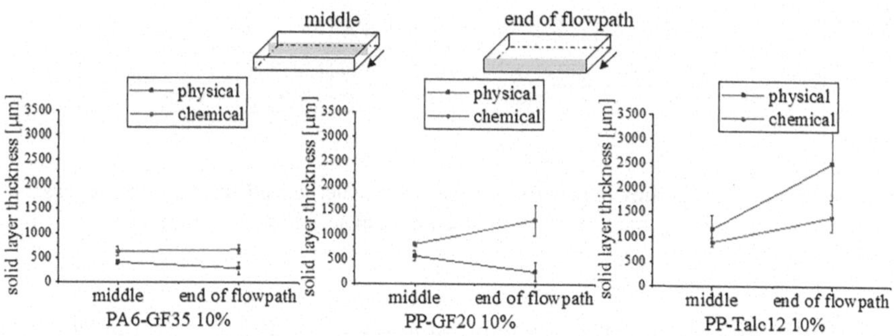

Fig. 4. Mean value of solid layer thickness dependent on material, foaming process and flow path position

The results show that the highest solid layer thickness can be found in physically foamed PP-Talc, deriving from the very large and in number small cells. Physically 10% foamed PA6-GF35 and PP-GF20 show similar values in the middle and at the end of the flow path. In contrast, using the chemical foaming process the thickness of the solid layer reaches even higher values for these materials. Comparing the thicknesses at the specific sampling points, the solid layer thicknesses for chemically foamed parts at the end of the flow path are higher than in the middle. In addition, using chemical foaming, higher solid layer thicknesses can be achieved than using the physical foaming process (except PP-talc).

The solid layer thickness for the maximum physically foamed parts are not determinable, because due to the high foaming degree, no continuous solid skin layer has formed. They are also not comparable with the chemically foamed because of the different degrees of foaming. However, 21% chemically foamed PP talc still has a solid layer thickness in the middle of the plate of 578 µm (SD: 73 µm) and 494 µm (SD: 161 µm) at the end of the flow path. Similar values reached 27% chemically foamed PA6-GF35. The solid layer thickness in the middle of the plate is 326 µm (SD = 86 µm) and 311 µm (SD = 52 µm) at the end of the flow path. These values are comparable to 10% physically foamed PA6-GF35 (middle: 410 µm (SD: 46 µm) and end of flow path (291 µm (SD: 122 µm)).

The achieved mechanical results for each material depending on the foaming process for tensile and bending strength shows Fig. 5.

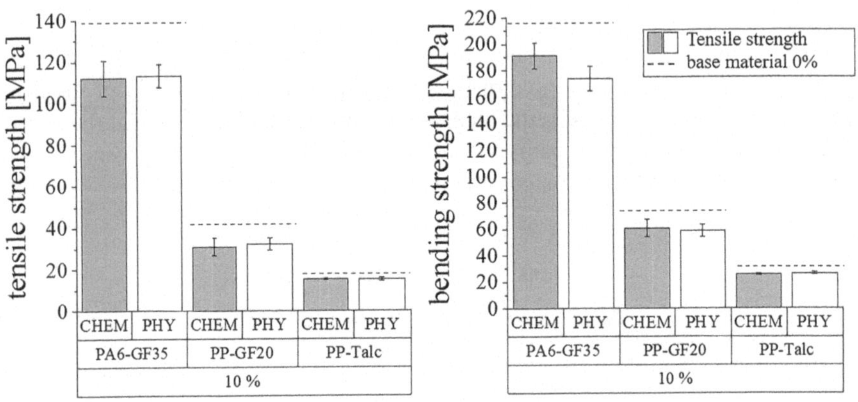

Fig. 5. Mechanical properties depending on the used material and the foaming process with 10% foaming degree compared to the non-foamed base material (dashed line)

The different basic material strength results from the fiber reinforcement and the respective base polymer. The fiber reinforcement increases the weld strength compared to unfilled material.

Despite the widely differing cell structures for the PP-material (seen in Fig. 3), there are no differences in tensile and bending strength using the chemical or physical foaming process. In contrast to that, a difference in the flexural strength of the foamed material can be seen in PA6-GF35, whereby the cell structures in the chemical and physical foaming process did not differ (Fig. 3). In general, the losses of the mechanical properties of all materials due to the foaming process are greater than the achieved density reduction. Thus PA6-GF35 loses 18%, PP-GF20 24% and PP-Talc12 about 17 % of its tensile strength due to a 10% weight reduction. The results from the bending tests show similar results for each material.

The mechanical properties of the maximum-foamed materials show that the mechanical strength of all materials decreases strongly as the degree of foaming increases. Physically highly-foamed PA6-GF35 (36%) lost 58% of its tensile strength compared to the non-foamed base material.

3.2 Mechanical and Optical Properties of Vibration Welded Parts

Based on the results of the structural analysis, the injection molded parts were joined by vibration welding. Therefore, the joint type (butt joint, T-joint), the welding parameters (pressure, welding path) and the degree of foaming (10%, max.%) were varied and investigated for all materials.

The results for the influence of the joint type are shown in Fig. 6. Here, a distinction is made between the results of the chemical (on the left side) and the physical foaming process (on right side) further divided according to the degree of foaming. The dotted line above the bar charts representing the base material strength as a reference of the non-foamed, joined base material. The presented results so far are all based on welding with standard parameters, mentioned above in Table 1.

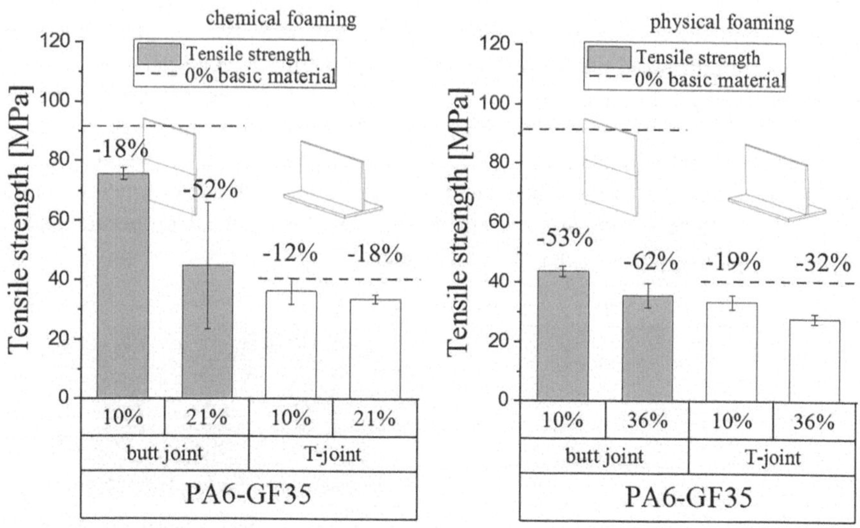

Fig. 6. Tensile strength of PA6-GF35 depending on joint type and foaming degree for chemical foaming process (left) and physical foaming process (right)

The results from the non- foamed and VIB-welded base material (dotted lines), show, that the joint type has a great influence on the welding strength. The achievable weld strength during butt joint welding is 92.6 MPa (SD: 4.12 MPa). In contrast to the

T-joint welding, where the maximum weld strength was 56% lower (40.9 MPa (SD: 4.92 MPa). An additional foaming of the material leads to further loss of mechanical strength. The differences between chemical and physical foamed material in dependence of the foaming degree is clearly visible in Fig. 6. A physically foamed PA6-GF35 loses 53% of its tensile strength in butt joint due to a 10% weight reduction. In which the chemically foamed counterpart is only reduced by 18% of its welding strength.

The loss of strength within T-joint welding in comparison to the basic material strength is less, independent of the foaming process and degree. The maximum achievable properties in the T-joint are similar for chemically and physically foamed components. Due to a higher weight reduction in the physical foaming process, the weld strength of physical foamed parts is lower than those of the chemical process.

Despite the same cell morphology of the physically and chemically foamed PA6 GF35, large differences in strength result during subsequent welding. 21% chemically foamed, butt-welded parts achieve comparable strength values as 10% physically foamed.

Figures 7 and 8 shows the appropriate microscopic images of the welds to the mechanical evaluations in Fig. 6.

The microcellular structure of physically and chemically foamed butt joints and T-joints of PA6-GF35 is subdivided further depending on the degree of foaming. A clearly visible increase in the number of cells with increasing degree of foaming can be seen.

Apart from changes in the number of cells and density, no qualitative statements regarding the weld quality in the microscopy of the welded parts are possible. Whether or not the solid skin layer has been punctured in consequence of the welding process, is difficult to assess with the help of just the microscope images.

Occasionally, some foam cells are visible in the joining level. The extent to which these lead to a decrease of the mechanical properties can not be determined by optical analysis alone. Here a comparison with the previous mechanical properties and the solid layer thickness measurement is to be made.

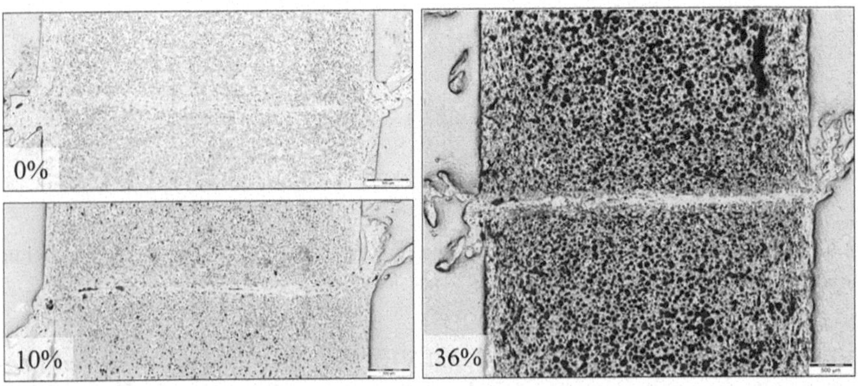

Fig. 7. VIB-welded (butt joint) of physically foamed PA6-GF35 in dependence of the foaming degree (welding parameter: $a = 0.9$ mm; $p_F = 0.5$ MPa; $s_F = 1$ mm)

Figure 8 compares microscopic images of chemically (on the left side) and physically (on the right side) welded parts using T-joint. A statement about the resulting mechanical characteristics is also not possible in this case. However, knowledge of the solid layer thicknesses in the center of the plate and at the end of the flow path, together with the welding parameters (joining path), makes it possible to conclude whether or not the weld has punctured the solid layer.

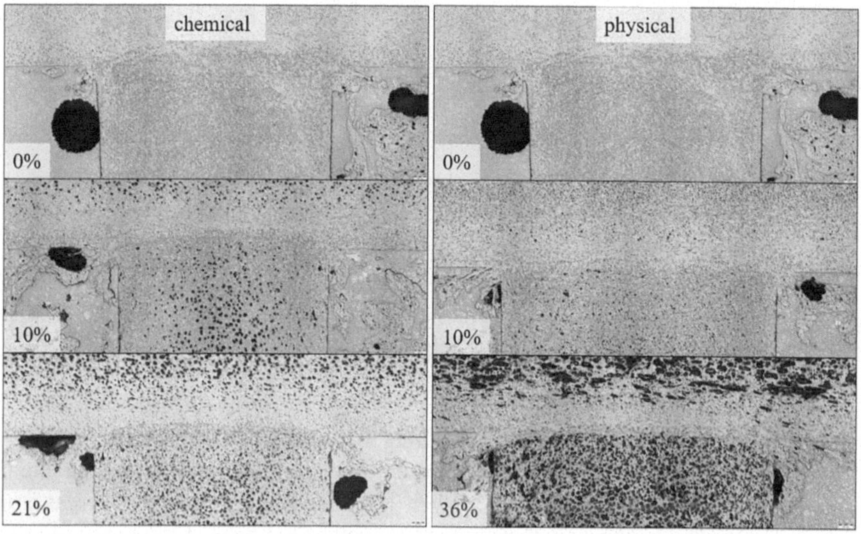

Fig. 8. VIB-welded (T joint) of chemically (left) and physically (right) foamed PA6-GF35 in dependence of the foaming degree (welding parameter: a = 0.9 mm; $p_F = 0.5$ MPa; $s_F = 1$ mm)

In order to investigate the influence of the joining pressure and the welding path, the welding parameters were varied in addition to the standard parameters. Figure 9 shows the effect of the variation on the tensile strength for chemically foamed (on the left side) and physically foamed (on the right side) parts.

As seen in the previous results, the chemical foaming process leads to a higher solid layer thickness within the parts. This fact alone leads to mechanical differences in the welding process.

The mechanical strength of welded, chemically foamed parts is significant higher than the physically ones, because the solid layer is twice as high (Fig. 4 chemically: 668 µm (SD: 89 µm); physically: 291 µm (SD: 122 µm)).

So in this case a welding of compact material is possible. The reachable mechanical properties decrease about 18% compared to the welded base material (standard welding parameters). A variation of the welding path up to 1.4 mm doesn't change the results. An increase of the joining pressure (from 0.5 to 2 MPa) decrease the maximum strength.

A significantly lower strength can be seen in the results from the physical foaming process (Fig. 9, on the right side). All chosen parameters punctured the solid layer

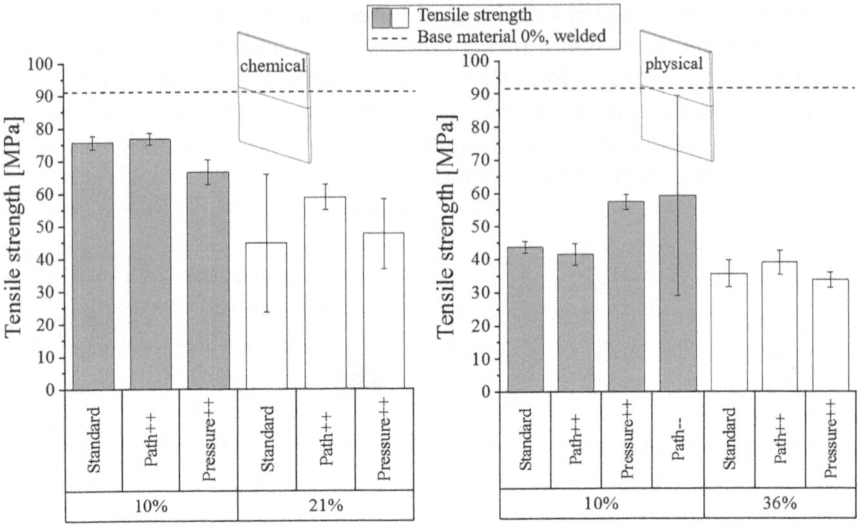

Fig. 9. Tensile strength for chemically (left) and physically foamed (right) PA6-GF35 within the VIB-process in dependence of parameter variation

thickness into the foamed core (except Path −−), so that the mechanical strength is reduced. In this case the shown values represent the strengths for welding outside the foamed core. Within a punctured solid layer, a higher joining pressure increases the tensile strength due to the compensation of the plate warpage. Under optimal conditions a maximum theoretical strength can be achieved by setting the joining path within the solid layer thickness (0.5 mm). Due to the plate warpage, the standard deviations are very high (Path−−). In addition, a 0.5 mm joining path is not industrially relevant and not realistic even for large components with warpage. Nevertheless, the results are intended to show the influence of welding inside and outside the solid skin layer.

4 Conclusion

On the way from manufacture by foam injection molding to welded component, a variety of parameters affect the foam structure.

The results have shown that there is a significant inhomogeneity of the characteristic three-layer structure within the parts regarding cell-size distribution and thickness of the solid skin layer for each material. Accordingly, the probability increases to weld into the (more unstable) foamed core.

The chosen foaming process, the type of blowing agent, material and above all the injection molding parameters have a decisive influence on the resulting cell morphology. The thickness of the solid layer has a significant influence for welding a foamed part. By varying the process parameters for the foam injection molding, the solid layer thickness can be selectively changed and the welding parameters can be adapted to the present material within the process limitations.

As presented in the results there is a significant difference between vibration welding of solid thermoplastics and their microcellular foamed counterparts. Regarding the welding process, the thickness of the solid skin layer and the joint type are of enormous importance for the mechanical characteristics. By parameter variation within the solid layer thickness, the mechanical strength can be further increased. When welding into the foamed core, it comes to reducing the strength. However, it is not always possible to weld outside the foamed core.

Due to a slightly warpage of the parts, an additionally adjustment of the joining path and pressure is necessary to get a consistent weld. In this case welding outside the foamed core is not possible if the solid layer thickness is less than the joining path. Especially in the industrial manufacturing process, small joining paths are not feasible due to process limitations, large component dimensions, warpage and high foaming degrees with low solid layer thicknesses.

Therefore the chemical foaming process provides higher solid layer thicknesses within the parts, with the result that the plate warpage can be compensated for with higher joining pressure or path.

The study underlined the importance of the process - material - structure - properties- relationships. By knowing the complex property relationships between influencing factors, it is possible to weld foamed components with good mechanical and optical properties.

Acknowledgements. The investigations were carried out in cooperation of the University of Bayreuth, Polymer Engineering Bayreuth.

The investigations were promoted from budgetary funds of the German Federal Ministry of Economic Affairs and Energy (BMWi) on the basis of a decision by the German Bundestag via the German Federation of Industrial Research Associations (AiF) We would like to record our appreciation of this support.

Furthermore, we thank the companies Lanxess Deutschland GmbH and LyondellBasell Industries for providing the material.

Gefördert durch: Bundesministerium für Wirtschaft und Energie aufgrund eines Beschlusses des Deutschen Bundestages

AiF iGF Industrielle Gemeinschaftsforschung

References

1. Altstaedt, V., Mantey, A.: Thermoplast-Schaumspritzgießen. Hanser, München (2010)
2. Okamoto, Kelvin T.: Microcellular Processing. Hanser, Cincinatti (2003)
3. Hopmann, C., Kreimeier, S., Schoengart, M.: Laser transmission welding of foamed thermoplastic injection moulded parts. In: PPS – Polymer Processing Society. IKV Aachen, RWTH Aachen University (2016)
4. Potente, H.: Schweißen von Thermoplasten mit zellularer Struktur. Final report (AiF – Project 13.595 N), University Paderborn (2006)

5. Kishbaugh, L., Kolshorn, U., Bradley, G.: Vibration and ultrasonic welding conditions and performance for glass fiber filled PA 6 and PA 6.6 injection moulded using the MuCell microcellular foaming process. In: Conference Proceedings, Blowing Agents and Foaming Processes (2007)
6. DIN EN ISO 6603-2:2002-04: Plastics – Determination of puncture impact behaviour of rigid plastics – Part 2: Instrumented puncture test (ISO 6603-2:2000). Beuth (2000)
7. DIN EN ISO 178:2019-08: Plastics – Determination of flexural properties (ISO 178:2019). Beuth (2019)
8. DIN EN ISO 527-1:2012-06: Plastics – Determination of tensile properties – Part 1: General principles (ISO 527-1:2012). Beuth (2012)
9. DVS-Richtlinie 2203: Prüfen von Schweißverbindungen an Tafeln und Rohren aus thermoplastischen Kunststoffen – Prüfverfahren – Anforderungen. DVS-Verlag, Düsseldorf (2003)

Prediction of the Bond Strength of Thermoplastics Welded by Laser Transmission Welding

Christian Hopmann, Simon Bölle(✉), and Lorenz Reithmayr

Institute for Plastics Processing, RWTH Aachen University, Aachen, Germany
simon.boelle@ikv.rwth-aachen.de

Abstract. Laser transmission welding is one of various welding techniques used to join thermoplastics. Low heat introduction into the welded parts and a high welding speed are the reasons why laser transmission welding established itself as a joining process in the plastics processing industry. To minimise defective parts and maximise productivity, it is essential to determine a set of ideal welding parameters that allow maximum bond strength at the lowest possible cycle times. To facilitate this process, simulation models provide detailed analysis without the need for destructive and costly part testing. To predict the weld strength of two thermoplastic parts joined by laser transmission welding, the Institute for Plastics Processing has developed a model that combines the simulated thermal properties of the material during and after welding with the molecular behaviour of plastic melts. Based on the results of the thermal modelling of the welding process a mathematical model describing the movement of polymer chains is used to calculate the resulting bond strength depending on material properties as well as heating and cooling rates. The temperature data of nodes situated at the interface of both welding partners are extracted from the simulation for every time increment of the simulation. The model, which is based on the reptation theory of polymer melts, is then used with these data and the bond strength is calculated. The results are validated by tensile tests on welded parts with the same input parameters used in simulation. In first results, the calculated bond strength shows a good agreement with the values measured in tensile tests. Occasional deviations can be explained by the fact that the material decomposition occurring in experimental welds is not considered in the thermal simulation and the reptation theory.

Keywords: Laser transmission welding · Process simulation · Tensile strength

1 Introduction and Objective

Parts and components made of engineering plastics are a basic component of modern commercial and industrial products [7, 28, 32]. The wide variety of achievable material properties and the large potential for continuous development are some of the main factors driving the expanding use of plastics in applications such as automotive,

C. Hopmann and R. Dahlmann (Eds.): *Advances in Polymer Processing 2020*,
pp. 247–257, 2020. https://doi.org/10.1007/978-3-662-60809-8_20

electrical and aerospace technologies [18, 22]. These new developments place increasingly high demands on manufacturing processes. The process of joining two or more components is one of the last steps of the manufacturing process but is critical for ensuring the correct functionality of the final part [4, 27].

Laser transmission welding of thermoplastics is one such joining process. During this process, a transparent welding partner is joined with an absorbing counterpart by heat energy provided by a laser beam directed at the weld seam [7, 19, 31]. The strength of the resulting weld is of decisive importance for the proper function of the part and the assembly. In addition to the basic strength of the welded materials, the weld strength is determined by a large number of process parameters, but cannot be estimated with sufficient accuracy only on the basis of these parameters. Hence, welded parts are usually designed in an empirical, iterative process in which the welding parameters as well as the part geometry are adapted according to guidelines and best practices. To cut down on this time-consuming and expensive approach, it is of major importance to predict the strength of the weld in dependency of the welding parameters. To this end, a numerical multi-scale approach for predicting weld strength and performance of parts joined by laser transmission welding is developed (Fig. 1) [16].

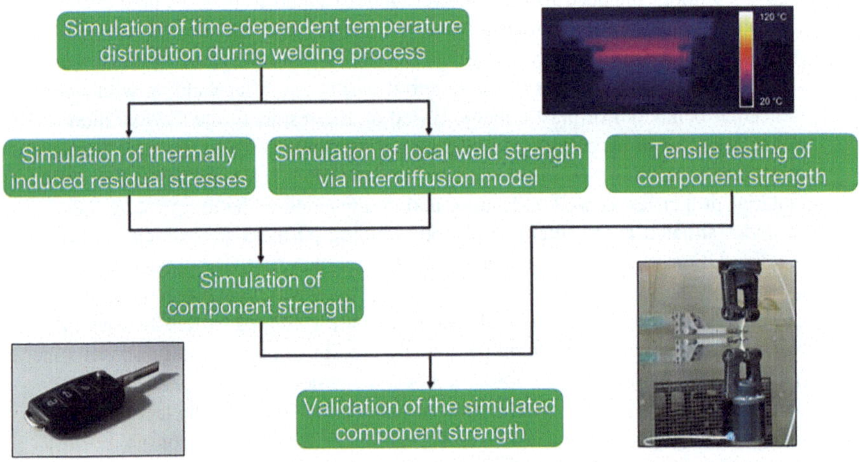

Fig. 1. Concept for the numerical description of the weld strength in welded components [16]

First, a 3-D model of the welding partners is developed and the temperature distribution during and after the welding process is simulated. This results in time- and location-dependent temperature profiles of the welded parts. Subsequently, both the thermally induced residual stresses as well as the resulting strength of the weld based on the temperature profile in the welding zone are calculated. The weld strength is initially modelled using equations arising from the reptation model proposed by *de Gennes* [8]. However, the strength of the bond is also influenced by residual stresses, which either add or detract from the calculated bond strength. By considering both the calculated bond strength and the simulated residual stresses, a model for predicting the tensile behaviour of the welded part is developed and validated in experimental tensile tests [16].

2 State of the Art

Laser transmission welding is a single-stage process in which the heating of the plastic and the joining process take place simultaneously. The optical properties of both materials are subjected to the condition that one welding partner must have a high degree of transmission in the laser wavelength range and the other a high degree of absorption. Before the welding process both components are positioned in the desired end position and the welding pressure is applied. The process sequence is shown schematically in Fig. 2. The laser beam passes through the transparent welding partner without significant heating. Only in the second welding partner, the laser beam is completely absorbed in the near-surface layer, whereby the laser energy is converted into heat energy and the plastic is melted. Due to heat conduction, the transparent welding partner is also plasticised indirectly in the area of the welding zone. The internal welding pressure resulting from the expansion of the plastic melt as well as from the externally applied welding pressure results in a material-locking connection of the components.

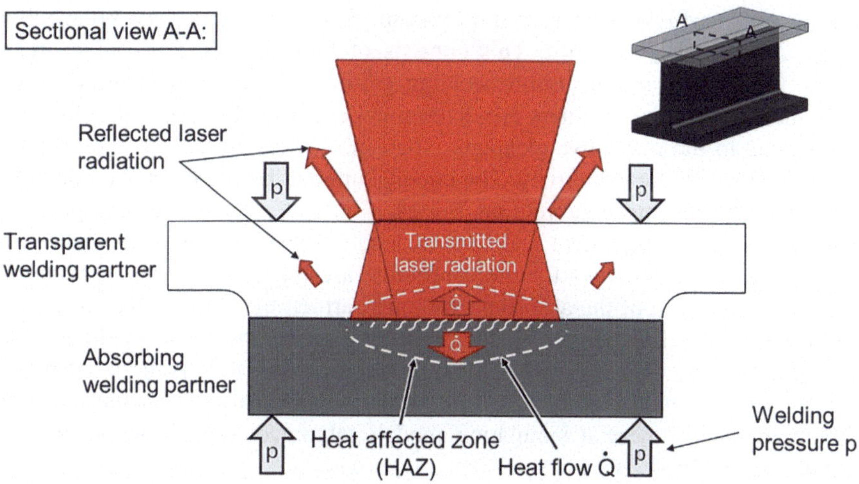

Fig. 2. Process principle of laser transmission welding [16]

Furthermore, different approaches for the numerical calculation of the heating process using the finite element method (FEM) are pursued [1, 3, 5, 6, 9–12, 17, 21, 23, 25, 26, 33, 34]. The listed works differ in the type of modelling as well as in the input parameters used. Both literature values and own measurements serve as input data, whereby the used measuring methods vary. Challenges arise with the temperature dependence of the material properties, the characterisation of the intensity distribution, the experimental validation of the calculated temperature distributions and the deviations between simulation results and experimentally determined results. First approaches for the numerical calculation of thermally induced residual stresses during laser welding with a moving spot, i.e. contour welding, are presented by *Potente et al.*, *Zoubeir and Elhem* and *Gupta and Pal* [14, 30, 34]. However, in their research, very high heating temperatures up to 700 °C were calculated [34], no information was given on the mechanical material model [14] or higher temperatures lead to lower residual stresses [30]. *Labeas et al.* introduce a simulation model for the calculation of residual stresses during contour welding of composite parts, in which only a simple, purely elastic material behaviour is implemented [24].

A simulation model with which both residual stresses and weld strengths can be predicted on the basis of a realistic numerical calculation of the temperature distributions in the welding process is currently non-existent.

3 Simulation Model for Laser Transmission Welding

3.1 Thermal Simulation

The simulation of the laser welding process is carried out with t-joint samples for the process variant of simultaneous welding for a polyamide 6.6 (PA 6.6) of the type Ultramid A3W from the manufacturer BASF SE, Ludwigshafen, Germany. For the laser-absorbing welding partner, samples with 0.3 wt.% carbon black are used. With the Abaqus FEM software from Dassault Systèmes SE, Vélizy-Villacoublay, France, a 3D model was built. This consists of two main sections which represent the transparent and absorbing welding partner. The thermal material properties of the two welding partners are determined experimentally for both materials and assigned to the respective elements. These properties are density, specific heat capacity and thermal conductivity. The energy supplied by the laser is modelled by means of a self-developed subroutine. For each time increment, the subroutine calculates the location-dependent and time-dependent heat flux supplied per volume element. The input variables for the subroutine are the optical material properties absorption coefficient of the absorbing welding partner, the extinction coefficient of the transparent welding partner, surface reflection at the absorbing welding partner and reflection factor in the transparent welding partner as well as the intensity distribution after radiating through the transparent welding partner. For more detailed information on the thermal simulation model, reference is made to the work of *Hopmann et al.* [16].

The simulation is validated by real welds performed with a fiber coupled diode laser type LDM 400-40 from Laserline GmbH, Mühlheim-Kärlich, Germany, with a wavelength of 940 nm. The laser couples its energy into an optical fiber, which leads to a line optics. The line optics widen the point-shaped laser beam and bring it into line form. Without the transparent welding partner influencing the laser beam, according to the manufacturer's specifications, the line has the dimensions $27 \times 1.5\ mm^2$ in the focal plane, which corresponds to the dimensions of the T-joints in the welding plane. The heat-affected zone (HAZ), which is defined in the simulation as the area in the weld cross-section that reaches the melt temperature of the material of 260 °C, is used for the validation of the thermal simulation. In real welds, the HAZ is measured by means of microscopic images in which the microstructure change becomes visible when the melt temperature is reached.

3.2 Thermomechanical Simulation

Based on the thermal simulation model, the mechanical stresses caused by the thermal expansion are calculated for each time step. For this purpose, an elastic-plastic material model is used for which the material data are determined in experimental tensile tests. This model describes reversible and irreversible mechanical effects that arise in response to forced deformations. The consideration of the irreversible plastic deformation allows the calculation of the remaining internal stresses after cooling of the welding partners [16]. For the thermomechanical simulation, the elements of the thermal 3D model are extended by the temperature-dependent, mechanical properties thermal expansion coefficient, Young's modulus, yield point and plastic stress-strain behaviour. The thermal expansion coefficient is measured as a function of temperature using the thermal-mechanical analysis device TMA 2940 from TA Instruments, New Castle, USA. The temperature-dependent Young's modulus, yield point and plastic stress-strain behaviour are determined with the universal testing machine type Z150 from Zwick GmbH & Co. Kg, Ulm, Germany.

Since an exact quantitative determination of residual stresses in the weld seam is not possible with current measurement methods, the validation is performed on the basis of the combined consideration of residual stresses and weld seam strength.

3.3 Calculation of the Weld Strength

The calculation of the weld strength is implemented according to the definition of the *degree of healing* $D_h(t)$ as per *Bastien and Gillespie* [2], which is based on the work of *de Gennes* [8]:

$$D_h(t) = \frac{\sigma}{\sigma_\infty} = \sum_{i=0}^{t/\Delta t} \left(\frac{{t_{i+1}}^{1/4} - {t_i}^{1/4}}{t_{rep}(T)^{1/4}} \right) \tag{1}$$

σ represents the weld strength in [MPa], σ_∞ the tensile strength of the base material in [MPa], t_i and t_{i+1} the start and end time of a time increment in the thermomechanical simulation in [s] and $t_{rep}(T)$ the temperature-dependent reptation time in [s]. The reptation time is defined by the approach of *Juhl, Christiansen and Jensen* [20]:

$$t_{rep}(T) = \frac{20M_e}{\pi^2 RT\rho(T)} \eta_0(T) \tag{2}$$

In this equation, M_e denotes the entanglement molecular weight of the polymer in [kg/mol], R the universal gas constant in [J/mol K], T the temperature in [K], $\rho(T)$ the temperature-dependent melt density in [kg/m^3] and $\eta_0(T)$ the temperature-dependent zero-shear viscosity, which is calculated via the Williams–Landel–Ferry (WLF) equation [29].

The equations are implemented in the numerical computing software MATLAB 2016a from MathWorks, Inc., Natick, USA. A routine was developed that calculates the zero-shear viscosity and density for every time increment of the welding process as a function of temperature based on the data resulting from the thermomechanical simulation model. Both values are used to calculate the temperature-dependent reptation time at every examined time increment. Subsequently, an individual degree of healing is calculated for each time increment. By summing all calculated individual degrees of healing, D_h is calculated, which is then multiplied by the tensile strength of the base material σ_∞ to calculate the final bond strength σ. In the last step, the residual stresses in the tensile test direction after the cooling of the welding partners are imported from the thermomechanical simulation and subtracted from the tensile strengths calculated with the reptation model.

The validation is based on tensile tests with welded T-joints. The tensile tests are carried out with a universal testing machine type Z10 from Zwick GmbH & Co. Kg, Ulm, Germany. The breaking force of the welded T-joints is measured and converted into a breaking stress using the microscopically determined weld area. The experimentally determined breaking stresses are then compared to the calculated weld strength.

4 Results

4.1 Thermal Simulation

Figure 3 shows the HAZ for different process points. On the left side, the microscopic image of an experimental weld is shown. This is compared to the temperature distribution in the section plane at the change-over point from the heating to the cooling phase from the simulation with the same process parameters. In sum, there is a very good agreement between the simulation and experimental welds for the considered process points.

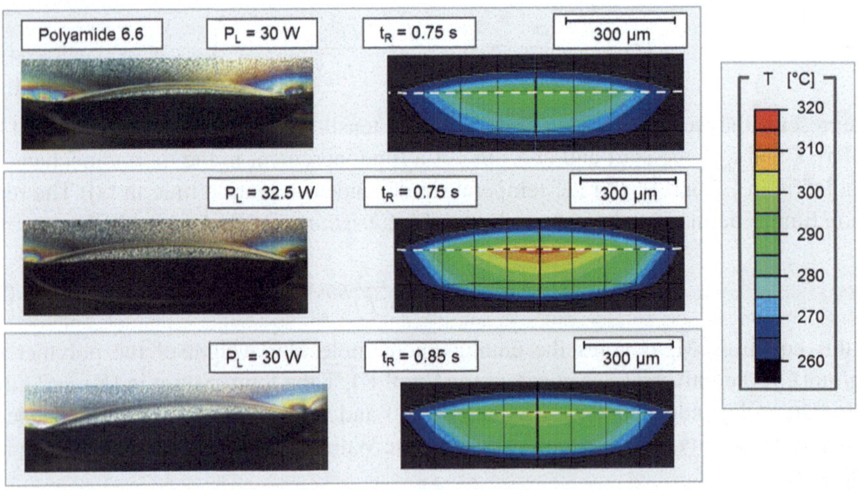

Fig. 3. Validation of the thermal simulation model based on the optical appearance of the HAZ

A comparison of the HAZ from the experiment and simulation for all considered process points can be found in Fig. 4. Starting from a weld with an irradiation time of 0.75 s and a laser power of 30 W, at which the welding partners just adhere, the laser power and irradiation time were increased gradually until degradation phenomena occur during the experimental welds (starting with a laser power of 35 W with a constant irradiation time of 0.95 s or an irradiation time of 1 s with a constant laser power of 27.5 W). In the simulation, temperatures above the degradation temperature of the material are also calculated for these process points. Since the degradation of material is not considered in the simulation, starting with these process points lead to differences between experimental and calculated heat-affected zones, which increase with higher energy input by the laser. However, for the process points where no degradation occurs the simulation shows a very good agreement with experimental welds.

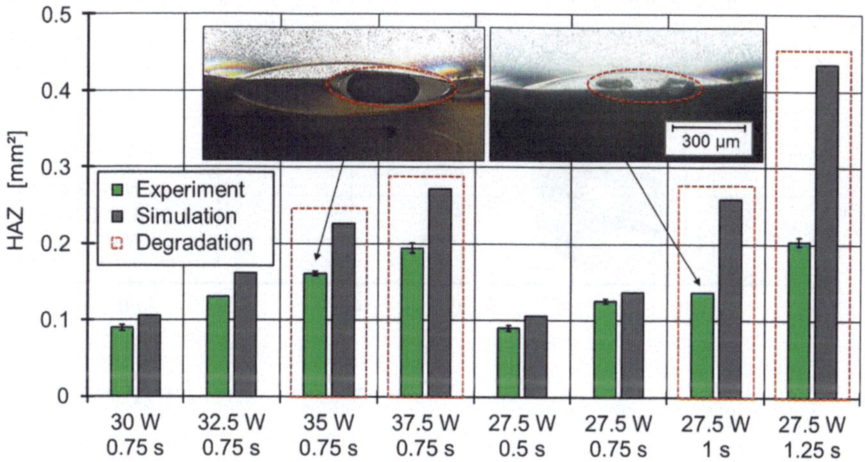

Fig. 4. Comparison of the HAZ from the simulations with experimental welds

4.2 Thermomechanical Simulation

Figure 5 shows the time course of the calculated, time-dependent and signed Von Mises stresses over heating and cooling phases in the welding process. The three curves show the progress of the stress in the middle node between the two welding partners for the laser powers of 30 W, 32.5 W and 35 W with an irradiation time of 0.75 s. In addition, the stress distributions in both joining partners are shown at different times in the welding process for a laser power of 32.5 W and an irradiation time of 0.75 s. Due to thermal expansion of the material during the heating process, compression stresses form in the weld seam, which are mechanically compensated in the environment by tensile stresses. When cooling begins, the deformed material and the formed compression stresses are reset. However, due to the partly irreversible deformation of the material, there is no complete recovery. This means that, after the material has cooled down to ambient temperature, tensile stresses form in the weld seam, which are compensated by compression stresses in the vicinity of the weld seam. The compression stresses occurring in the weld seam during heating and the resulting tensile stresses after cooling increase as expected with increasing laser power and thus increasing temperature and thermal expansion. Even if no validation of the pure residual stresses in the weld seam is possible with current means, the experimentally determined residual stresses of injection moulded components with values between 5–10 MPa provide a rough indication [13, 15]. Since slightly higher heating and cooling rates are present in laser transmission welding, the calculated residual stresses of 15–25 MPa for PA 6.6 are generally of a realistic order of magnitude.

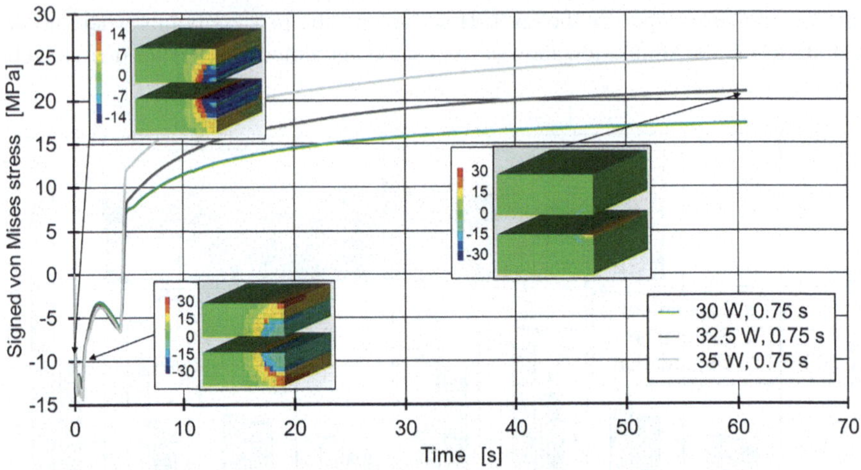

Fig. 5. Signed Von-Mises stresses for an irradiation time of 0.75 s with laser powers of 30 W, 32.5 W and 35 W

4.3 Calculation of the Weld Strength

Figure 6 shows the comparison of the weld strength from the simulations with experimental welds for different process points. The horizontal, green line marks the tensile strength of the base material. Increasing laser power and irradiation times generally lead to a longer time of the material being in the molten state, while at the same time shorter reptation times are present. This results in increasing calculated weld strengths in the simulation. The degradation phenomena already described in Sect. 4.1 lead to a decrease or no further increase in the experimentally determined weld strength at the corresponding process points. Since these phenomena are not taken into account in the reptation model, this effect is not apparent in the calculated simulation values. Particularly at the process point with an irradiation time of 1.05 s and a laser power of 30 W, a clear deviation between experimental and simulative results can be seen. Even if the calculation results from the simulation slightly underestimate the weld strength in real welds, there is a relatively good agreement between simulation and experiment.

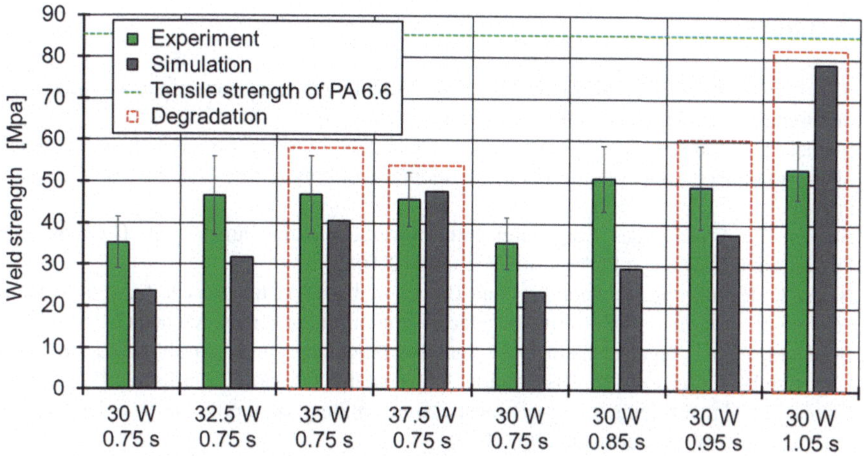

Fig. 6. Comparison of the weld strength from the simulations with experimental welds

5 Conclusion and Outlook

In the presented study, a 3D model for the thermal simulation of laser transmission welding was developed. For polyamide 6.6, the comparison of heat-affected zones in the simulation with heat-affected zones determined via microscopic images of experimental welds shows that the thermal model leads to realistic calculation results. Based on the thermal model, a further simulation model was developed to calculate the thermally induced residual stresses caused by the welding process. For this purpose, temperature-dependent elastic-plastic mechanical material behaviour was used, which leads to results in a realistic order of magnitude. In parallel, a routine for the calculation of the weld strength was developed on the basis of the thermal simulation model, which considers the degree of healing according to the reptation model combined with calculated residual stresses. In general, the calculated strengths for the considered material show a good agreement with the strengths of experimental welds. However, since the decomposition of the material is not taken into account in the simulation model, deviations from real welds occur at process points where degradation phenomena are present.

In forthcoming research, the model will first be transferred to other materials in addition to polyamide 6.6. Furthermore, a coupled model for flow simulation is to be developed in order to take into account the melt flow during the welding process.

Acknowledgement. The depicted research has been funded by the Deutsche Foschungsgesellschaft (DFG) as part of the research project "Integrative calculation of the weld strength of plastics parts based on an interdiffusion model presented for laser transmission welding." We would like to extend our gratitude to the DFG.

References

1. Acherjee, B., Kuar, A.S., Mitra, S., Misra, D.: Modeling of laser transmission contour welding process using FEA and DoE. Opt. Laser Technol. **44**(5), 1281–1289 (2012)
2. Bastien, L.J., Gillepie Jr., J.W.: A non-isothermal healing model for strength and toughness of fusion bonded joints of amorphous thermoplastics. Polym. Eng. Sci. **31**(24), 1721–1730 (1991)
3. Becker, F.: Einsatz des Laserdurchstrahlschweißens zum Fügen von Thermoplasten. Dissertation, Universität Paderborn, Paderborn (2003)
4. Beiss, T.: Einführung, Technologie- und Branchenüberblick. In: Proceeding of Kunststoffe erfolgreich verbinden – Innovative Fügetechnologien für die Praxis. Aachen (2016)
5. Bonefeld, D.: Eigenspannungen, Spaltüberbrückung und Strahloszillation beim Laserdurchstrahlschweissen. Dissertation, Universität Paderborn, Paderborn (2012)
6. Chen, M.: Gap bridging in laser transmission welding of thermoplastics. Dissertation, Queen's UniversityOntario, Ontario (2009)
7. Coelho, J.P., Abreu, M.A., Pires, M.C.: High-speed laser welding of plastic films. Opt. Lasers Eng. **34**(10), 385–395 (1991)
8. De Gennes, P.G.: Reptation of a polymer chain in the presence of fixed obstacles. J. Chem. Phys. **55**(1), 572–579 (1971)
9. Fargas, M., Wilke, L., Meier, O., Potente, H.: Analysis of weld seam quality for laser transmission welding of thermoplastics based on fluid dynamical processes. In: Proceedings of the 65th Annual Technical Conference (ANTEC). Cincinatti (2007)
10. Fiegler, G.: Ein Beitrag zum Prozessverständnis des Laserdurchstrahlschweißens von Kunststoffen anhand der Verfahrensvarianten Quasi-Simultan- und Simultanschweißen. Dissertation, Universität Paderborn, Paderborn (2007)
11. Frick, T.: Untersuchung der prozessbestimmenden Strahl-Stoff-Wechselwirkungen beim Laserstrahlschweißen von Kunststoffen. Dissertation, Friedrich-Alexander-Universität Erlangen-Nürnberg, Erlangen (2007)
12. Grewell, D., Benatar, A.: Semiempirical, squeeze flow, and intermolecular diffusion model. II. Model verification using laser microwelding. Polym. Eng. Sci. **48**(8), 1542–1549 (2008)
13. Guevara-Morales, A., Figueroa-Lopez, U.: Residual stresses in injection molded products. J. Mater. Sci. **43**(13), 4399–4415 (2014)
14. Gupta, S.K., Pal, P.K.: Analysis of through transmission laser welding of nylon6 by finite element simulation. Manag. Prod. Eng. Rev. **9**(4), 56–69 (2018)
15. Hastenberg, C.H.V., Wildervanck, P.C., Leenen, A.J.H., Schennink, G.: The measurement of thermal stress distributions along the flow path in injection-molded flat plates. Polym. Eng. Sci. **32**(7), 506–515 (1992)
16. Hopmann, Ch., Bölle, S., Kreimeier, S.: Modeling of the thermally induced residual stresses during laser transmission welding of thermoplastics. Weld. World **63**, 1–13 (2019)
17. Ilie, M., Kneip, J.C., Mattei, S., Nichici, A., Roze, C., Girasole, T.: Through-transmission laser welding of polymers – Temperature field modeling and infrared investigation. Infrared Phys. Technol. **51**(1), 73–79 (2007)
18. Jänecke, M.: Leichtbau mit technischen Textilien. Kunststoffe **105**(2), 26–30 (2015)
19. Jones, I.: Laser welding of plastic components. Assem. Autom. **22**(2), 129–135 (2002)
20. Juhl, T.B., Christiansen, J.D., Jensen, E.A.: Investigation on high strength laser welds of polypropylene and high-density polyethylene. J. Appl. Polym. Sci. **1289**(5), 2679–2685 (2013)
21. Klein, H.M.: Laserschweißen von Kunststoffen in der Mikrotechnik. Dissertation, RWTH Aachen, Aachen (2001)

22. Kreimeier, S.: Thermische Simulation des Laserdurchstrahlschweißprozesses von teilkristallinen Thermoplasten. Dissertation, RWTH Aachen, Aachen (2017)
23. Lakemeier, P., Schoeppner, V.: Simulation-based investigation of the temperature influence during laser transmission welding of thermoplastics. In: Proceedings of the 75th Annual Technical Conference (ANTEC), Anaheim (2017)
24. Labeas, G.N., Moraitis, G.A., Katsiropoulos, Ch.V: Optimization of laser transmission welding process for thermoplastic composite parts using thermo-mechanical simulation. J. Compos. Mater. **44**(1), 113–130 (2010)
25. Mayboudi, L.S.: Heat transfer modelling and thermal imaging experiments in laser transmission welding of thermoplastics. Dissertation, Queen's University Ontario, Ontario (2008)
26. Mayboudi, L.S., Birk, A.M., Zak, G., Bates, P.J.: Infrared observations and finite element modeling of a laser transmission welding process. J. Laser Appl. **21**(3), 111–118 (2009)
27. Messner, R.W.: Joining composite materials and structures: some thought-provoking possibilities. J. Thermoplast. Compos. Mater. **17**(1), 51–75 (2004)
28. N.N.: PlasticsEurope Annual Review 2017–2018. Annual report, Plastics Europe AISBL, Brussels (2018)
29. Osswald, T.A.: Rudolph, N.: Polymer rheology – fundamentals and applications, 1st edn. Hanser, München (2015)
30. Potente, H., Wilke, L., Ridder, H., Mahnken, R., Shaben, A.: Simulation of the residual stresses in the contour laser welding of thermoplastics. Polym. Eng. Sci. **48**(4), 767–773 (2008)
31. Reinl, S., Rau, A.: Laserkunststoffschweißen in der industriellen Serienproduktion. Laser Mag. **28**(2), 17–21 (2011)
32. Sparks, J.A.: Low cost technologies for aerospace applications. Microprocess. Microsyst. **20**(8), 449–454 (1997)
33. Wang, C., Yan, T., Liu, H., Zhong, H.: Temperature field and fluid field simulation of laser transmission welding polycarbonate. In: Proceedings of the 76th Annual Technical Conference (ANTEC), Orlando (2018)
34. Zoubeir, T., Elhem, G.: Numerical study of laser diode transmission welding of a polypropylene mini-tank: temperature field and residual stresses distribution. Polym. Test. **30**(1), 23–34 (2011)

Permeation Properties of Laser-Sintered Polyamide 12 Sheets in Comparison to an Extruded Polyamide 12 Film

Anna Liebrich[1,3,4(✉)], Horst-Christian Langowski[1,3,4], Regina Schreiber[2,4], and Bernd R. Pinzer[2,5]

[1] Chair of Food Packaging Technology, Technical University of Munich, 85350 Freising, Germany
anna.liebrich@ivv.fraunhofer.de
[2] University of Applied Sciences Kempten, 87435 Kempten, Germany
[3] Fraunhofer Institute of Process Engineering and Packaging IVV, 85354 Freising, Germany
[4] Kompetenzzentrum für angewandte Forschung in der Lebensmittel- und Verpackungstechnologie (KLEVERTEC), 87437 Kempten, Germany
[5] Laboratory for Optical 3D Metrology and Computer Vision, University of Applied Sciences Kempten, 87435 Kempten, Germany

Abstract. Laser sintering of polymers is widely used for the production of individual products and small-batch series. However, the qualification of laser-sintered polymeric components for new application fields, e.g. in the food and packaging industry, is still limited due to missing knowledge on physicochemical material properties. This work investigates the mass transfer of low molecular weight substances through laser-sintered polyamide 12 sheets in comparison to an extruded polyamide 12 film. Analysis of structural material properties reveals significant differences between both materials depending on the production processes. Despite their apparent porosity, laser-sintered sheets show lower permeation coefficients for water vapor and oxygen compared to extruded films. This might be related to the higher crystallinity of the laser-sintered vs. the extruded material, arising from the slow cooling rate of the polymer in the laser sintering process.

This research brings knowledge of the different permeation behavior of laser-sintered and extruded polyamide 12 in relation to the structural properties.

Keywords: Laser sintering · Permeation properties · Porosity · Crystallinity · Crystal structure · Polyamide 12

1 Introduction

Additive manufacturing (AM) technologies offer the potential to fabricate individualized products with increased part complexity at low production volumes. Laser sintering (LS) of polymers is an AM technique that is established for direct manufacturing

C. Hopmann and R. Dahlmann (Eds.): *Advances in Polymer Processing 2020*, pp. 258–269, 2020. https://doi.org/10.1007/978-3-662-60809-8_21

of products and small-batch series in many industries, mainly automotive, aerospace and biomedical sectors [1, 2]. In the food and packaging industry, plastics are commonly used for products manufactured in large quantities for which LS parts are not suitable. Nevertheless, there is a potential for using LS parts primarily for the individualization of products, for the production of small-batches and also for various grippers. As an example, current interest exists in the confectionery industry for the production of casting molds. However, the application of the technology remains limited due to insufficient knowledge of structural properties and functional performance of LS materials and components [3]. In particular, a deeper understanding of the physicochemical material behavior is missing so far. Until now, there is little information available on the mass transport in LS polymers and the resultant barrier properties towards gases and vapors.

In the LS process, a thermoplastic polymer powder is selectively fused by means of laser energy. As the production technique works free of pressure and the surrounding powder is used to hold the melt, LS polymers present different structural properties compared to traditional manufacturing technologies like injection molding [4, 5]. Typical material characteristics include porosity as well as an inhomogeneous large spherulitic crystal structure of the polymer matrix [3]. The mass transport of low molecular weight substances through polymers, besides their chemical structure, strongly depends on the physical material properties such as density or crystallinity [6]. Therefore, transferring material parameters from e.g. extruded polymer films might only be possible to a limited extent. Moreover, only a little research has been done on the permeation properties of polyamide 12 (PA12), which is the mostly applied LS powder material [7].

The main objective of the research was to examine the barrier properties against gases and vapors of LS PA12 in comparison to extruded PA12. Water vapor and oxygen permeation experiments were carried out on disc-shaped samples produced by both manufacturing techniques using the same PA12 powder as starting material. For an improved understanding of the transport behavior, the structural properties of different materials were analyzed by means of X-ray micro tomography (CT), differential scanning calorimetry (DSC) and polarized light microscopy.

2 Basic Permeation Theory

The transport of gases and vapors through a polymeric layer is basically described by the model of solution-diffusion mechanism which involves the adsorption and solution of the penetrating substances at the surface of higher concentration, their diffusion through the polymer layer under the effect of concentration gradients and their desorption on the side of lower concentration [8].

The permeation rate, i.e. the flux density J [mol m^{-2} s^{-1}] in the steady-state is given by

$$J = Q(p_1 - p_2) = \frac{DS(p_1 - p_2)}{d} = \frac{P(p_1 - p_2)}{d} \tag{1}$$

where Q is the permeability [mol m−^{2} s^{-1} Pa−^{1}], p_1 and p_2 are the upstream and downstream pressures [Pa], d is the thickness of the polymer substrate [m], D is the diffusion coefficient [m^2 s−^{1}], S is the Henry's law solubility coefficient [mol m^{-3} s−^{1} Pa^{-1}] and P is the permeation coefficient [mol m m−^{2} s^{-1} Pa−^{1}] [9, 10].

3 Materials and Methods

3.1 Production of Samples

Laser Sintering. Disc-shaped sheets with a nominal diameter of 70 mm and a nominal thickness of 1000 μm were produced on a EOS (EOS Electro Optical Systems GmbH, Krailling, Germany) P110 LS machine using PA2200 PA12 powder supplied by EOS GmbH with a mixing ratio 50/50 between virgin and recycled powder (particle volume median diameter = 57 μm). The sheets were built in vertical direction (z-direction) and positioned close to each other in the building chamber (nominal distance = 3 mm) in order to eliminate process-related variabilities in the material properties caused by e.g. the inhomogeneous temperature distribution across the building platform [11]. The standard process parameters recommended by the LS machine manufacturer EOS GmbH for PA2200 PA12 powder at 100 μm layer thickness were applied. The scanning strategy was an alternating x-y scanning pattern including two contour lines and enclosed raster lines ("hatch lines"). Due to the fixed hatch grid of the LS system, the sheets were produced with different scanning patterns depending on their positioning within the building chamber. Figure 1 displays magnified sections of the different x-y scanning patterns applied. After the building process, the sheets were finished by bead blasting in order to remove the remaining powder material.

Cast Film Extrusion. In order to produce starting material for the cast film extrusion process PA2200 powder was converted to small granules using a co-rotating twin-screw compounder ZK 25 T (screw diameter (D) = 25 mm, screw length = 24 D) with a rotational speed of 100 rpm and a mass temperature of 210 °C, followed by a strand pelletizer CSG 171 T (both COLLIN Lab & Pilot Solutions GmbH, Maitenbeth, Germany). The granules were dried at 80 °C for 24 h to remove any residual moisture before processing on a lab-scale cast film line CR 136/350 with an extruder E 30 P (screw diameter (D) = 30 mm, screw length = 30 D, nozzle width = 300 mm) (COLLIN Lab & Pilot Solutions GmbH, Maitenbeth, Germany). The extrusion parameters were a rotational speed of 90 rpm and a mass temperature of 210 °C. The chill roll unit was kept at a haul-off speed of 0.5 m/min resulting in a final film thickness of approximately 500 μm. For the permeation experiments, disc-shaped samples with a diameter of 70 mm were manually cut out of the produced film.

3.2 Analysis of the Material Structure

X-ray Micro Tomography (CT). CT was used to investigate the porosity of the LS sheets. As the extruded material showed a macroscopically homogeneous structure and no pores were detectable, it was excluded from the CT analysis. One LS sheet

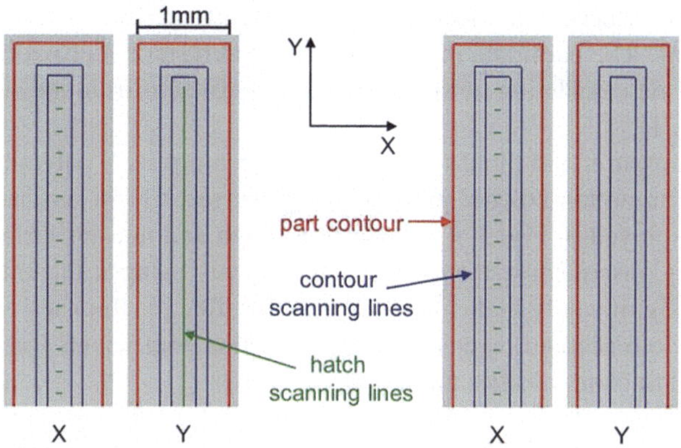

Fig. 1. Magnified sections of different x-y scanning patterns used for laser sintering of sheets with a nominal thickness of 1000 μm in vertical direction.

was randomly selected out of the entire batch produced and a rectangular sample with dimensions (25 × 10) mm^2 was manually cut out from the middle of the sheet. CT scans were performed using a Scanco μCT 40 desktop device with an acceleration voltage of 70 keV, a current of 57 μA and 1000 projections recorded per 180° sample rotation. These settings resulted in a voxel size of $(6\ \mu m)^3$. Post-processing of the raw data and subsequent data analysis was conducted following the method described in [12]. The presented results, including overall porosity and pore number density of the sample, as well as the equivalent pore diameter of individual pores, refer to a volume of interest (VOI) of 8.384 mm^3 (= (500 × 500) voxels times the thickness of the sample). This allows the assumption that values are representative of the entire volume of the sheet [12].

Differential Scanning Calorimetry (DSC). The melting and crystallization transitions of different sample materials were studied by means of DSC using a DSC 3+ STARe System (Mettler-Toledo GmbH, Gießen, Germany) under nitrogen atmosphere. Heating and cooling cycles were performed between 23 °C and 230 °C with a rate of 10 K/min. For each material investigated, the DSC analysis was conducted on three individual samples with masses of (8 ± 0.5) mg. Regarding the LS sheets, powder particles attached to the surfaces of the samples were carefully removed with a scalpel tool. The melting enthalpy of the samples was derived from the first heating cycle using a linear baseline from 120 °C to 200 °C. The crystallinity was determined as the ratio between the melting enthalpy of the samples and a theoretical melting enthalpy of 209.3 J g^{-1} for 100% crystalline PA12 [13].

Polarized Light Microscopy. The crystal structure of different materials was examined by means of transmission light microscopy under polarized light. Pictures were taken from thin sections of 5 μm thickness using a Leitz Diaplan with a camera type DFC 295 (Leica Microsystems, Wetzlar, Germany).

3.3 Permeation Experiments

Thickness Measurements. Thickness measurements were conducted with a precision thickness gauge FT3 (Hanatek, St. Leonards-on-Sea, UK) at 23 °C and 50% r.h.

For each sample, a harmonic mean value was calculated from five measurements at random locations. The harmonic mean has been chosen since it provides the appropriate average for the calculation of the permeation coefficient from the steady-state flux density (see Eq. 1).

Due to the pressure- and mold-less production technique LS polymer material generally presents unmolten powder particles at the surfaces [14]. As can be seen from the CT slice displayed in Fig. 2, this leads to a rough and uneven surface structure of the LS sheets. Consequently, thickness measurements using a thickness gauge give results that are significantly higher compared to the effective thicknesses of the sheets. This was taken into account by calculating permeation coefficients for all LS samples with the gauged thickness values minus 120 μm [15].

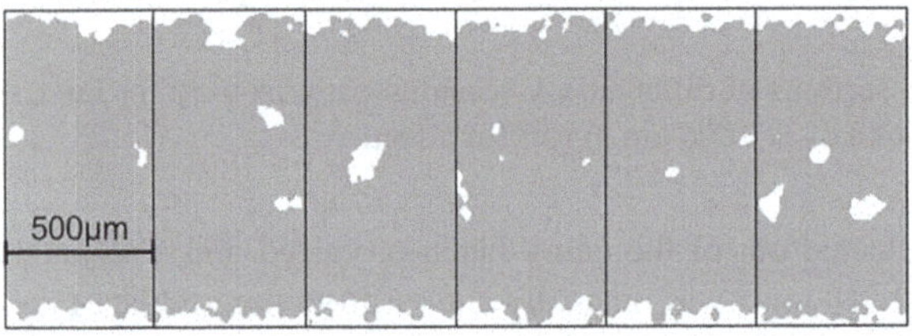

Fig. 2. CT slice of a laser-sintered polyamide 12 sheet with a nominal thickness of 1000 μm built in vertical direction.

Sample Preparation. In order to eliminate leakages and undesired side diffusion effects during permeation measurements, the LS samples, containing rough surfaces, were tightly embedded between two sheets of a self-adhesive aluminum foil with a circular cut-out area of 25 cm^2. In addition, a two-component adhesive was applied along the border of the cut-out area to ensure that no gases could pass through the interface between aluminum foil and the sheets (Fig. 3). The extruded samples were prepared accordingly for comparable measurement conditions.

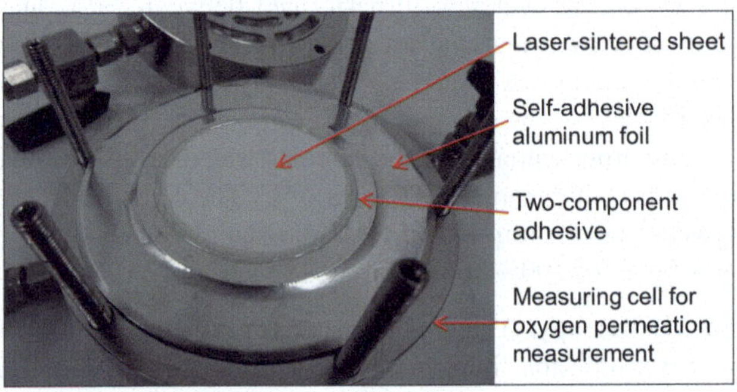

Fig. 3. Sample preparation for permeation measurements.

Water Vapor Permeation Measurements. The water vapor transmission rate (WVTR) was measured gravimetrically in accordance with DIN 53122-1 at 23 °C and a humidity gradient of 85 → 0% r.h. For normalization to a pressure difference of 1 Pa, the results were divided by the water vapor partial pressure at 85% r.h. (= 2391.8 Pa).

Oxygen Permeation Measurements. The oxygen permeability was determined with an optical measurement method according to DIN 53380-5 at 23 °C and 50% r.h. using measurement equipment from PreSens (Precision Sensing GmbH, Regensburg, Germany). To reduce the amount of oxygen dissolved in the samples, the LS sheets were stored in the measuring cells under nitrogen atmosphere several days before starting with the standard measurement procedure [16].

4 Results

4.1 Porosity

Figure 4a shows the local distribution of different pore sizes in the LS sheet as revealed by CT analysis. It can be recognized that the pores are mainly located in the inner region of the sheet surrounded by a dense skin. This might primarily arise from

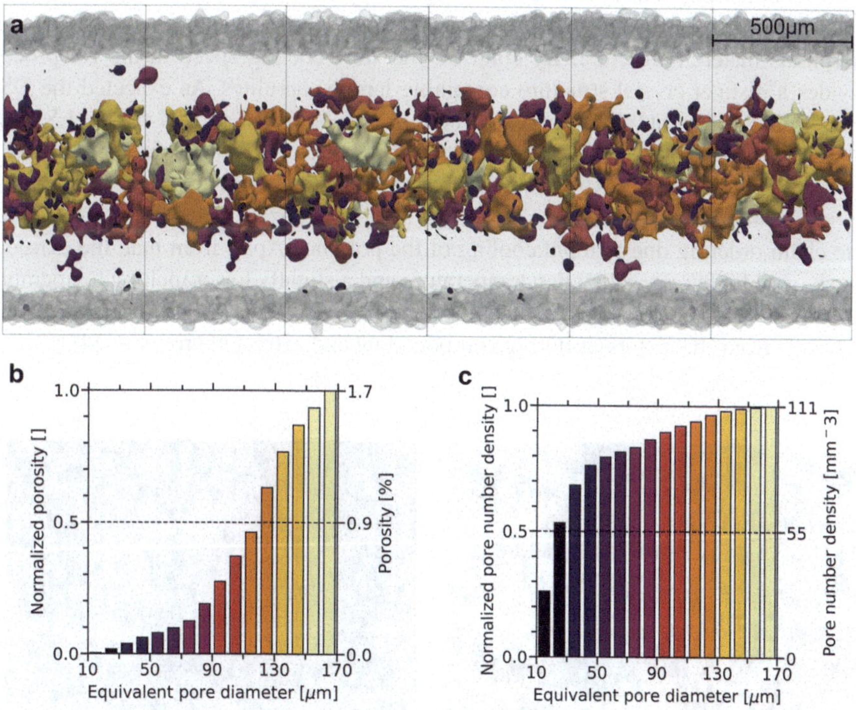

Fig. 4. Semi-transparent view of visualized CT data through a 3 mm (= 500 voxels) wide cross-section **a**), cumulative pore size distributions of porosity **b**) and pore number density **c**) within this section, for a laser-sintered polyamide 12 sheet with a nominal thickness of 1000 µm built in vertical direction. (Overall porosity: the ratio of pore volume over the entire sample volume, Pore number density: the number of single pores divided by the sample volume, Equivalent pore diameter: the diameter of an (imaginary) sphere of the same volume as the pore [12]).

the scanning strategy used to produce the sheets including different exposure parameters for contours and hatch scanning (see Fig. 1). Apart from that, the local porosity distribution might further be influenced by different temperature gradients and thus different cooling rates comparing different areas of the sheet [12, 17].

As it can be derived from Fig. 4b, c, the investigated LS sample shows an overall porosity and a pore number density of detected pores of 1.7% and 111 pores per mm^3, respectively. Pores are in a size range up to an equivalent pore diameter of 167 μm (= 0.0024 mm^3 pore volume) with a volume median of 123 μm and a number median of 29 μm.

4.2 Crystallinity and Crystal Structure

From microscopic pictures displayed in Fig. 5, it becomes visible that the LS sheets and the extruded film show significantly different crystal structures which might be linked to the different processing conditions of the material. The dark spots observed in both pictures could be small particles of titanium dioxide and silica that are added to the powder to improve whiteness and flowability [18–20]. In the cast film extrusion process the molten polymer is cooled down rapidly by chill rolls immediately after being ejected from the extrusion nozzle. This gives low crystal growth leading to a more amorphous material structure. By contrast, the slow cooling speed in the LS process provides a distinct crystal structure containing large spherulites. As expected the DSC analysis gives different crystallinities for both materials which are 38% for the LS PA12 and 22% for the extruded PA12 (see Table 1). From DSC curves displayed in Fig. 6, it can be observed that the extruded material provides an exothermic peak at 161 °C in the first heating run which can be described as recrystallization and signalizes incomplete chain ordering due to rapid cooling of the polymer. Apart from that, the extruded film shows a higher crystallization temperature and a slightly lower melting temperature compared to the LS sheet indicating a shorter molecular chain length that most probably results from shear stresses during compounding and extrusion processes [4].

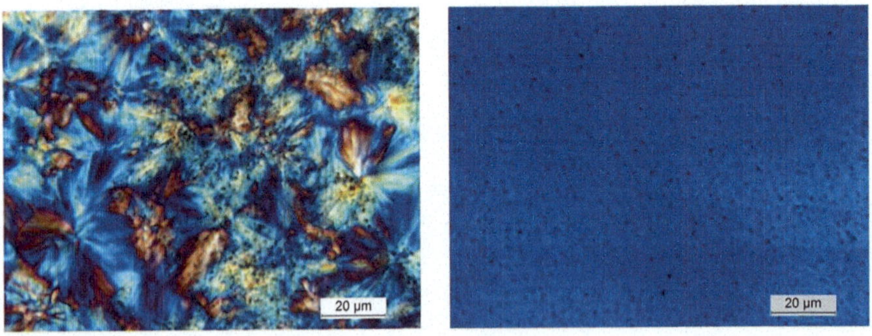

Fig. 5. Thin section polarized light microscopy pictures from samples of laser-sintered (left) and extruded (right) polyamide 12.

Compared to the LS and the extruded material, the PA12 powder presents a higher melting temperature and higher melting enthalpy in the first heating run which can be explained by a special modification of the crystalline morphology of the polymer during powder production by solution-precipitation process [21]. From Fig. 6, it can be recognized that the DSC curve of the LS sample shows a decent shoulder at a temperature corresponding to the melting temperature of the powder. Since unmolten powder particles adherent to the surfaces of the LS sheets were removed before DSC measurements, this points out that partially molten powder cores are present in the inner polymer structure of the LS material [22].

Table 1. DSC data of different polyamide 12 materials. Values are the average of three individual measurements at the 95% level of confidence.

	PA12 powder	Laser-sintered PA12	Extruded PA12
Recrystallization temperature [°C]	–	–	161.0 ± 0.4
Melting temperature [°C]	186.6 ± 0.4	180.9 ± 0.9	178.3± 1.1
Crystallization temperature [°C]	147.2 ± 0.4	147.7 ± 3.1	150.2 ± 1.2
Melting enthalpy [J g^{-1}]	110.9 ± 1.2	80.2 ± 2.8	46.4 ± 0.5
Crystallinity [%]	53.0 ± 0.6	38.3 ± 1.3	22.1 ± 0.2

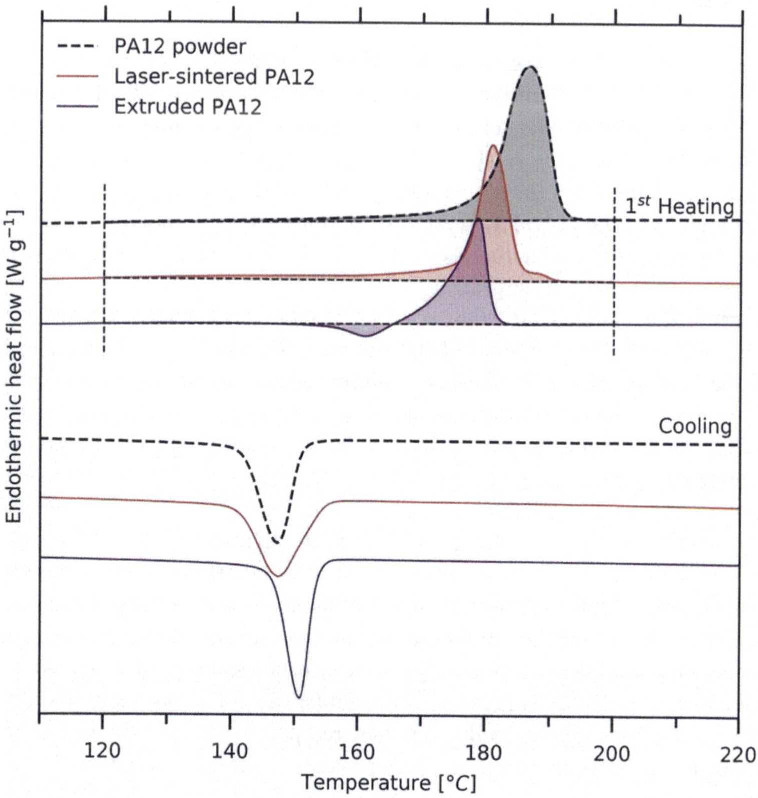

Fig. 6. DSC curves for samples of different polyamide 12 materials for the 1st heating and cooling run.

4.3 Water Vapor and Oxygen Permeation Properties

Table 2 summarizes permeation results for the LS sheets and the extruded film. Values mainly match with permeation data found in the literature for conventionally manufactured PA12 [23, 24]. Both materials present a ratio of water vapor to oxygen permeation coefficients of $\approx 1.9 \times 10^3$ while permeation coefficients for the extruded film are factor 1.6 higher compared to the LS sheets. No significant differences in the permeation properties were found between LS sheets produced with different scanning patterns which is in line with results reported for the mechanical material behavior of thin tensile test samples built in vertical orientation [25].

Table 2. Permeation data of different polyamide 12 materials. Values are the average of three individual measurements at the 95% level of confidence.

	Laser-sintered PA12	Extruded PA12
Water vapor permeation coefficient [mol m m^{-2} s^{-1} Pa^{-1}]	$(8.0 \pm 1.5) \times 10^{-14}$	$(1.3 \pm 0.1) \times 10^{-13}$
Oxygen permeation coefficient [mol m m^{-2} s^{-1} Pa^{-1}]	$(4.3 \pm 0.6) \times 10^{-17}$	$(6.9 \pm 0.4) \times 10^{-17}$

5 Discussion

LS and extruded PA12 show significantly different material structures resulting from the different processing techniques. Despite apparent porosity, the LS sheets show lower permeation coefficients for water vapor and oxygen compared to the extruded film. This might be primarily related to the higher crystallinity of the LS sheets vs. the extruded film, arising from the slow cooling in the LS process. At temperatures below the melting point of the polymer, crystalline regions are in general much less accessible to most penetrants, thus acting as excluded volumes and impermeable barriers for sorption and diffusion processes [26].

Regarding the material structure of the LS sheets, CT analysis reveals porosity mainly located in the inner region of the sheets, while only a few pores are detected close to the borders. However, the pore number density distribution shows very high numbers of small pores indicating that additional pores close to the detection limit of the CT analysis could also be present in significant amounts. Apart from that, the presence of unmolten powder cores in the material structure signalizes incomplete melting of the polymer due to insufficient energy input during sintering suggesting poor consolidation of the polymer [7]. This could lead to the hypothesis that LS sheets present a higher permeable material structure in which mass transfer would be mainly determined by the flow of substances through a continuous network of pores and microcavities rather than follow rules of solution-diffusion mechanism. However, the permeation behavior does not confirm this assumption for any of the samples investigated. Contrary to expectations, the LS sheets present barrier properties for water vapor and oxygen in the same order as for conventionally manufactured polymers typically used in technical and packaging applications (see Fig. 7). A possible explanation could be, that the LS sheets show a material behavior similar to a laminate structure with a more permeable porous inner layer surrounded by two denser layers of greater permeation resistance.

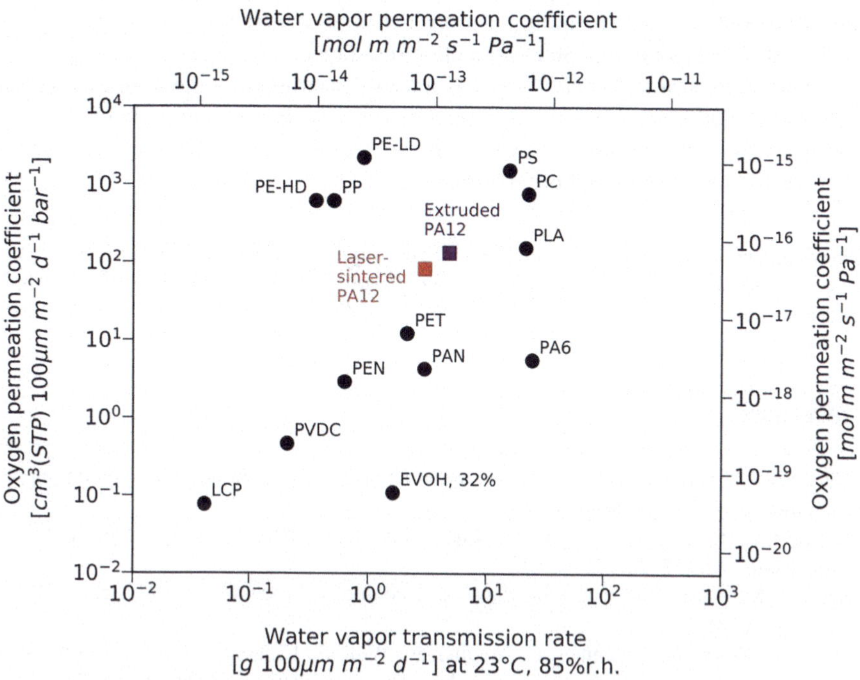

PA6	Polyamide 6	PLA	Polylactic acid
PAN	Polyacrylonitril	PP	Polypropylene
PC	Polycarbonate	PS	Polystyrene
PE-HD	High-density polyethylene	PVDC	Polyvinylidene chloride
PE-LD	Low-density polyethylene	EVOH	Ethylene-vinylalcohol copolymer (percentage: fraction of ethylene)
PEN	Polyethylene naphtalate	LCP	Liquid crystal polymer
PET	Polyethylene terephtalate		

Fig. 7. Water vapor and oxygen permeation properties of laser-sintered and extruded polyamide 12 in relation to conventionally manufactured polymers typically used in technical and packaging applications, at 23 °C (modified from [8]).

6 Conclusion

This work analyzed the permeation properties in relation to the material structures of LS and extruded PA12. Based on the main findings, the following conclusions can be drawn:

- LS and extruded PA12 exhibit significantly different material structures resulting from different production processes.
- Despite the porosity, LS PA12 shows lower permeation coefficients for water vapor and oxygen compared to extruded PA12 which might be related to the higher crystallinity arising from the slow cooling process.
- The permeation behavior of LS sheets does not indicate the presence of a continuous network of pores or microcavities. Based on the CT results, the transport behavior of LS sheets could be comparable to the permeation through a

multi-layered structure including a more permeable porous inner region enclosed by two denser layers of greater permeation resistance.

- Permeation experiments could serve as a useful instrument to analyze the inherent material characteristics of LS polymers and to verify their structural integrity.
- Future work is needed to study the permeation properties of LS sheets depending on their thickness and build orientation.

Acknowledgments. The authors would like to thank Monika Gessler from EOS Electro Optical Systems GmbH (Krailling, Germany) for her helpful advice and for providing the LS samples for this research.

References

1. Gebhardt, A., Hötter, J.-S.: Additive Manufacturing: 3D Printing for Prototyping and Manufacturing. Hanser, Munich (2016)
2. Goodridge, R., Tuck, C., Hague, R.: Laser sintering of polyamides and other polymers. Prog. Mater. Sci. **57**, 229–267 (2012)
3. Schmid, M.: Laser Sintering with Plastics: Technology, Processes, and Materials. Hanser, Munich (2018)
4. Wörz, A., Wudy, K., Drummer, D., Wegner, A., Witt, G.: Comparison of long-term properties of laser sintered and injection molded polyamide 12 parts. J. Polym. Eng. **38**, 573–582 (2018)
5. Van Hooreweder, B., Moens, D., Boonen, R., Kruth, J.-P., Sas, P.: On the difference in material structure and fatigue properties of nylon specimens produced by injection molding and selective laser sintering. Polym. Test. **32**, 972–981 (2013)
6. Brandrup, J., Immergut, E.H., Grulke, E.A., Abe, A., Bloch, D.R.: Polymer Handbook, 4th edn. Wiley, New York (1999)
7. Bourell, D.L., Watt, T.J., Leigh, D.K., Fulcher, B.: Performance limitations in polymer laser sintering. Phys. Procedia **56**, 147–156 (2014)
8. Langowski, H.C.: Permeation of Gases and Condensable Substances through Monolayer and Multilayer Structures. In: Piringer, O.G., Baner, A.L. (eds.) Plastic Packaging. Interactions with Food and Pharmaceuticals, pp. 297–347. Wiley-VCH, Weinheim (2008)
9. Barrer, R.M.: Diffusion in and Through Solids. University Press, Cambridge (1941)
10. Crank, J.: The Mathematics of Diffusion, 2nd edn. Clarendon Press, Oxford (1975)
11. Wegner, A.: Theory on the continuation of melting processes as basic requirement for a robust processing in laser sintering of thermoplastics (Diss.), Universitätsbibliothek Duisburg-Essen (2015)
12. Liebrich, A., Langowski, H.C., Schreiber, R., Pinzer, B.R.: Porosity distribution in laser-sintered polymeric thin sheets as revealed by X-ray micro tomography. Polym. Test. **76**, 286–297 (2019)
13. Gogolewski, S., Czerntawska, K., Gastorek, M.: Effect of annealing on thermal properties and crystalline structure of polyamides. Nylon 12 (polylaurolactam). Colloid Polym. Sci. **258**, 1130–1136 (1980)
14. Launhardt, M., Wörz, A., Loderer, A., Laumer, T., Drummer, D., Hausotte, T., Schmidt, M.: Detecting surface roughness on SLS parts with various measuring techniques. Polym. Test. **53**, 217–226 (2016)

15. Liebrich, A., Langowski, H.C., Schreiber, R., Pinzer, B.R.: Permeation properties of laser-sintered polyamide 12 sheets in relation to their material structure, to be published in Additive Manufacturing
16. Müller, K., Scheuerer, Z., Florian, V., Skutschik, T., Sängerlaub, S.: Comparison of test methods for oxygen permeability: optical method versus carrier gas method. Polym. Test. **63**, 126–132 (2017)
17. Rüsenberg, S., Schmidt, L., Schmid, H.: Mechanical and physical properties—A way to assess quality of laser sintered parts. In: Proceedings of the 22nd International Solid Freeform Fabrication Symposium, pp. 239–251 (2011)
18. Verbelen, L., Dadbakhsh, S., Van den Eynde, M., Kruth, J.-P., Goderis, B., Van Puyvelde, P.: Characterization of polyamide powders for determination of laser sintering processability. Eur. Polym. J. **75**, 163–174 (2016)
19. Meyer, K.R., Hornung, K.H., Feldmann, R., Smigerski, H.J.: Method for polytropically precipitating polyamide powder coating compositions where the polyamides have at least 10 aliphatically bound carbon atoms per carbonamide group. US Patent 4,334,056, 1982
20. Scholten, H., Christoph, W.: Use of a nylon-12 for selective laser sintering. US Patent 6,245,281, 2001
21. Schmid, M., Amado, A., Wegener, K.: Materials perspective of polymers for additive manufacturing with selective laser sintering. J. Mater. Res. **29**, 1824–1832 (2014)
22. Zarringhalam, H., Majewski, C., Hopkinson, N.: Degree of particle melt in Nylon-12 selective laser-sintered parts. Rapid Prototyp. J. **15**, 126–132 (2009)
23. Domininghaus, H.: Kunststoffe: Eigenschaften und Anwendungen (Plastics: Properties and Applications), 7th edn. Springer, Berlin (2008)
24. McKeen, L.W.: Permeability Properties of Plastics and Elastomers. William Andrew Publishing, Oxford (2012)
25. Tasch, D., Mad, A., Stadlbauer, R., Schagerl, M.: Thickness dependency of mechanical properties of laser-sintered polyamide lightweight structures. Addit. Manuf. **23**, 25–33 (2018)
26. Michaels, A., Parker Jr., R.: Sorption and flow of gases in polyethylene. J. Polym. Sci. **41**, 53–71 (1959)

Viscoelastic Modelling of Polymer Melts and Rubber Compounds

Walter Friesenbichler[1](✉), S. Stieger[1], R. Kerschbaumer[2], G. Berger-Weber[1], A. Neunhaeuserer[1], and E. Mitsoulis[3]

[1] Institute of Injection Moulding of Polymers, Montanuniversitaet, 8700 Leoben, Austria
Walter.Friesenbichler@unileoben.ac.at
[2] Polymer Competence Center Leoben GmbH, 8700 Leoben, Austria
[3] School of Mining Engineering and Metallurgy, National Technical University of Athens, 157 80 Zografou, Greece

Abstract. Thermoplastic melts and rubber compounds are viscoelastic fluids. They show a complex flow behavior, which is influenced by various factors such as polymer type, molar mass distribution, recipe, filler-filler network and in some cases wall slippage. Most of the state-of-the-art simulation software packages use viscous material models for the calculation of the flow field as well as pressure and temperature distribution, neglecting the viscoelastic nature of polymers. This simplification may lead to an underestimated pressure demand in injection molding simulation.

This contribution presents how to correctly measure viscosity data (shear and extensional viscosity) for thermoplastics and rubber compounds taking into account the pressure dependency of the viscosity and the influence of viscous dissipation in capillary rheometry at higher shear rates. Moreover, a guideline on how to best fit rheological data with the viscoelastic K-BKZ/Wagner model is outlined. Comparing CFD simulation results to experimental data, only the K-BKZ/Wagner model is able to correctly predict pressure losses of contraction flow dominated geometries. Examples will be given for NBR and PP-PNC.

Keywords: Polymer melt · Viscoelasticity · K-BKZ/Wagner model · Injection pressure

1 Introduction

Thermoplastics and rubber compounds exhibit a viscoelastic material behavior. When simulating the flow of such complex fluids, the true material behavior is usually not taken into account. Instead, generalized Newtonian fluid flow models are used to solve the conservation equations (mass, momentum and energy). This simplification may lead to deviations between experiment and simulation, especially when predicting the filling pressure in rubber injection molding. Fasching [1] showed for a NBR rubber compound with a hardness of ShoreA 75, that the injection pressure is clearly

C. Hopmann and R. Dahlmann (Eds.): *Advances in Polymer Processing 2020*, pp. 270–282, 2020. https://doi.org/10.1007/978-3-662-60809-8_22

underestimated using commercial injection molding simulation software for a thick-walled hollow tube (Fig. 1). In this study the screw antechamber, machine nozzle, sprue and the part was considered applying a viscous Carreau-WLF model. Moreover, the dissipation produced in injection molding process was much higher than predicted by the simulation. While the measurement of bulk material ejected into air exhibit temperature rises of 12–30 K compared to the set melt temperature, Cadmould 3D-F® assumes no significant temperature rise resulting from viscous heating.

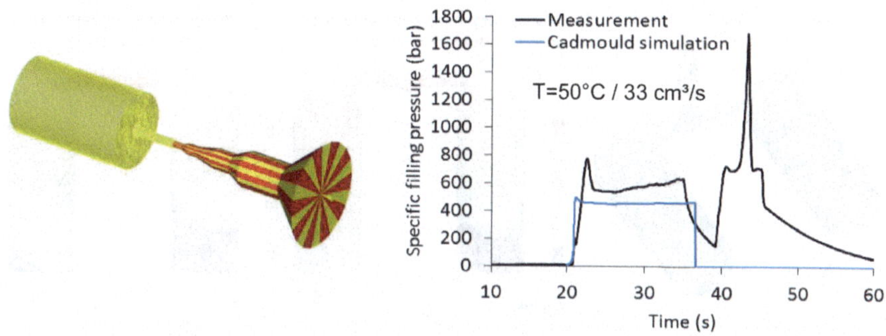

Fig. 1. CADMOULD 3D-F® geometry model of a thick-walled hollow tube (left) [1]; measured vs. predicted injection pressure (right) [19]

Additionally, purely viscous rheological models are not able to predict empirically proven flow phenomena, such as extrudate swell and vortices [2]. However, an exact numerical description of these viscoelastic flow phenomena is fundamentally important when designing extrusion tools, runner systems or complex injection molds, and is consequently subject of numerous research projects.

For unfilled polymers, various research groups had great success when using the Kaye-Bernstein-Kearsley-Zapas (K-BKZ) model to predict viscoelastic dominated flow phenomena. Ansari et al. [3, 4] managed to precisely predict the pressure drops in various capillaries both for a LDPE and an HDPE melt. Konaganti et al. [5] predicted extrudate swell for an HDPE melt well; he achieved remarkable results when including a crystallization kinetics model. Another benchmark problem was solved by Mitsoulis [6], who was able to numerically reproduce the experimentally observed flow patterns of a branched LDPE and a linear HDPE melt in an abrupt 20:1 contraction. Recently, Mitsoulis et al. [7] proved the applicability of the K-BKZ model to a wall slipping SBR rubber compound. For the first time, CFD simulations were carried out using a viscoelastic integral-type fluid flow model for a rubber compound, which provided an excellent prediction of measured pressure drops in an orifice die and in various injection molding dies. In [8] Mitsoulis et al. showed for a polypropylene nanocomposite with 5 wt.% organically modified montmorillonite (Nanofil 5) and 5 wt.% compatibilizer (SCONA TPPP 2112 GA) that viscous modelling failed to predict the real injection pressure. The measured pressure values for injecting through a short hyperbolical nozzle (L/D = 1.7, D = 1.5 mm) at a shear rate level of 500,000 s^{-1} was only matched when carrying out viscoelastic CFD simulations and considering the pressure dependence of viscosity.

Considering the pressure dependence of the viscosity is especially important when predicting the pressure demand for thin-walled parts or ones with long flow distances. Figure 2 [9] shows the geometrical model of a thin-walled bushing and the comparison of measured filling pressure to numerical predictions with and without considering the pressure dependence of the viscosity.

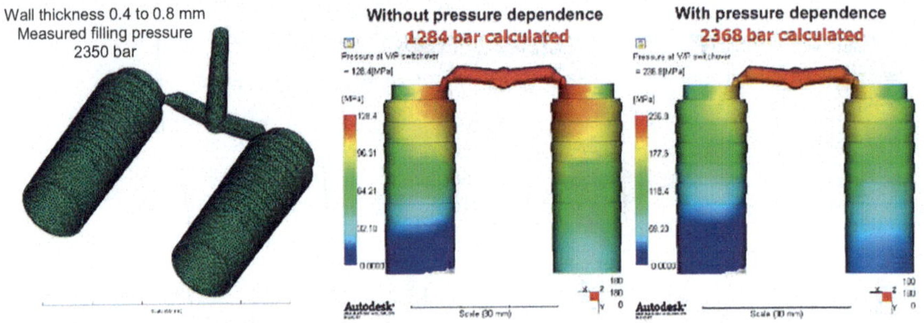

Fig. 2. Autodesk® Moldflow Insight geometry model and measured injection pressure for a thin-walled bushing (left); comparison of calculated injection pressure without (1284 bar, center) and with pressure dependent viscosity (2368 bar, right) for syndiotactic polystyrene [9]

The measured filling pressure of 2350 bar was only predicted correctly when taking the pressure dependence of the viscosity into account.

This study outlines how to fit the K-BKZ/Wagner model best for a highly filled rubber compound. Furthermore, measured pressure drops in an orifice die and one capillary are compared to viscoelastic (K-BKZ/Wagner), viscous (power-law) and viscoplastic (Herschel-Bulkley) CFD simulation results.

2 Experimental

2.1 Material

In the present study, an industrially used, carbon black filled Acrylonitrile Butadiene Rubber (NBR) with a hardness of ShoreA 75 was tested. At room temperature, the density is 1.21 g/cm^3. The compound was provided by our company partner SKF Sealing Solutions Austria GmbH, Judenburg, Austria. However, the exact composition must remain confidential.

2.2 Oscillatory and Capillary Rheology

Dynamic tests were carried out using a Rubber Process Analyzer (D-MDR3000) from MonTech Werkstoffprüfmaschinen GmbH. All measurements were performed in the linear viscoelastic region under sinusoidal shear load. The phase shift of the sample's response gives information on the viscoelasticity of the tested material. The storage modulus G′ represents the ideally elastic, the loss modulus G″ the purely viscous part of the overall viscoelastic material behavior of polymers. Time-temperature superposition (TTS) allows the construction of a master curve for any reference temperature within the window of experiments. Thus, frequencies which are not accessible under isothermal test conditions may be taken into account in linear viscoelastic modelling. The selected temperature spectrum in this study was 100 K (+60 °C to +160 °C).

The steady-state shear viscosity was determined on a high pressure capillary rheometer of the company GÖTTFERT Werkstoff-Prüfmaschinen GmbH, type RG 50. The inlet pressure loss was determined using a tapered orifice die. This procedure is recommended by Perko et al. [10] for rubber compounds. Moreover, pressure drops in three different round dies (diameter = 1 mm) were measured (L/D = 5/10/20). The true shear rate results from the established correction by Weißenberg and Rabinowitsch.

Moreover, the steady-state extensional viscosity may be estimated from inlet pressure losses observed with the orifice die. Established models were proposed by Cogswell [11], Obendrauf [12] and Binding [13]. Perko et al. [12] showed that the extensional viscosity after Binding correlates best with measured values obtained with Sentmanat's Extension Rheometer (SER) [14]. Consequently, the steady-state extensional viscosity after Binding was exclusively used in this study.

2.3 Simulation

All CFD simulations presented in this study were performed using the commercial software package Polyflow® (developed by ANSYS Inc., Canonsburg, PA, USA) in the version 19.2. The geometry of the tapered orifice die and the inlet flow in an abrupt capillary are illustrated in Fig. 3. During injection the entangled polymer chains will align in flow direction. In the corners a secondary flow may occur depending on the rheological properties of the fluid.

A two-dimensional (2D) axisymmetric finite element model (FEM) of the orifice die (L/D = 0.2/1) and one abrupt capillary (L/D = 20/1) was built.

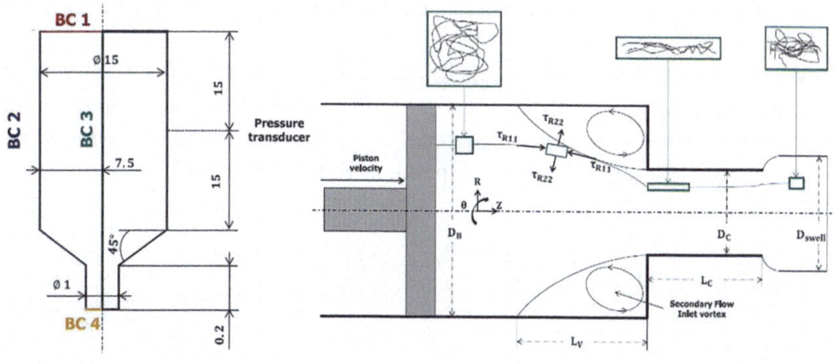

Fig. 3. CFD simulation setup of the orifice die (left) and inlet flow of an abrupt capillary (right)

The governing equations for incompressible fluids under steady-state, isothermal flow conditions may be written as [15]:

$$\nabla \cdot \bar{v} = 0 \tag{1}$$

$$-\nabla p + \nabla \cdot \bar{\bar{\tau}} = 0 \tag{2}$$

where $\bar{v}$ is the velocity vector, p is the pressure and $\bar{\bar{\tau}}$ is the stress tensor. Any influence of inertia and gravity was neglected due to the high viscosity of the NBR. The viscosity η of purely viscous fluids depends only on the rate of deformation $\bar{\bar{D}}$, but not on the magnitude of the imposed deformation (strain). The viscous stresses of non-Newtonian, incompressible and inelastic fluids are given as:

$$\bar{\bar{\tau}} = 2\,\eta(\dot{\gamma}, T, p, c)\,\bar{\bar{D}} \tag{3}$$

There are numerous mathematical models available, which are able to reflect the dependency of the viscosity on the shear rate $\dot{\gamma}$, the temperature T, the pressure level p and even the level of conversion c. In the present study, we are only considering shear-rate dependency. Highly filled rubber compounds exhibit a shear thinning behavior over a broad shear rate region with no Newtonian plateau. A model widely used in rubber rheology, which is able to reflect these characteristics is the simple power-law model [15]:

$$\eta(\dot{\gamma}) = K\,(\lambda \cdot \dot{\gamma})^{n-1} \tag{4}$$

where K is the consistency index, λ is the natural time (reciprocal of the reference shear rate) and n is the flow exponent. Since previous studies indicate the presence of a yield stress for highly filled rubber compounds, we used as a second model, an analytic expression of the viscoplastic Herschel-Bulkley model [15] implemented in Polyflow®.

$$\eta(\dot{\gamma}) = \tau_0 \left(\frac{1 - \exp\left(\frac{-3 \cdot \dot{\gamma}}{\dot{\gamma}_c}\right)}{\dot{\gamma}} \right) + K \left(\frac{\dot{\gamma}}{\dot{\gamma}_C} \right)^{n-1} \tag{5}$$

where τ_0 is the yield stress, $\dot{\gamma}_c$ the critical shear rate, K the consistency index and n the flow exponent.

However, viscous fluid flow models assume inelasticity, which is incorrect for any polymer, but especially for rubber compounds. They exhibit a strong degree of elasticity even in an uncured state. We considered the viscoelastic nature of the tested rubber compound using the integral and time dependent K-BKZ model [15]:

$$\bar{\bar{\tau}} = \frac{1}{1-\theta} \int_{-\infty}^{t} M\left(t - t'\right) H\left(I_{C^{-1}}, I_C\right) \left[\bar{\bar{C}}_t^{-1}\left(t'\right) + \theta \cdot \bar{\bar{C}}_t\left(t'\right) \right] dt' \tag{6}$$

with the material constant θ, the present time t, the past time t', the memory function $M(t-t')$, the damping function $H(I_{C^{-1}}, I_C)$, the Finger strain tensor $\overline{\overline{C}}_t^{-1}(t')$ and the Cauchy-Green tensor $\overline{\overline{C}}_t(t')$. The material constant θ is given by

$$\frac{N_2}{N_1} = \frac{\theta}{1-\theta} \tag{7}$$

where N_1 and N_2 are the first and second normal stress differences. Due to difficulties in measuring these characteristics especially for highly filled polymer systems the material constant θ is often set to zero in numerical studies [7, 8]. We followed this approach modelling the NBR rubber compound.

The linear viscoelastic material behavior is described through the memory function. It reflects the entanglement network of macromolecular chains. In unloaded condition, they exhibit a coiled-like structure of high entropy. Applying strain, polymers respond with a contrary force, since the macromolecular chains try to retain the past state of high entropy. However, applied strain leads over time to a restructuring of the entanglement network, which results in a decrease of the contrary force (relaxation). One exponential function is not able to reflect the complex relaxation mechanisms of polymers. Consequently, a series of Maxwell elements (prony series) is typically used to discretize the continuous relaxation spectrum (Eq. 8).

$$M(t-t') = \sum\nolimits_{i=1}^{N} \frac{g_i}{\lambda_i} \cdot \exp\left(-\frac{t-t'}{\lambda_i}\right) \tag{8}$$

with i representing the ith mode of the total number of modes N, the relaxation moduli g_i and relaxation times λ_i. In order to describe the pronounced nonlinear viscoelastic material behavior of rubber compounds, a damping function proposed by Wagner [16] was used. It reflects the dependency of rheological properties on the magnitude of the imposed deformation (strain).

$$H(I_{C^{-1}}, I_C) = \exp(-\alpha\sqrt{\beta I_{C^{-1}} + (1-\beta)I_C - 3} \tag{9}$$

Where $I_{C^{-1}}$ and I_C are the first invariants of the Finger strain tensor and Cauchy-Green tensor, respectively. α and β are material constants.

Polyflow® uses an iterative strategy to solve the integral constitutive equations of the K-BKZ type [15]. The stress tensor in the momentum Eq. (2) is replaced by an extra stress tensor $\overline{\overline{\tau}}_E$:

$$\overline{\overline{\tau}}_E = \omega \cdot \overline{\overline{\tau}} + (1-\omega) \cdot 2\,\eta_E \cdot \overline{\overline{D}} \tag{10}$$

where $\overline{\overline{\tau}}$ are viscoelastic stresses calculated from Eq. (6), η_E is the evolutive (Newtonian) viscosity and ω is a control parameter between 0 and 1. For $\omega = 0$, the flow problem is purely viscous (Newtonian), while $\omega = 1$ represents a full solution of the original viscoelastic Eq. (6). The evolutive viscosity is calculated using linear viscoelastic material data:

$$\eta_E = \sum_{i=1}^{N} g_i \cdot \lambda_i \tag{11}$$

All boundary conditions used in CFD simulations are listed in Table 1 with corresponding surfaces illustrated in Fig. 3. At the inflow (BC 1) we imposed a volumetric flow rate Q, which can be calculated as:

$$Q = \frac{\pi D_b^2 v_P}{4} \tag{12}$$

where D_b is the barrel diameter of the capillary rheometer and v_p is the set velocity of the piston. The apparent shear rate at the capillary wall is written as:

$$\dot{\gamma}_a = \frac{32Q}{\pi D^3} \tag{13}$$

with the diameter D of the capillary die. In case of an orifice die it is the smallest diameter. At the capillary and barrel wall (BC 2) a no-slip boundary condition is set. Moreover, we assumed a vanishing tangential force (fs) and a zero normal velocity at the axis of symmetry (BC 3).

Table 1. Boundary conditions (BC) in CFD simulations.

Boundary conditions (BC)[a]	Description
BC 1	Fully developed normal velocity profile (inflow)
BC 2	Normal (v_n) and tangential velocity (v_s) = 0 (wall)
BC 3	Tangential force (f_s) and normal velocity (v_n) = 0 (axis of symmetry)
BC 4	Normal force (f_n) and tangential velocity (v_s) = 0 (viscous/viscoplastic outflow)
BC 4	Tangential velocity (v_s) = 0, fully developed normal velocity profile (same volumetric flow rate as inflow), pressure = 0 at the exit barrel wall node (viscoelastic outflow)

[a] Corresponding surface is illustrated in Fig. 3

Finally, a distinction was made between viscous/viscoplastic and viscoelastic models at the outflow (BC 4). Using the viscous and viscoplastic models, a vanishing normal force (f_n) together with a zero tangential velocity was applied. For the viscoelastic fluid flow simulations, a zero tangential velocity together with a fully developed normal velocity profile (same volumetric flow rate as inflow) was applied at the outflow, since we found a zero force BC to massively overestimate the pressure level at the exit. Additionally, a pressure value of zero was assigned at the exit barrel wall node.

Experimental data indicates, that pressure and temperature dependencies of the viscosity effectively cancel each other out. Consequently, isothermal flow simulations were performed, omitting both dependencies.

3 Results and Discussion

3.1 Rheological Characterization

Due to the high filler content, rubber compounds are complex thermo-rheological systems showing filler-filler and filler-polymer interactions, which result in the well-known Payne effect [17, 18]. Due to the filler network the rubber compounds exhibit a power-law behavior even at very low shear rates and no Newtonian plateau. Fasching [1] proposed pre-shearing conditions for dynamic testing, which lead to a maximum breakdown of the filler network with minimized damage to the polymer chains. In steady-state rheological testing, the observed properties always reflect a specific deformation history. Consequently, Fasching's procedure results in rheological data much closer to those conditions, which we are aiming to describe numerically.

In a first step the rheological behavior of NBR was characterized by means of oscillatory rheometry. Comparing the complex viscosity to the steady-state shear viscosity proves the inapplicability of the Cox-Merz rule to the tested NBR rubber compound at 100 °C (Fig. 4). Pre-shearing the NBR rubber compound following Fasching's guideline only reduces the gap between steady-state and complex viscosity but fails closing it.

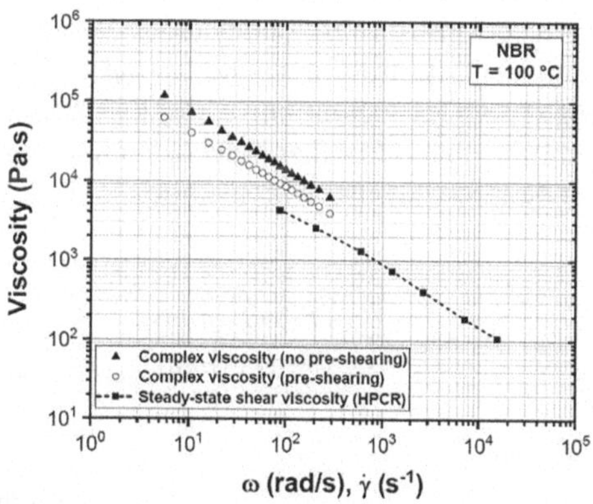

Fig. 4. Complex and steady-state shear viscosity as a function of frequency and shear rate for NBR at 100 °C (HPCR High Pressure Capillary Rheometer)

In capillary flow experiments no stick-slip was observed at the surface of extruded strands and no pressure fluctuations were present.

Plotting the pressure losses as a function of the capillary length, (Fig. 5, right), proves a high linearity indicating that temperature and pressure dependencies of the viscosity effectively cancel each other out.

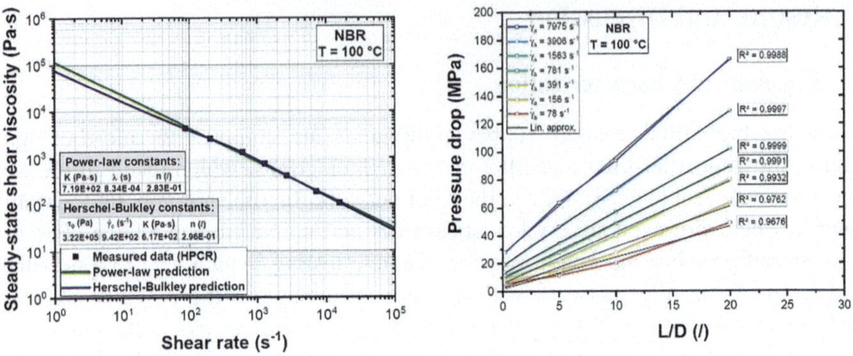

Fig. 5. Viscous and viscoplastic modelling of the NBR at 100 °C (left); Bagley-plot at the reference temperature (right) (HPCR High Pressure capillary Rheometer)

3.2 Viscoelastic Modelling

TTS allows covering a frequency range exceeding just one isothermal dynamic shear test. A broad frequency range of storage and loss moduli is of great importance fitting viscoelastic fluid flow models. Consequently, oscillatory tests using Fasching's concept of pre-shearing were performed over a temperature range of 100 K (60–160 °C). In a first step the rheological data from dynamic testing at 6 different temperatures were mastered applying TTS. The definition of the loss factor tan(δ) results in an invariance to vertical shifting b_t and is therefore recommended for mastering complex thermo-rheological systems (Eq. 14) with the horizontal shifting factor a_t and the angular frequency ω.

$$\tan(\delta) = \frac{b_t \cdot G''(a_t\omega)}{b_t \cdot G'(a_t\omega)} \tag{14}$$

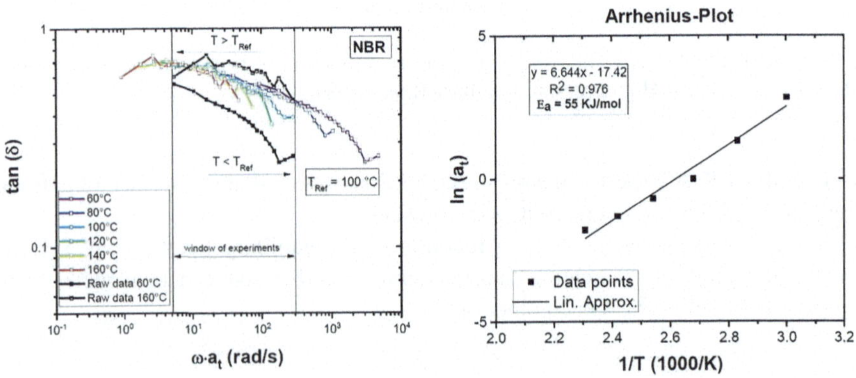

Fig. 6. Mastering of loss factor tanδ (left); Arrhenius plot (right)

This procedure allowed covering a frequency range of almost 4 decades (up to approx. 10^4 rad/s) at the reference temperature of 100 °C (Fig. 6, left). In order to check the physical plausibility of this mastering, the horizontal shifting factors were plotted over the inverse temperature (Fig. 6, right). A coefficient of determination of >97% proves linear correlation and indicates that the mastering is correct.

In ANSYS Polyflow®, the continuous relaxation spectrum can only be considered through a maximum of 8 Maxwell elements. Nevertheless, a comparison of measured data to model predictions (Fig. 7, left) proves that this number is already enough to describe the linear viscoelastic behavior of polymers well.

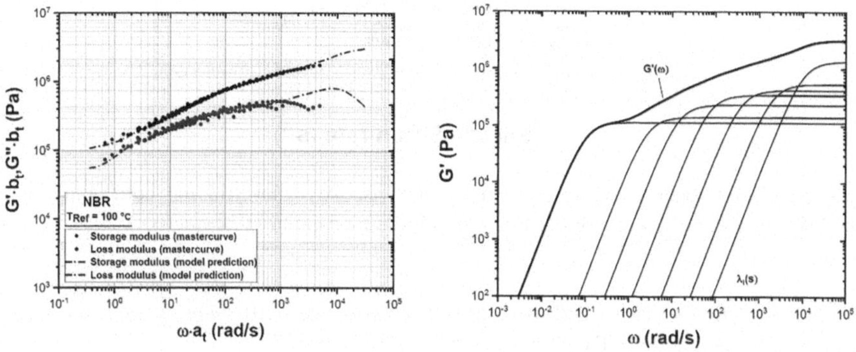

Fig. 7. Mastered G′/G″ data for NBR at 100 °C and model predictions (left); G′ curves for the 8 chosen Maxwell elements (right)

Modelling the nonlinear viscoelastic material behavior, rheological data from capillary rheometry were used. In case of a pure shear flow, the damping function (Eq. 9) depends only on the material parameter α. Thus, this model constant was fitted using steady-state shear viscosity data. Now, only the material parameter β remains as an unknown quantity in the system, which was fitted using steady-state extensional viscosity data after Binding [13].

A comparison of model predictions and measured data (Figs. 7, left and 8) proves that also highly filled rubber compounds can be described with viscoelastic fluid flow models well following the outlined procedure. All K-BKZ/Wagner constants of this fit are listed in Table 2.

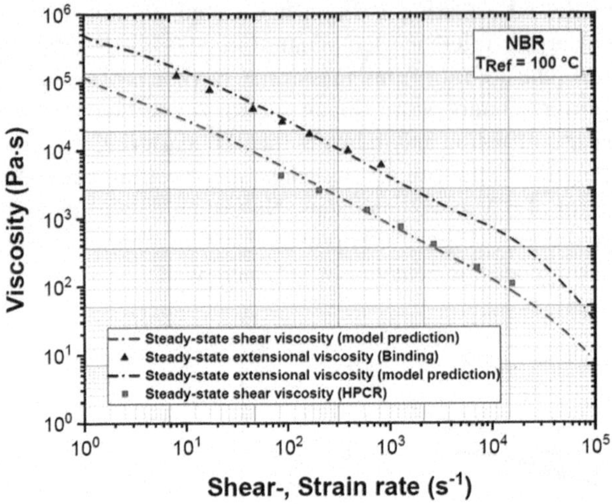

Fig. 8. Measured shear viscosity and extensional viscosity after Binding both compared to K-BKZ/Wagner model predictions (HPCR High Pressure capillary Rheometer)

Table 2. Relaxation spectrum and material constants for the NBR rubber compound obeying the K-BKZ/Wagner model at the reference temperature of 100 °C

Modes	λ_i (s)	g_i (Pa)	θ (/)	α (/)	β (/)
1	1.00E−04	1.35E+06	0	0.34	0.22
2	5.18E−04	5.41E+05			
3	2.68E−03	4.21E+05			
4	1.39E−02	3.45E+05			
5	7.20E−02	2.33E+05			
6	3.73E−01	1.41E+05			
7	1.93E+00	2.19E−01			
8	1.00E+01	1.13E+05			

3.3 Numerical Results

Finally, the influence of viscoelastic modelling on CFD simulation was investigated. The 15:1 contraction flow of an orifice die (L/D = 0.2/1) was analyzed as well as a shear dominated capillary flow (L/D = 20).

As expected, using a viscous or viscoplastic fluid flow model leads to a distinctive underestimation of measured pressure drops (power law and Herschel-Bulkley - Fig. 9, left). Mean deviations increase with rising apparent shear rate levels indicating that pressure losses resulting from extension get more and more dominant.

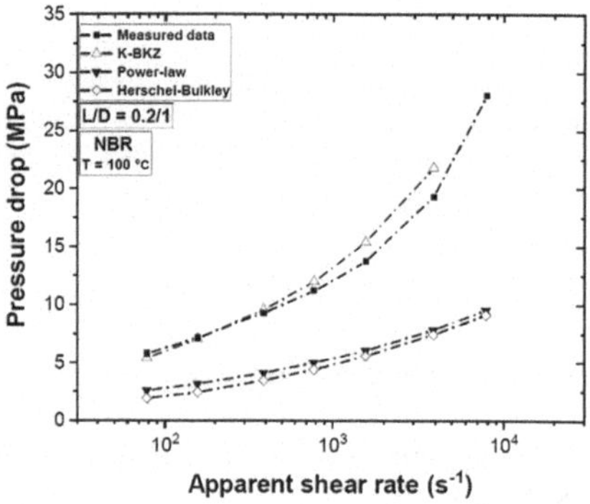

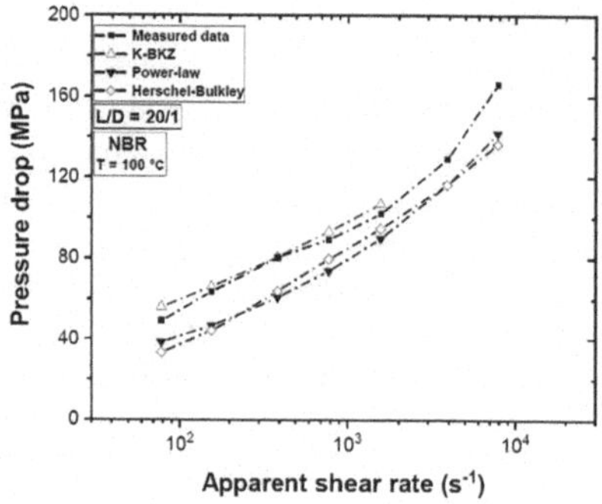

Fig. 9. Comparison of measured and predicted pressure drops (K-BKZ, Power-law, Herschel-Bulkley) for an orifice die (left) and a capillary with L/D = 20

Consequently, both the viscous and the viscoplastic models are not able to describe a contraction flow dominated geometry correctly. However, the viscoelastic K-BKZ/Wagner model ("K-BKZ" - Fig. 9) is able to predict measured pressure drops of the orifice die well. The influence of the contraction flow to the overall pressure drop is minimized for the L/D = 20 capillary. As a result, the viscous and viscoplastic models match measured pressure data better. Nevertheless, the K-BKZ/Wagner model provides also a better pressure prediction for the shear dominated geometry.

4 Conclusion

The pressure driven contraction and capillary flow of a highly carbon-black filled NBR rubber compound was studied both experimentally and numerically. First, almost perfect linearity of the pressure as a function of the capillary length was observed, proving that temperature and pressure dependencies of the viscosity effectively cancel each other out in capillary rheometry.

Next, dynamic testing taking into account the sensitivity of rheological data to the history of deformation was performed over a temperature range of 100 K. Applying TTS allowed modelling of linear viscoelastic data over a frequency range of more than 4 decades.

Carrying out CFD simulations, only the K-BKZ/Wagner model was able to predict pressure drops in a contraction flow dominated orifice die (L/D = 0.2/1) correctly.

Acknowledgements. This research work was supported by the Austrian Research Promotion Agency (FFG) as part of the "RubExject II" project (corresponding project number 855873) and the company partners SKF Sealing Solutions Austria GmbH, Judenburg, Austria, IB Steiner, Spielberg, Austria and ELMET Elastomere Produktions- und Dienstleistungs-GmbH, Oftering, Austria. The authors further thank company PKT Praezisions-kunststofftechnik Buertlmair Gesellschaft m.b.H. and Prof. T. Lucyshyn, Dr. Ivica Duretek, MSc Stephan Schuschnigg and Dr. Matthias Haselmann for their respective contributions.

Literatur

1. Fasching, M.: Robust processing in rubber injection molding using advanced simulation methods and material data. Dissertation, Montanuniversitaet Leoben (2015)
2. Mitsoulis, E.: 50 Years of the K-BKZ Constitutive Relation for Polymers. ISRN Polym. Sci. 1–22 (2013)
3. Ansari, M., Hatzikiriakos, S.G., Mitsoulis, E.: Slip effects in HDPE flows. J. Non-Newtonian Fluid Mech. (2011)
4. Ansari, M., Zisis, T., Hatzikiriakos, S.G., Mitsoulis, E.: Capillary flow of low-density polyethylene. Polym. Eng. Sci. **52**, 649–662 (2012)
5. Konaganti, V.K., Derakhshandeh, M., Ebrahimi, M., Mitsoulis, E., Hatzikiriakos, S.G.: Non-isothermal extrudate swell. Phys. Fluids **28**, 123101 (2016)
6. Mitsoulis, E.: Effect of Viscoelasticity in Fountain Flow of Polyethylene Melts. Int. Polym. Proc. **24**, 439–451 (2009)
7. Mitsoulis, E., Battisti, M., Neunhäuserer, A., Perko, L., Friesenbichler, W., Ansari, M., Hatzikiriakos, S.G.: Flow behaviour of rubber in capillary and injection moulding dies. Rubber Compos. **46**, 110–118 (2017)
8. Mitsoulis, E., Battisti, M., Neunhäuserer, A., Perko, L., Friesenbichler, W.: Flow behavior of PP-polymer nanocomposites in injection molding hyperbolical dies. Adv. Polym. Technol. 1–11 (2018)
9. Lucyshyn, T.: Simulation report for a thin-walled bushing for company PKT Praezisionskunststofftechnik Buertlmair Gesellschaft m.b.H. Institute of Polymer Processing, Montanuniversitaet Leoben (2016)
10. Perko, L., Fasching, M., Friesenbichler, W.: Model for the prediction of bulk temperature changes and pressure losses in rubber compounds flowing through conical dies: An engineering approach. Polym. Eng. Sci. **55**, 701–709 (2014)
11. Cogswell, F.N.: Converging flow of polymer melts in extrusion dies. Polym. Eng. Sci. **12**, 64–73 (1972)
12. Perko, L., Friesenbichler, W., Obendrauf, W., Buchebner, V., Chaloupka, G.: Elongational viscosity of rubber compounds and improving corresponding models. Adv. Prod. Eng. Manag. **8**, 126–133 (2013)
13. Binding, D.: An approximate analysis for contraction and converging flows. J. Non-Newtonian Fluid Mech. **27**, 173–189 (1988)
14. Sentmanat, M.L., Rheol, M.L.: Miniature universal testing platform: from extensional melt rheology to solid-state deformation behavior. Acta **43**, 657 (2004)
15. ANSYS Inc.: ANSYS Polyflow User's Guide: Release 18.2: Canonsburg, PA, USA (2017)
16. Wagner, M.H.: A constitutive analysis of uniaxial elongational flow data of low-density polyethylene melt. J. Non-Newtonian Fluid Mech. **4**, 39–55 (1978)
17. Leblanc, J.: Rubber–filler interactions and rheological properties in filled compounds. Prog. Polym. Sci. **27**, 627–687 (2002)
18. Payne, A.R.: The dynamic properties of carbon black-loaded natural rubber vulcanizates. Part I. J. Appl. Polym. Sci. **6**, 57–63 (1962)
19. Friesenbichler, W., Berger, G., Fasching, M.: Simulation of Rubber Injection Molding – Challenges and Limitations. Conference Proceedings of the 7th bi-annual International Conference of Polymer & Molds Innovations (PMI). Gent 2016. pp. 179–183

A Study on the Determination of Virtual Process and Quality Data in Injection Molding Simulation

Armin Kech(✉) and Patricia Maris-Haug

Robert Bosch GmbH, CR/APP, Robert-Bosch-Campus 1, 71272 Renningen, Germany
armin.kech@de.bosch.com

Abstract. The current applied research study is embedded into the topic of connected industries and industry 4.0 at Robert Bosch GmbH by investigating the potential usage of process simulation to assist process optimization and hence make it more efficient.

This work presents the state-of-the-art regarding a comparison between process simulation results on one hand and measured sensor data as well as measured part weight on the other hand. The investigations were carried out at the example of a multi-pin connector device. For the study, commercially available software was used, such as AUTODESK MOLDFLOW AMI® and Moldex3D®.

During the injection molding trials using a short glass fiber-reinforced PBT grade in a double cavity mold, the in-mold cavity pressure was recorded with flush-mounted pressure sensors. After processing with varying process parameters, the weight of the samples and two dimensions of the samples were measured. Ten different settings were used for the investigation in order to determine the current gap between simulation results and measured data. Furthermore, the deviations of currently used model approaches and measured material data were discussed to improve the understanding of the gathered results.

Keywords: Injection molding · Simulation · Virtual processing

1 Introduction

This study aims on the evaluation whether commercially available software codes can be used to train a neural network (NN) which is used for process monitoring, optimization and control in injection molding. Currently the training of NN requires a huge amount of experimental data acquisition on the injection molding machine and in measuring labs afterwards. If it would be possible to substitute part of this training by simulated data it would be highly efficient, concerning the avoidance of waste and energy, and the reduction of time as well as costs. The simulation of real parts including the runner and cooling system is state-of-the-art and some players with

C. Hopmann and R. Dahlmann (Eds.): *Advances in Polymer Processing 2020*,
pp. 283–290, 2020. https://doi.org/10.1007/978-3-662-60809-8_23

their software codes are on the market. Two of them were used in this study. A typical product that is manufactured in-house at Bosch was used in combination with a DoE approach to determine experimental data.

2 Experimental and Numerical Setup

The material used was a PBT-GF30 from DuPont [1]. It was injection molded using an injection molding machine (370 S) from Arburg [2] with a screw diameter of 30 mm. The objective of these trials was to generate a database to be used for process monitoring and control [3] and to provide data for a comparison of simulated and real data concerning weight, pressure prediction and some pre-defined dimensions. The part used for this study is a 26 pin connector which is manufactured in-house at Bosch, see Fig. 1. The part is a two cavity part with a cold runner between the two cavity halves and a pressure sensor installed in one cavity, flush mounted.

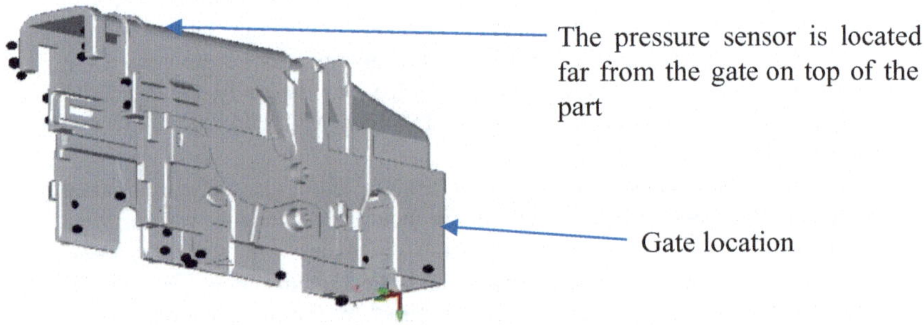

Fig. 1. 26 pin connector

The simulation was performed using AUTODESK MOLDFLOW AMI® 2019 and Moldex3D® R17.

2.1 Design of Experiments (DoE) in Injection Molding

A Sobol DoE scheme was defined for three varying parameters, which are packing pressure, mold temperature and injection speed. The melt temperature, the switch-over point and the holding as well as the cooling times were kept constant for all settings. It was used a hydrolysis stabilized PBT with 30 wt.% short glass fiber reinforcement. In total 312 settings were defined, each one with only one single part but with multiple levels for the three parameters to be varied. Table 1 shows the used parameters with its definitions and upper (max.) and lover (min.) bounds.

Table 1. Bounds for the varied parameters in the DoE

Parameter	Packing pressure	Mold temperature	Injection Speed
Abbreviation	pNach	TWz	Vein
Unit	[bar]	[°C]	[cm^3/s]
Max.	700	120	70
Min.	300	50	20
Step width	1	1	1

2.2 Simulation Setup of the Problem

The different simulation setups were modelled in both software packages. First, see Fig. 2, the system of a two cavity mold was modelled in AUTODESK MOLDFLOW AMI® by a discretization of the CAD data provided for the part. The cold runner with the gate of the part was modelled as beam elements, while the part was discretized with 3D liner tetrahedral elements with 12 elements over thickness. By applying an appearance number of two for the part and the small runner leading to the gate as well as the gate itself, a two cavity situation was simulated. As there was no information on the cooling channels layout, an arbitrary layout was used in the simulation.

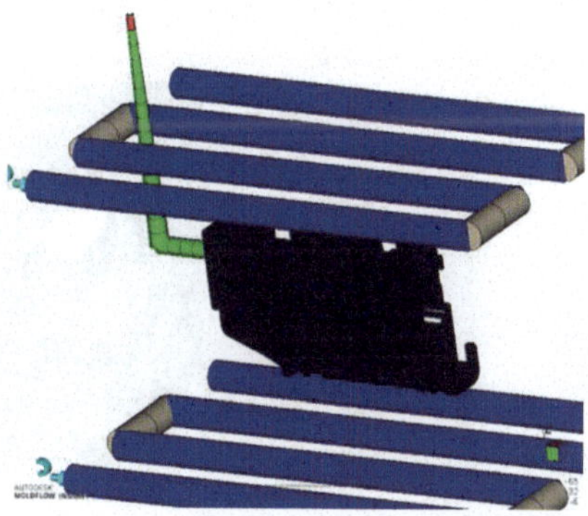

Fig. 2. Model of 26 pin connector in MOLDFLOW Synergy®. The cooling channels are given in blue with connectors at the outside of the drillings, the cold runners is green and the injection point is given by the yellow cone on a single hot runner element

Second, see Fig. 3, both cavities were modelled in Moldex3D Designer®. In contrary to MOLDFLOW, both cavities were modelled connected by the runner, which, again, was modelled with beam elements. As for MOLDFLOW the cooling system is not the actual, but an arbitrary one.

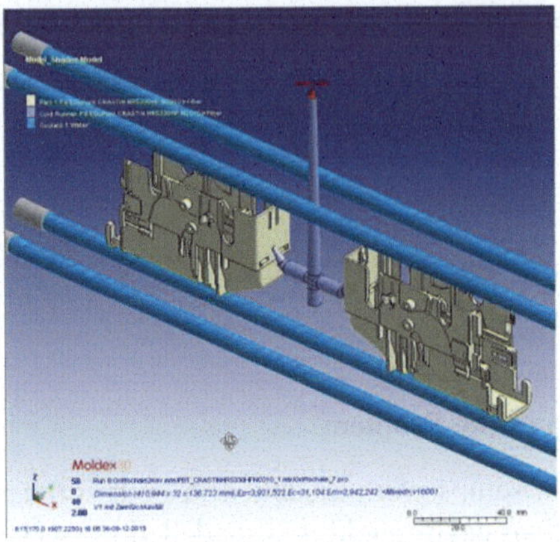

Fig. 3. Model of 26 pin connector in Moldex3D Designer®. The cooling channels are given in blue, the cold runners is violet and the melt entrance is marked with a red area

There is a difference in the discretization approach. While MOLDFLOW uses only tetrahedral elements in 3D, Moldex3D has a Boundary Layer Mesh (BLM) with hexagonal elements on the outer surface and tetrahedral elements only in the center of the part, see as an example Fig. 4.

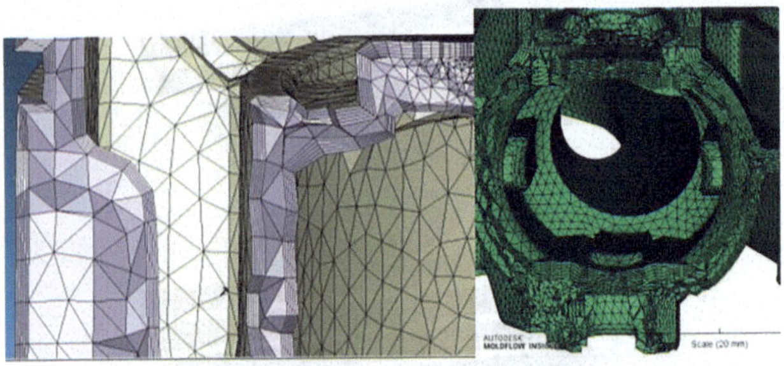

Fig. 4. Typical BLM mesh in Moldex3D (left) and tetrahedral mesh in Moldflow AMI(right)

Ten settings from the Sobol DoE scheme were simulated with the following settings for the varying parameters, see Table 2.

Table 2. Settings used in simulation

Setting	pNach	Twz	Vein
V0	650	120	26
V1	324	120	55
V2	576	120	61
V10	682	120	40
V30	696	120	47
V50	480	120	63
V70	338	120	30
V90	438	85	50
V150	414	85	40
V257	690	50	55

During the injection molding trials the sensor values were registered. Therefore, they could be used for the setup of the simulations. The actual movement of the screw was used to set the switch-over point as a function of screw position or injection time. The actual hydraulic pressure data was used to define the packing phase for each setting and the cycle time, see Fig. 5.

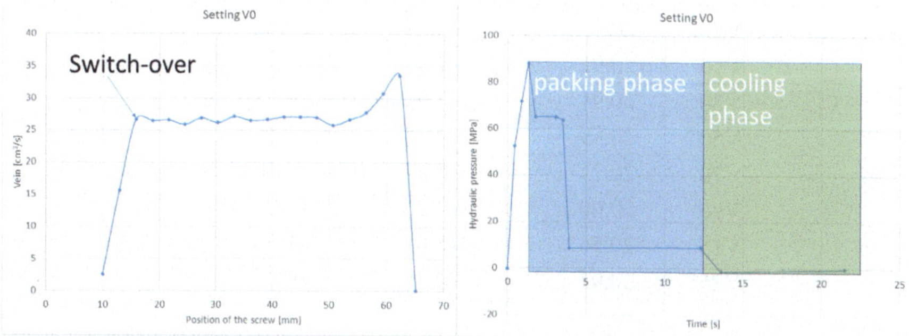

Fig. 5. Injection speed over screw position (left) and hydraulic pressure over time (right) for V0

3 Results and Discussion

Directly after injection molding the samples were weighted and the transient pressure was recorded for the pressure transducer used.

3.1 Results from Simulations

In Table 3 the comparison between measured and simulated weight is shown. For the Software B results an average between the weights in both cavities, which were nearly identical, was taken.

Table 3. Comparison between measured and simulated results on weight in [g]

Setting	Measurement	Software A	Software B
V0	20.95	21.60	20.90
V1	20.41	21.43	20.76
V2	20.77	21.45	20.84
V10	20.92	21.53	20.91
V30	20.91	21.55	21.02
V50	20.59	21.43	20.91
V70	20.77	21.49	20.87
V90	20.60	21.58	20.77
V150	20.56	21.59	20.88
V257	20.93	21.84	20.98

Table 4 shows the deviation in [%] between measured data and simulated ones.

Table 4. Difference between measured and simulated data for weight in [%]

Setting	Measurement	Software A	Software B
V0	20.95	+3.10	−0.24
V1	20.41	+5.00	+1.71
V2	20.77	+3.27	+0.34
V10	20.92	+2.92	−0.05
V30	20.91	+3.06	+0.53
V50	20.59	+4.08	+1.55
V70	20.77	+3.47	+0.48
V90	20.60	+4.76	+0.83
V150	20.56	+5.01	+1.56
V257	20.93	+4.35	+0.24

The pressure recorded and the pressure simulated are given in Table 5 for four of the ten settings.

Table 5. Comparison of recorded pressure and simulated pressure for four different settings, V0 (top, left), V1 (top, right), V10 (bottom, left) and V70 (bottom, right)

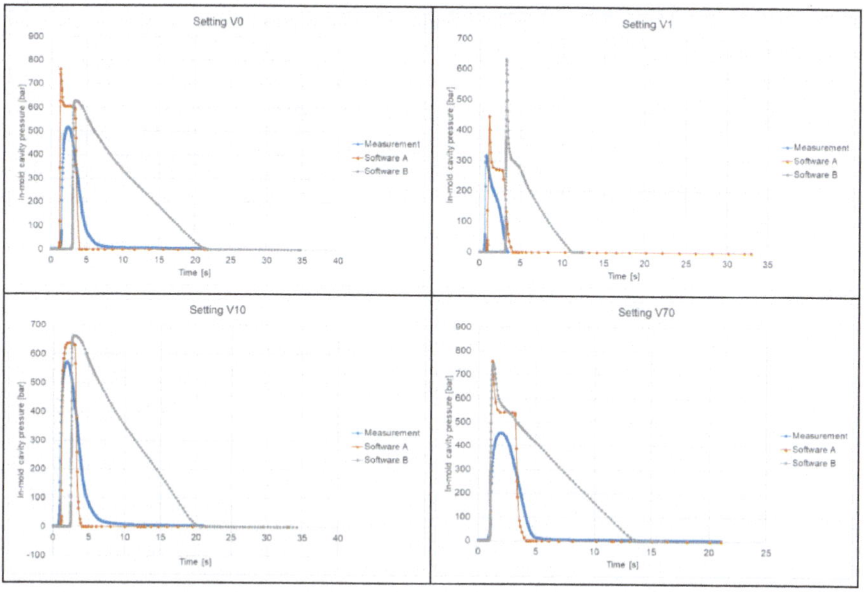

3.2 Discussion of Results

The weight is very well predicted, an index that the used material data were decently characterized and the geometry of the cavity model was used in the simulations. Nevertheless, the tendencies between the different settings are less well predicted but still usable and the results for Software B are slightly better.

The situation is different for the predicted transient pressures. The predicted max. pressure is always significantly higher than the recorded data. However, Software A shows a better agreement concerning the shape and the drop of the pressure than Software B does. It has to be mentioned, that the new option that takes into account visco-elastic effects in the polymer melt could not yet be applied on the data.

4 Conclusion

With these evaluation it is shown, that the weight can be predicted well with both software codes used in this study. The predicted pressure is still to close on the setup data for the hydraulic pressure with no pressure drop compared to the set max. value for packing pressure. In a next step the data has to be evaluated in that sense that it has to be analyzed whether there is a correlation between the predicted pressure data and the measured data. For this purpose AI/ML methods are currently discussed. Nevertheless, it is also obvious that without improvement on the predicted pressure data the usage for training an optimization method for injection molding production is rather limited.

References

1. DuPont Homepage: http://www.dupont.de/produkte-und-dienstleistungen/plastics-polymers-resins/thermoplastics/articles/crastin-hydrolosis-resistant-plastic.html
2. Arburg Homepage: https://www.arburg.com/fileadmin/redaktion/mediathek/technische_daten/arburg_allrounder_370s_multi-component_td_523673_en_gb.pdf
3. Lehtonen, P.: Verbundprojekt: Rationelle Energieeinsparung und Minimierung des Materialeinsatzes beim Spritzgießen (REMIS), Teilvorhaben: Modellierung und statistische Analyse: Schlussbericht, https://doi.org/10.2314/GBV:503687332

Strength Development in Overmolded Structures

Thijs Donderwinkel[1(✉)], Martin van Drongelen[2], and Sebastiaan Wijskamp[1]

[1] ThermoPlastic composites Research Center (TPRC), Palatijn 15, 7521 PN Enschede, The Netherlands
thijs.donderwinkel@tprc.nl

[2] Chair of Production Technology, Faculty of Engineering Technology, University of Twente, 7500 AE Enschede, The Netherlands

Abstract. Overmolding of thermoplastic composites is a technology in which a thermoplastic composite is thermoformed and subsequently injection overmolded. Although the feasibility of the process is increasingly demonstrated, it is acknowledged that there is a lack of proper design tools that can be used for a right-the-first-time design strategy. Here, a modelling strategy is proposed for the prediction of the bond strength between a composite insert and an injected polymer. The development of the interface strength is affected by the process history as well, where the temperature and polymer chain mobility play an important role. In the model, the melting behavior of the polymer interface is described using the temperature evolution on the interface combined with experimentally determined polymer melting kinetics via flash differential scanning calorimetry (DSC). Dedicated test geometries were developed and manufactured to evaluate the bond strength under different loading conditions. Short beam strength experiments were used to study the flow length dependency of the interface strength and were correlated with the predicted melting evolution on the interface. The outcome was critically reviewed leading to preliminary guidelines for design, materials and processing as well as routes to further mature this technology.

Keywords: Overmolding · Interface strength · Degree of melting

1 Introduction

1.1 Overmolding

Overmolding of thermoplastic composites is a hybrid process that combines thermoforming of a continuous fiber reinforced thermoplastic blank with an injection or compression molding process. The process allows for complex parts with high structural performance due to the continuous fibers that, in the ideal case, are positioned along the load paths in the part. Further advantages are the potential for high level of function integration, net shape processing and large series production in an automated

C. Hopmann and R. Dahlmann (Eds.): *Advances in Polymer Processing 2020*,
pp. 291–300, 2020. https://doi.org/10.1007/978-3-662-60809-8_24

process. A schematic representation of the overmolding process is shown in Fig. 1. The process consists of the melting and forming of a blank, which is consequently over injected with a polymer that is compatible with the polymer matrix of the composite blank. This allows, for example, the addition of reinforcing ribs to increase the geometrical stiffness.

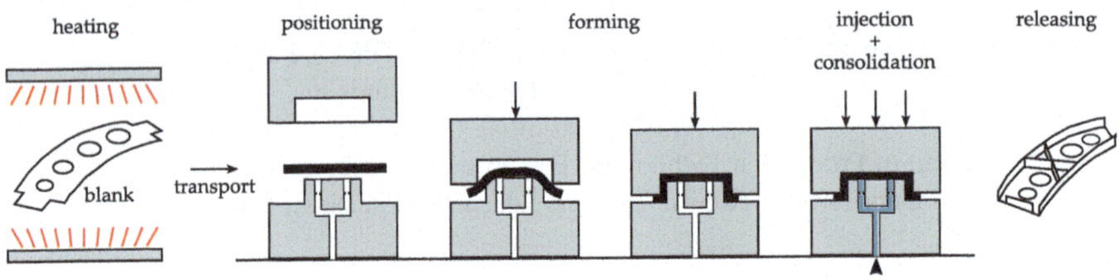

Fig. 1. Schematic representation of the overmolding process.

The interface strength between composite blank and injected polymer is critical for the performance of the overmolded part. Currently, commercial design tools are not available to predict this interface strength. Consequently, mold design and process optimization are done by trial-and-error. The objective of the current study is to develop and experimentally validate a predictive model for the interface strength between the injected polymer and composite blank. This model is coupled with commercial injection molding and structural analysis software in order to simulate the overmolding process as part of the mold design procedure.

1.2 Interface Strength Development

An unfilled or short fiber reinforced polymer is injected on to the laminate, referred to as insert, once the molds are fully closed. The time available for bonding during overmolding is limited, since the mold is kept at a low temperature to ensure rapid cooling and a short cycle time. The thermal energy of the injected polymer melt is used to locally heat up the interface with the insert and thereby developing a bond via reptation of amorphous polymer chains, as shown in Fig. 2. A packing pressure is applied to let the material cool down and solidify while pressurized.

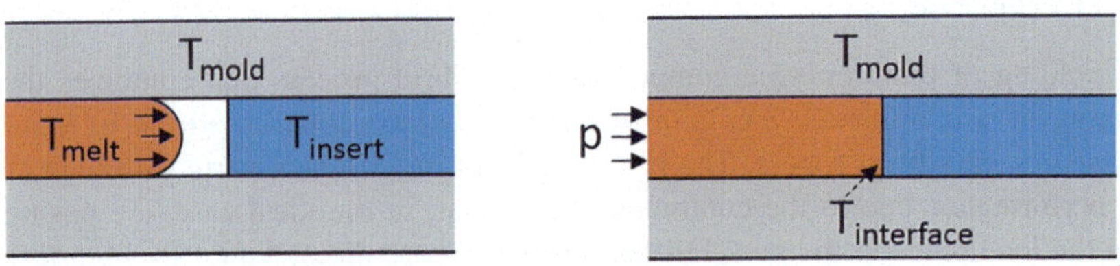

Fig. 2. Left: polymer injection; Right: packing under pressure *p*. The temperature of the mold (T_{mold}), injected polymer (T_{melt}) and insert (T_{insert}) determine the temperature of the interface ($T_{interface}$).

In literature, various authors [1, 2] have observed a sudden increase in interface performance upon melting for semi-crystalline polymers. This behavior was explained by the dissolving of the immobile crystalline fraction near the melting temperature, which also frees the rigid amorphous fraction [1]. Subsequently polymer chains can move across the interface and form a bond, so-called healing.

A non-isothermal healing model was introduced by Yang and Pitchumani [3] for amorphous polymers. The model uses polymer reptation times to describe the evolution of the healing process. For semi-crystallites reptation times can be in the order of milliseconds above the melting temperature [4], which leads to instantaneous healing. It was therefore suggested by Bouwman et al. [5] to describe the evolution for healing via the melting trajectory instead of an onset temperature [6–8]. The so-called degree of melting was introduced, which represents the fraction of molten crystallites with respect to the original state. The presented model assumes a direct relation between the temperature and the degree of melting based on input from a DSC experiment.

Effects such as superheating and recrystallization can arise during heating, which are heating rate dependent as was experimentally observed using DSC [9, 10] and flash DSC [10–13] experiments. Moreover the melting behavior of a semi-crystalline polymer is affected by its crystallization history, as observed by Furushima et al. [11] for PEEK. The critical heating rate to suppress reorganization of imperfect crystallites decreases with increasing crystallization temperature. This was explained by the formation of more stable crystallites at higher crystallization temperatures. Similar behavior was observed for polyamide-6 [12] and isotactic polypropylene [13].

In the current work the influence of rate dependency and crystallization history will be studied of the Victrex AE™ 250 polyaryletherketone (PAEK) polymer using flash DSC. Subsequently a generic model is introduced to predict the evolution of the degree of melting during fusion bonding processes. Experimentally-determined melting curves are inserted into the model to calibrate the kinetics of the melting behavior. This allows for the introduction of rate and crystallization history dependency. Ultimately the predicted degree of melting is compared to strength values from short beam strength experiments.

2 Experimental Study

Flash DSC (Mettler-Toledo Flash DSC 1) allows for a wide temperature range and high heating and cooling rates (−4000 °C/s up to 40,000 °C/s) and is thereby suited for emulating the process conditions during overmolding. A sensor chip was prepared to measure the response of the PAEK polymer.

The effect of the crystallization history is emulated before the specimen is subjected to heating. First the material is heated rapidly to dissolve the crystalline fraction and subsequently cooled to the isothermal tooling temperature, which is maintained for 300 s. Via this method, the stamp forming process prior to overmolding is emulated and allows for a fully crystallized specimen with values of 26.2%. The melting behavior of the PAEK polymer was studied for heating rates between 10 °C/s and 5000 °C/s after isothermal crystallization. Moreover the effect of the isothermal crystallization temperature was evaluated.

A second order polynomial baseline is applied to the heat flow signal to obtain the melting peak for each rate. Subsequently the resulting melting peak is normalized by the associated heating rate for each signal. After horizontal alignment, the heat capacity of each melting peak is obtained as displayed in Fig. 3.

The evolution of the melting peak changes with increasing heating rate, where a horizontal shift is observed in the temperature associated with the melting peak. Moreover the double melting peak observed for rates up to 300 °C/s disappears for higher rates. The presence of the double melting peak is observed in literature as well for polyetheretherketone (PEEK) polymers [9–11]. This behavior is explained by the melting of imperfect crystallites formed during cooling, followed by reorganization of the polymer chains via crystallization and subsequent melting of these stable crystallites during heating. The isothermal crystallization temperature affects the onset and peak temperature of the melting peak, which indicates that more stable crystallites are formed at the higher crystallization temperature.

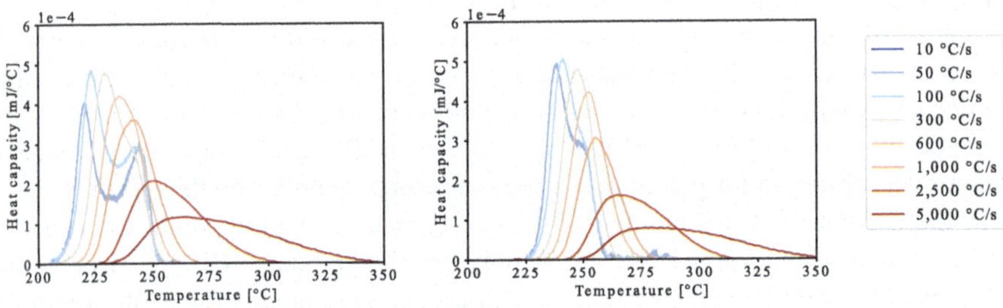

Fig. 3. Heat capacity during heating of the PAEK specimen using various heating rates after isothermal crystallization at 200 °C (left) and 220 °C (right).

A second order polynomial baseline is applied to the heat flow signal to obtain the melting evolution for each rate. The evolution can be expressed as the degree of melting X_m, which is found by integrating the heat flow absorbed during the melting process via Eq. 1.

$$X_m = \frac{1}{Q_T} \int_0^T \frac{\partial Q}{\partial T} dT \tag{1}$$

Here Q, T and Q_T represent the heat flow, temperature and total heat of fusion respectively.

The degree of melting is calculated for each heating rate from the experiment, which results in Fig. 4. A positive shift in the onset of melt can be observed in the figure with increasing heating rate, whereas the gradient of the curves decreases with higher rate. These effects can be attributed to the lag in the polymer response, since the crystalline fraction has less time to respond to the increasing temperature.

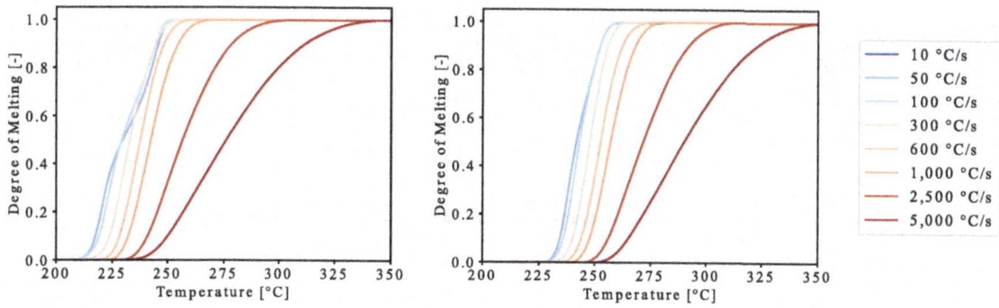

Fig. 4. Degree of melting evolution of the PAEK specimen using various heating rates after isothermal crystallization at 200 °C (left) and 220 °C (right). The curve associated with 10 °C/s is omitted for clarity.

3 Numerical Study

Compared to previous work [5] a next step was taken by including heating rate dependency in the model. The experimental data presented in the previous section describes the evolution of the degree of melting for constant heating rates. Via this data the development of the molten state can be evaluated for arbitrary temperatures and heating rates within the range of the experimental data via a two-dimensional interpolation scheme. An explicit update via Forward-Euler was introduced to describe the evolution of degree of melting for an arbitrary temperature evolution via Eq. 2,

$$X_m^{k+1} = \frac{\partial X_m}{\partial t}^k \Delta t + X_m^k \tag{2}$$

where k denotes the current time increment and the current time step $\Delta t = t^{k+1} - t^k$.

The model obtains the melting rate at the current degree of melting value and heating rate via Eq. 3 using the experimental data and interpolation scheme,

$$\frac{\partial X_m}{\partial t}^k = \frac{\partial X_{m,exp}}{\partial t}\left(X_m^k, \frac{\partial T}{\partial t}^k\right) \tag{3}$$

where subscript *exp* denotes the experimental origin of the data. The temperature is not included in the kinetics of an arbitrary increment.

The model allows for an accurate description of the growth regime of the experimental degree of melting, since the experimental increase is significant with respect to the numerical incrementation. A visual representation of the algorithm for three subsequent increments is shown in Fig. 5.

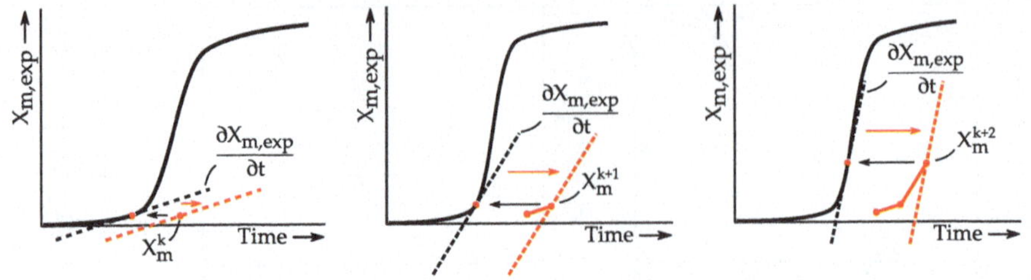

Fig. 5. Explicit update of the degree of melting for subsequent increments k, $k + 1$ and $k + 2$.

4 Results and Discussion

The performance of the numerical model will be evaluated in the current section, where it will be compared with the experimental data. First, the experimental temperature data from the constant heating rate experiments will be inserted in the model. Subsequently, the model will be used to predict the degree of melting evolution for an overmolded s-rib geometry and correlate the results with mechanical test data.

4.1 Degree of Melting Modeling for Constant Heating Rates

The temperature evolution of the flash DSC experiments is inserted in the numerical model to predict the degree of melting. Since the model uses the flash DSC data as input, the resulting degree of melting reproduces the experimental curves as shown in Fig. 6. A linear two-dimensional interpolation scheme has been used to describe the melting rate for arbitrary temperatures and heating rates. A step size of 1 °C and 0.001 for the temperature and degree of melting discretization is used respectively.

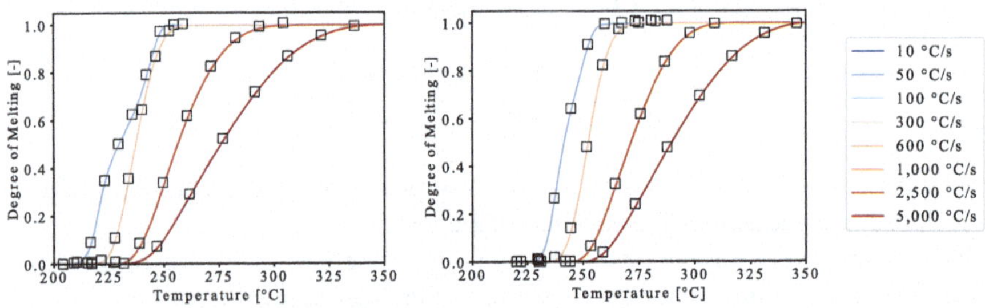

Fig. 6. Experimental and modeled degree of melting evolution of the PAEK specimen using various heating after isothermal crystallization at 200 °C (left) and 220 °C (right). Various increments predicted by the numerical model are denoted by the square symbols. The curve associated with 50, 600, 2500 and 5000 °C/s are shown exclusively for clarity. Moreover the amount of displayed data points has been reduced to 10 per model.

The numerical model is able to accurately represent the trends of the experimental data including the onset and growth. The model does however overestimate the experimental curves in the late regime, however the error is below 1% for all rates. The late regime has a minimal variation of the gradient, which is consequently less accurately captured by the model since a small amount of increments is available in this domain.

4.2 Degree of Melting Modeling for Non-constant Heating Rates

Additionally an experiment was performed to study the melting behavior of the PAEK polymer under process conditions. Two custom heating curves, shown in the left image of Fig. 7, were programmed in the flash DSC to emulate the temperature evolution during overmolding process for a center and edge locations in a rib. Prior to applying the heating curve, the specimen was isothermally crystallized at 220 °C for 300 s to allow for full crystallization. The resulting degree of melting is displayed in the right image of Fig. 7.

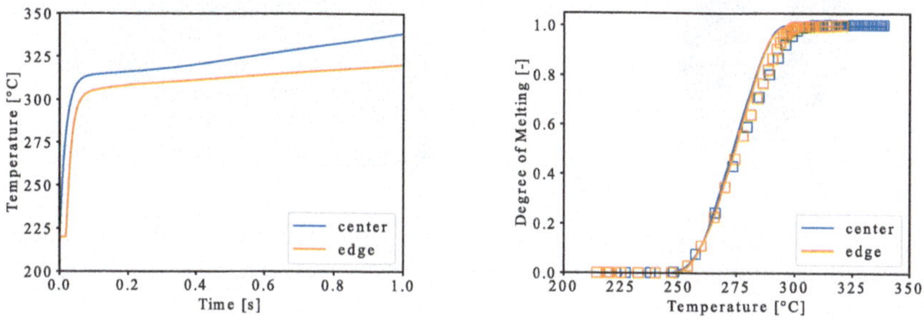

Fig. 7. Left: temperature evolution of two locations in a rib during overmolding; Right: experimental (solid line) and modeled (square symbols) degree of melting evolution of the PAEK specimen for two locations during the overmolding process. The amount of displayed data points has been reduced to 50 per model.

The temperature history was inserted into the numerical model to predict the degree of melting evolution during overmolding for the two locations. The onset and early regime of melting is captured accurately by the model for both locations, whereas the growth is slightly underestimated. The explicit scheme continues to increase the degree of melting up to 1.005 in the late regime, as was observed for constant heating rates too.

4.3 Correlation with Structural Performance

Short beam strength experiments were performed on specimens created using the overmolding process. A s-rib geometry was developed to allow for the investigation of the strength development over the flow path length, shown in the left image of Fig. 8. The composite insert is comprised of Toray Cetex® TC1225 unidirectional tape with a cross-ply layup of eight plies and subsequently overmolded with Victrex® PEEK 90HMF40 for the introduction of ribs. The tool temperature during stamp forming of the composite inserts was 220 °C. During the subsequent injection molding stage, the polymer material is injected at 400 °C and the tool temperature was varied between 200 °C and 230 °C. The orientation of the evaluated laminates was $[[0,90]_s]_s$ and

$[[90,0]_s]_s$. The specimen dimensions were 28.5 mm × 8.2 mm × 4.8 mm and were produced via milling. The position of each specimen is shown in Fig. 8. The resulting short beam shear strength is displayed in the left image of Fig. 9 for specimens with the top ply perpendicular to the flow direction.

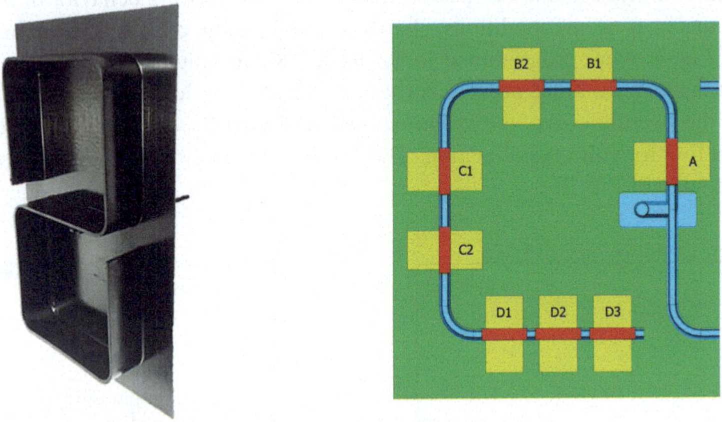

Fig. 8. Left: s-rib geometry after two-step overmolding process; Right: specimen positions for short beam strength testing. The location of the in-flow is represented by the blue rectangular section.

The numerical temperature evolution in the s-rib geometry is obtained for each interface node on the insert side from an injection molding simulation performed in Autodesk® Moldflow. Subsequently, the temperature history is inserted in the degree of melting model to obtain the evolution of the interface strength for each specimen. The resulting average degree of melting per specimen is shown in the right image of Fig. 9 for the two tool temperatures.

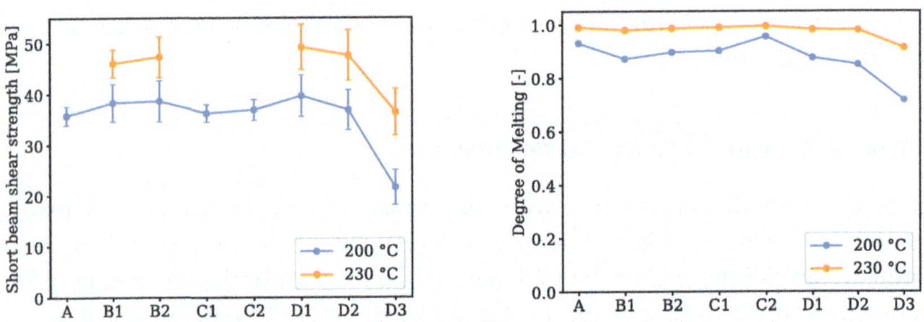

Fig. 9. Flow length dependency of the strength evolution after overmolding. Two tool temperatures are displayed at 200 °C (blue) and 230 °C (orange). Left: short beam shear strength results of specimens with the top ply fiber orientation perpendicular to the overmolded rib. Right: average degree of melting per specimen as predicted by the numerical model.

The evolution of the short beam shear strength specimen displays a near constant value of 35.5 MPa up to specimen D2 for a tool temperature of 200 °C. A steep decrease in strength is observed at specimen D3. A similar trend with a higher average value is observed in the strength values of the specimens produced using a tool temperature of 230 °C.

The degree of melting shows a near constant evolution along the flow path with a decrease for specimen D3 as well, which is the last specimen to be filled during processing. Flow renewal is shorter at this location consequently. Moreover, heat is lost to the tooling perpendicular to the flow direction, which is not present for the other specimens. A higher tool temperature results in a smaller heat loss and consequently more heat is available to melt the polymer material at the interface, which results in a higher degree of melting as shown by Fig. 9.

5 Conclusions and Outlook

The overmolding process is a promising technology for the manufacturing of complex thermoplastic composite parts with integrated functionality. Design tools are needed to enable a right-the-first-time design of new parts. A numerical method for the prediction of the bond strength development at the interface during overmolding was proposed, which uses input from flash DSC experiments on the melting behavior. The model is able to give a prediction of melting at the interface between a melted injection polymer and a solid thermoplastic composite insert. A correlation with the structural performance was presented using short beam shear strength experiments. The evolution of strength displays minimal variation along the flow path except for a steep decrease in the specimen near the end of the rib. Tooling temperature affects the strength development significantly, which was displayed by the numerical model as well. Further work focuses on the quantitative prediction of the interface performance via the integration in structural analysis software.

Acknowledgements. The authors gratefully acknowledge the support from the industrial and academic members of the ThermoPlastic composites Research Center (TRPC), as well as the project partners Autodesk, KraussMaffei, Safran, Magneti Marelli and National Composites Centre.

References

1. Boiko, Y.M., Guérin, G., Marikhin, V.A., Prud'homme, R.E.: Healing of interfaces of amorphous and semi-crystalline poly(ethylene terephthalate) in the vicinity of the glass transition temperature. Polymer **42**(21), 8695–8702 (2001)
2. Lamethe, J.F., Beauchêne, P., Léger, L.: Polymer dynamics applied to PEEK matrix composite welding. Aerosp. Sci. Technol. **9**, 233–240 (2005)
3. Yang, F., Pitchumani, R.: Healing of thermoplastic polymers at an interface under nonisothermal conditions. Macromolecules **35**, 3213–3224 (2002)
4. Grouve, W.J.B., Warnet, L.L., Rietman, B., Visser, H.A., Akkerman, R.: Optimization of the tape placement process parameters for carbon-PPS composites. Compos. Part A **50**, 44–53 (2013)

5. Bouwman, M.M., Donderwinkel, T.G., Krämer, E.T.M., Wijskamp, S., Costa, F.: Overmolding – An integrated design approach for dimensional accuracy and strength of structural parts. In: Beckwith, S.W., Brusso, J., Busel, J., Fullwood, D., Hayden, M., Howell, D. (eds.) CAMX Conference 2016, pp. 387–401. CAMX, Anaheim (2016)
6. Lee, W.I.L., Springer, G.S.: A model of the manufacturing process of thermoplastic matrix composites. J. Compos. Mater. **21**(11), 1017–1055 (1987)
7. Giusti, R., Lucchetta, G.: Modeling the adhesion bonding mechanism in overmolding hybrid structural parts for lightweight applications. Key Eng. Mater. **611–612**, 915–921 (2014)
8. Colak, Z., Sonmez, F., Kalenderoglu, V.: Process modeling and optimization of resistance welding for thermoplastic composites. J. Compos. Mater. **36**(6), 721–744 (2002)
9. Lee, Y.: Physical properties of poly(ether ether ketone). Doctoral dissertations 1896 – February 2014 (739), University of Massachusetts, Amherst (1988)
10. Marsh, J.J.: Characterisation of crystallization and melting in thermoplastic polymers using chip calorimetry. PhD thesis, University of Birmingham, Birmingham (2016)
11. Furushima, Y., Toda, A., Rousseaux, V., Bailly, C., Zhuravlev, E., Schick, C.: Quantitative understanding of two distint melting kinetics of an isothermally crystallized poly(ether ether ketone). Polymer **99**, 97–104 (2016)
12. Furushima, Y., Nakada, M., Ishikiriyama, K., Toda, A., Androsch, R., Zhuravlev, E., Schick, C.: Two crystal populations with different melting/reorganization kinetics of isothermally crystallized polyamide 6. J. Polym. Sci. Part B Polym. Phys. **54**(20), 2126–2138 (2016)
13. Baeten, D.: Fast scanning chip calorimetry combined with time resolved X-ray diffraction : a new view on polymer crystallization and melting. PhD thesis, KU Leuven, Leuven (2016)

Simulation of Solidification of a Nucleated Isotactic Polypropylene in a Quiescent Condition

Hamed Nokhostin(✉) and Christian Hopmann

Institute for Plastics Processing (IKV), Seffenter Weg 201,
52074 Aachen, Germany
hamed.nokhostin@ikv.rwth-aachen.de

Abstract. Nucleating agents play an important role as additives in the production of injection-moulded components from semi-crystalline thermoplastics. To date, however, no work has been published in the scientific literature, which simulates the influence of nucleating agents on the formation of the microstructure of semi-crystalline thermoplastics. This research gap is to be closed within the framework of the research.

The aim of previous researches at the Institute for Plastics Processing (IKV) was to predict the microstructure of injection-moulded components using an in-house developed model to describe the crystallisation of semi-crystalline thermoplastics, which was implemented into software, called “SphaeroSim”. Although, it was able to simulate only a homogeneous pure melt without any additive. Within this work, the SphaeroSim is further developed to consider the additives and to distinguish between alpha- and beta nucleating agents. To consider the nucleating agents, the possibility to set up predefined nuclei is added to the model. Finally, the growth rate of beta crystals is measured on the hot stage at different temperatures to realise the consideration of beta nucleating agents. Finally, the solidification process on the hot stage is simulated and compared to the experimental results.

Keywords: Crystallisation · Semi-crystalline thermoplastics · Multi-scale simulation

1 Introduction

1.1 Importance of Nucleating Agents

Nucleating agents play an important role as additives in the production of injection-moulded components from semi-crystalline thermoplastics. On one hand, they decrease the cycle time during processing because the component solidifies faster. On the other hand, nucleating agents have a major effect on the nucleation rate and the crystallinity of the polymer matrices. Critical processes occur during the crystallisation process. The decrease in the temperature of the plastic melt leads to

C. Hopmann and R. Dahlmann (Eds.): *Advances in Polymer Processing 2020*,
pp. 301–312, 2020. https://doi.org/10.1007/978-3-662-60809-8_25

crystallisation, usually with the formation of alpha crystals in isotactic polypropylene, which are the most stable crystals with high tensile strength but low toughness [1]. Despite their low thermal stability, the beta crystals exhibit greater toughness. The addition of nucleating agents contributes to the formation of beta crystals, which leads to the equilibrium of the properties through rapid solidification or better order of the microstructure, resulting in the improved mechanical performance of the component. Besides, depending on the application of the component, the addition of well-chosen nucleating agents helps to form small spherulites that do not interfere with light transmission, so that components can be manufactured with high clarity [2].

To take into account the influences of nucleating agents on the formation of the microstructure and the material properties of plastics, experimental investigations have been carried out to date. Their first and most important influence is the crystallisation rate, which increases significantly with the addition of nucleating agents. However, the presence of nucleating agents does not change the crystallisation mechanism [3, 4]. Besides, the crystallisation temperature is also increased by the addition of small amounts of nucleating agents. For example, the crystallisation temperature of semi-crystalline polypropylene increases to about 120 °C [5] due to the addition of only 0.2 wt.% beta-nucleating agents. Important beta-nucleating agents are silica, hydroxides, calcium carbonate and amides.

Microscopic images show that the spherulites of isotactic polypropylene (iPP) grow very slowly to of 100–200 μm. The presence of nanoscale beta-nucleating agents (a styrene-butadiene-rubber based nucleating agent) leads to a higher number of nuclei and a dramatic decrease in spherulite diameter (about 3 μm), resulting in a very fine and homogeneous distribution of crystals [5]. However, the addition of beta-nucleating agents is the most effective and accessible method to obtain a high amount of beta-modification in iPP. Although beta-modification can improve the toughness of the material, the softer behaviour of beta-iPP also brings a significant decrease in strength and stiffness [1, 6, 7]. However, it is possible to compensate for the loss of strength with other additives such as ultra-high molecular weight polyethylene (UHMWPE) [8]. Besides, the impact strength can be increased by more than 100% using beta-nucleating agents [9].

To date, no scientific literature has been published that simulates the influence of nucleating agents on the formation of the microstructure of an injection-moulded component made of semi-crystalline thermoplastics. This research gap is to be closed within the framework of this research.

1.2 A Multiscale Simulation Chain to Predict Microstructure Dependent Effective Elastic Properties of Injection-Moulded Polypropylene

Previously, we have introduced a multi-scale integrative simulation chain on an ICME platform [11] to calculate the effective mechanical properties of an injection-moulded part, by considering the microstructure and the resulting inhomogeneity. It consists of three self-developed simulation tools, which overall work on three characteristic scales. The chain starts with the injection moulding simulation, where filling, cooling and holding pressure of a part can be numerically simulated. It is a macroscale simulation, where three main equations are solved: conservation of mass, conservation of momentum and energy, and multiphase flow. All material properties are modelled independence of a macroscale crystallisation degree computed with a simple Nakamura approach [12]. The results are transferred to another tool – SphaeroSim – to describe the solidification process on a mesoscale. The goal is to simulate the spherulitic microstructure of the part. An event-based model is adopted to develop SphaeroSim, which describes the solidification process via the nucleation and growth model. For implementation, a Cellular Automaton model is used, in which the phase state of small melt volumes depends on the state of its direct neighbour volumes [13]. Third and last step of the simulation chain is the calculation of the effective properties of the previously determined inhomogeneous microstructure. For the homogenisation, a two-step asymptotic approach is used. In the first step, the structure of a single crystalline is homogenised on the nanoscale. It is composed of pure amorphous and crystalline phases of the material, which has to be characterised in terms of mechanical properties. It is one of the biggest challenges for the homogenisation, since it is difficult to separate and create purely amorphous and crystalline phases. Direct measurements are almost impossible. Therefore, the simulation chain was extended by a molecular dynamics simulation approach to calculate the mechanical properties of purely crystalline and amorphous phases. This enables the consideration of different materials or modifications of the base material. In this way, the presented simulation chain can be used in a more agile and individual way. In the second step of the homogenisation, the properties of the spherulites have to be calculated using the data from the calculated microstructure. Therefore, the mesoscale mesh of the solidified microstructure is used. The simulation chain is illustrated in Fig. 1.

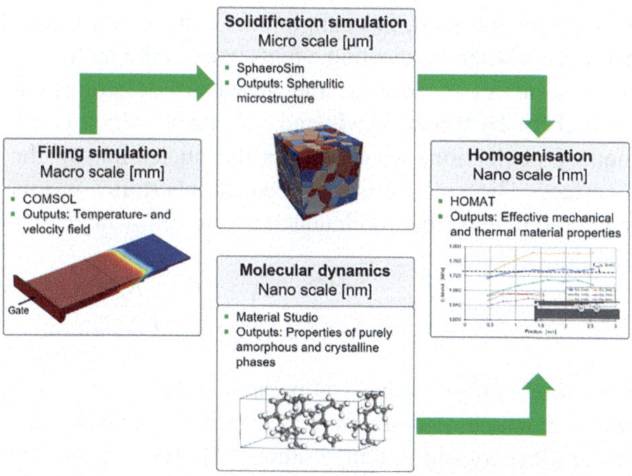

Fig. 1. The developed multi-scale integrative simulation chain

2 A New Model to Predict the Microstructure of Semi-Crystalline Thermoplastics

In SphaeroSim, a microscale crystallisation model was implemented. The model is based on the nucleation and growth approach. First, a nucleation model was derived from Ziabicki's formalism [22]. In contrast to Ziabicki's work, the Gibbs energy was retained as an independent model parameter for as long as possible, since it offers the possibility of mapping process-dependent influences by the melt flow. To map the non-linear viscoelastic behaviour of the plastic melts, relaxation functions based on the reptation theory were additionally introduced into the model description. For the calculation of crystal growth, the standard model according to Hoffman, Davis and Lauritzen was used [23]. In the course of the model implementation, the crystallisation model was implemented in SphaeroSim using the cellular automata. In our implemented approach, the problem was first transformed into a state formulation, i.e. each element is assigned a unique state. In the implementation, an event-based approach was used. An event describes a point in time at which state changes of the element can happen. Nucleation and growth events are distinguished. Besides, it is possible to map flow-dependent and flow-independent as well as thermal and athermal nucleation using SphaeroSim [12, 14].

Adding special nucleating agents to the iPP may result in the presence of both alpha and beta modifications in one product. Since they have different properties, it is important to distinguish between them in the predicted microstructure. Our crystallisation model was firstly able to consider only one crystal modification, whose unit cell had a basic geometric structure with the form of a parallelepiped. A new nucleation model as well as a new growth model should be implemented in SphaeroSim to consider the beta modification. Although, within the scope of this work we only focus on the implementation of the growth model. To eliminate the effect of beta nucleation, we decided to study the nucleated iPP, since the homogenous nucleation, which happens in the pure parts of the melt, is negligible in comparison to the heterogeneous nucleation, which occurs where nucleating agents exist in the melt.

The implemented model to describe the growth of spherulites is based on the widely used approach of Hoffman, Davis and Lauritzen, which consider the growth process as secondary nucleation, which means the nucleation on the surface of an existing crystal surface. The radial growth speed of spherulites in this model is only influenced by temperature and can be calculated by the following equation.

$$v_G = v_0 exp\left(-\frac{U^*}{R(T - T_\infty)}\right) exp\left(-\frac{K_g}{T\Delta G_q}\right) \quad (1)$$

v_0 and K_g correspond to a constant material parameter, U^* to an activation energy and T_∞ to a temperature, at which no more molecular movement takes place. T_∞ is usually 30–60 °C below the glass transition temperature [15]. The required material parameters are already determined experimentally in our previous works [16].

To consider the beta modification, the implemented growth model was extended to consider both alpha and beta modifications, so that two different growth speeds can be calculated for each time and temperature, which can be assigned to the related crystal modification. Nucleating agents were also modelled as predefined nuclei, which can be either alpha or beta. It should be mentioned that the existence of both modifications as nucleating agents is also possible. Since in the numerical implementation of the crystallisation model volumetric elements are used, a simplification is taken into account: We assume that the sizes of the nucleating agents are as big as one element, which in SphaeroSim can be one micrometre, minimum.

3 Set of Experiments to Calibrate the Implemented Model for Simulation

To simulate the nucleated iPP, the growth speed was experimentally measured using isotactic polypropylene 505P from Sabic BV, Bergen Op Zoom, Netherlands. NJSTAR NU-100 from New Japan Chemical Co., Ltd., Osaka, Japan and talc powder from Dr. Lohmann Diaclean GmbH, Dortmund, Germany were used as beta- and alpha-nucleating agents, respectively. A twin-mixer of type 350/350-E from Brabender GmbH & Co. KG, Duisburg, Germany was used to prepare the compounds (Table 1).

Table 1. Combination of the prepared compounds

Num.	Abbreviation	Material	Nucleating agent (NA)	Weight% of NA
1	Ref. I	iPP 505P	–	
2	Ref. II	iPP 505P	–	
3	Combi. I	iPP 505P	Talc	0.5
4	Combi. II	iPP 505P	Talc	0.1
5	Combi. III	iPP 505P	NJSTAR	0.5
6	Combi. IV	iPP 505P	NJSTAR	0.1
7	Combi. V	iPP 505P	NJSTAR	0.002
8	Combi. VI	iPP 505P	NJSTAR	0.0002

To prove the presence of beta modification in the material, DSC measurements were carried out. Since alpha and beta modifications have different melting temperatures [17], it was expected that two peaks could be observed by heating the material [18, 19]. Therefore, all eight mixed materials were examined using a Q1000 type DSC from TA Instruments, Belgium.

To measure the growth speed of spherulites, experiments on the hot stage were carried out under the polarisation microscope, in which a very thin layer of iPP was heated up to 180 °C and then cooled down to a certain temperature. The cooling process should happen as fast as possible, since the measurements have to be done at a constant temperature and not during cooling. If it is not fast enough, the sample solidifies before growth speed can be measured. From the beginning of the plateau temperature, images were captured every 10 s. Later the radius of the single spherulite was measured and the growth speed was determined for each temperature. Then by a parameter fitting the required material parameters were calibrated for the beta modification.

To later verify the results of the new implementation, a new variable was defined as an output for SphaeroSim, which is a number with two digits (Table 2). The first digit distinguishes homogeneous (normal) from heterogeneous (nucleated) nuclei, which can be either zero for homogeneous or one for heterogeneous nuclei. The second digit distinguishes different crystalline modification, which can be either zero for alpha modification or one for beta modification in this context.

Table 2. Values and related meanings for the new defined output

Value	Description
00	Normal alpha nuclei
01	Nucleated alpha nuclei
10	Normal beta nuclei
11	Nucleated beta nuclei

4 Results

The DSC-results show that the two references contain only alpha modifications (see Fig. 2). This is also the case with the combinations with talc powder. However, with the second nucleating agent (NJSTAR) the two modifications are clearly visible at the same time. It can be seen that the proportion of nucleating agent in Combi. VI is so low, that no beta modification can be observed.

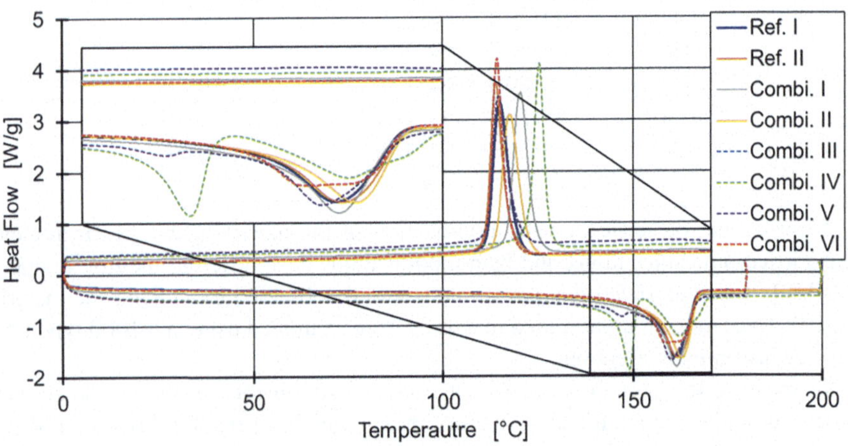

Fig. 2. The heating and cooling curves for prepared compounds.

Since alpha and beta spherulites have different structures at the molecular level, it was expected to see these differences using a polarisation microscope, since the lamellar structure of the spherulite can be observed only with a polarisation filter. In the literature, there are images in which beta spherulites are distinguished from alpha [17–20]. To verify this claim, thin sections of the Combis III, IV and V were prepared and examined under the polarisation microscope on a FTIR-600 heating stage from Linkam Scientific Instruments Ltd., Epsom, UK. However, the microstructures of Combi. III and IV were so fine that no single spherulites were visible. With the Combi. V, spherulites were clearly observed under the microscope (see Fig. 3).

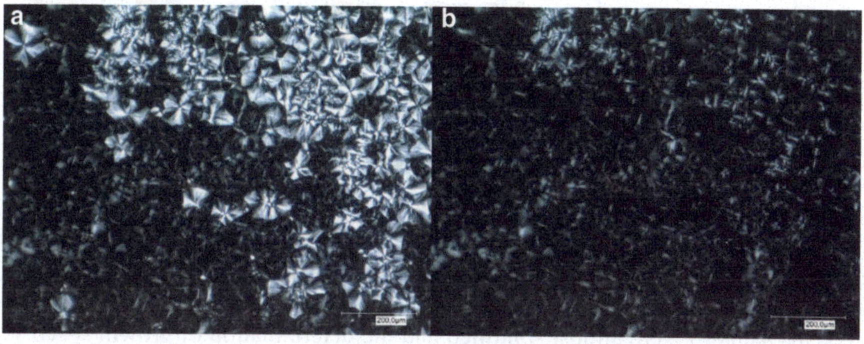

Fig. 3. Microscopic image of Combi. V: (**a**) at room temperature (**b**) at a temperature of 150 °C.

Figure 3a shows the contrast between two groups of spherulites with different brightness. According to the literature, the bright spherulites are beta-modifications and the darks alpha-modifications. The melting temperatures of the two modifications were determined from the DSC results:

- Alpha: 158.87 °C
- Beta: 144.5 °C.

To verify the observation, the thin section was again heated up to 150 °C on a hot stage. As expected, the beta spherulites melted when this temperature was reached, which shows that the light spherulites correspond to beta modifications (Fig. 3).

Additional hot stage experiments were carried out with the Combi. V to measure the growth speed. Five different cooling rates (50, 60, 70, 80 and 90 K/min) with four plateau temperatures (122, 127, 132 and 137 °C) were defined for these experiments. The results can be seen in Fig. 4. The growth speed could not be measured at 132 °C with 80 K/min, since the solidification was too fast.

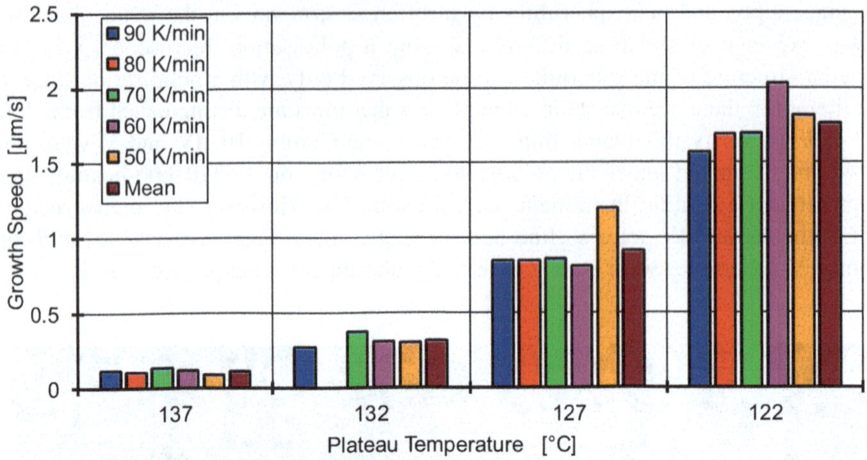

Fig. 4. The measured growth speed of the beta-spherulites.

Comparing our results with data from the literature (see Fig. 5), it can be seen that both are in a good agreement. However, as the plateau temperature decreases, our results deviate more. It might be due to the fact, that after reaching the lower temperatures, the samples were almost solidified and the determination of growth speed was based on only three captured images.

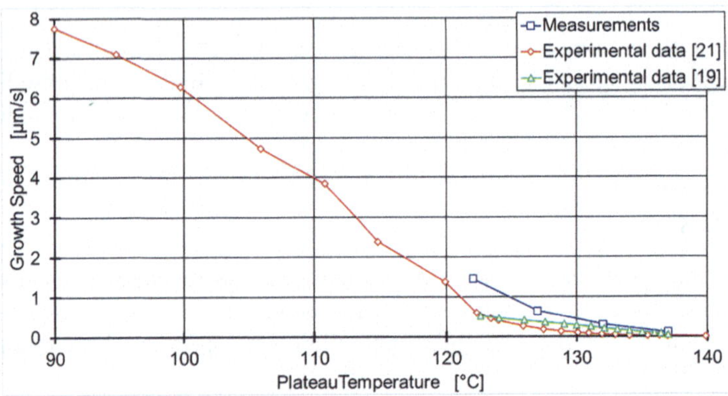

Fig. 5. Comparison of measured growth speed with literature values.

The required material parameters were determined for the growth model and inserted into the SphaeroSim. As can be seen in Fig. 6, the colored points represents the nucleating agents, which are implemented as predefined nuclei and exist from the beginning of the simulation.

Fig. 6. The presence of nucleating agents at the beginning of the simulation.

Simulation of the solidification process (see Fig. 7) shows us that in the defined simulation area (300 * 300 * 1 μm) with 180 predefined nuclei only three new nuclei were created from the melt. It proves that our assumption was correct and nucleating agents dominantly govern the nucleation process and the homogenous nucleation can be neglected.

Fig. 7. Simulated solidification process of Combi. V.

Finally, the predicted microstructure is compared with a captured microscopic image with the same temperature history (see Fig. 8). The highest spherulite diameter from the simulation is 51 μm. From the microscopic image, this value was around 62 μm, which shows a very good agreement between simulation and experiment. Although, it should be mentioned, that the nucleating agents are not homogenously distributed within the iPP, since the mixing quality by twin mixer is not ideal.

Fig. 8. Comparison between the spherulite diameters of the predicted and the captured microstructure for Combi. V with the cooling rate of 90 K/min and the plateau temperature of 122 °C.

5 Conclusion

SphaeroSim, simulation software for crystallisation of semi-crystalline thermoplastics, was further developed within the scope of this work to consider both crystal modifications of alpha and beta. The model was experimentally calibrated for beta modification of iPP employing hot stage measurements. The results show a need for more investigations regarding this matter and a measurement of the growth speed at lower temperatures is required. The hot stage experiment was simulated, where temperature sinks from 180 °C to four different plateaus and stops until the sample solidifies. The results show a good agreement with experiments, comparing the diameters of the spherulites. Finally, it can be said that SphaeroSim can predict a nucleated semi-crystalline thermoplastic under consideration of different crystal modifications.

In future work, an additional nucleation model should be implemented to SphaeroSim to simulate the solidification process under consideration of different crystal forms and distinguish between them. To achieve that, experiments should be carried out to determine the required material parameters for both alpha and beta modifications of iPP. Furthermore, studies to determine the thermal and mechanical properties of each modification should be done, to calculate the effective properties of an injection-moulded part with a combination of both phases more accurately.

Acknowledgment. This dedicated research has been funded by the *Deutsche Forschungsgemeinschaft* (DFG) as a part of the project "Simulation of the development of the microstructure of injection-moulded semi-crystalline thermoplastics by means of a multi-scale approach under consideration of shear-induced crystal forms (alpha and beta)" (Project-ID: HO 4776/53-1). We would like to extend our thanks to the DFG.

References

1. Tordjeman, P., Robert, C., Marin, G., Gerard, P.: The effect of α, β crystalline structure on the mechanical properties of polypropylene. Eur. Phys. J. **4**(4), 459–465 (2001)
2. Wypych, G.: Handbook of Nucleating Agents. ChemTec Publishing, Canada (2016)
3. Palza, H., Vera, J., Wilhelm, M., Zapata, P.: Spherulite growth rate in polypropylene/silica nanoparticle composites: effect of particle morphology and compatibilizer. Macromol. Mater. Eng. **296**(8), 744–751 (2011)
4. Xu, T., Zhang, A., Zhao, Y., Han, Z., Xue, L.: Crystallization kinetics and morphology of biodegradable poly(lactic acid) with a hydrazide nucleating agent. Polym. Test. **45**, 101–106 (2015)
5. Petchwattana, N., Covavisaruch, S., Sripanya, P.: Effect of nano-scaled styrene butadiene rubber based nucleating agent on the thermal, crystallization and physical properties of isotactic polypropylene. J. Alloys Compd. **582**, 190–195 (2014)
6. Karger-Kocsis, J.: How does "phase transformation toughening" work in semicrystalline polymers? Polym. Eng. Sci. **36**(2), 203–210 (1996)
7. Varley, R.J., Dell'Olio, M., Yuan, Q., Khor, S., Leong, K.H., Bateman, S.: Different β nucleants and the resultant microstructural, fracture, and tensile properties for filled and unfilled ISO polypropylene. J. Appl. Polym. Sci. **128**(1), 619–627 (2013)
8. Zhang, Y., Zhang, L., Liu, H., Du, H., Zhang, J., Wang, T., Zhang, X.: Novel approach to tune mechanics of β-nucleation agent nucleated polypropylene: role of oriented β spherulite. Polymer **54**(21), 6026–6035 (2013)
9. Kersch, M., Schmidt, H.W., Altstädt, V.: Influence of different beta-nucleating agents on the morphology of isotactic polypropylene and their toughening effectiveness. Polymer **98**, 320–326 (2016)
10. Michler, G.H., Balta-Calleja, F.J.: Nano- and Micromechanics of Polymers. Hanser, München (2012)
11. Laschet, G., Spekowius, M., Spina, R., Hopmann, C.: Multiscale simulation to predict microstructure dependent effective elastic properties of an injection molded polypropylene component. Mech. Mater. **105**, 123–137 (2017)
12. Spina, R., Spekowius, R., Hopmann, C.: Multiphysics simulation of thermoplastic polymer crystallization. Mater. Des. **105**, 455–469 (2016)
13. Wienke, S., Spekowius, M., Dammer, A., Mey, D.A., Hopmann, C., Müller, M.S.: Towards an accurate simulation of the crystallization process in injection molded plastic components by hybrid parallelization. Int. J. High Perform. Comput. Appl. **28**(3), 356–367 (2013)
14. Spina, R., Spekowius, M., Hopmann, C.: Analysis of polymer crystallization with a multiscale modeling approach. Key Eng. Mater. **611**, 928–936 (2014)
15. Lamberti, G.: A direct way to determine iPP density nucleation from DSC isothermal measurements. Polym. Bull. **52**(6), 443–449 (2004)
16. Spekowius, M.: New Microscale Model for the Description of Crystallization of Semi-Crystalline Thermoplastics. Verlag Mainz, Aachen (2017)
17. Lotz, B.: α and β phases of isotactic polypropylene: a case of growth kinetics `phase reentrency' in polymer crystallization. Polymer **39**(19), 4561–4567 (1998)
18. Lovinger, A.J., Chua, J.O., Gryte, C.C.: Studies on the α and β forms of isotactic polypropylene by crystallization in a temperature gradient. J. Polym. Sci. Polym. Phys. Ed. **15**(4), 641–656 (1977)
19. Zhang, B., Chen, J., Zhang, X., Shen, C.: Formation of β-cylindrites under supercooled extrusion of isotactic polypropylene at low shear stress. Polymer **52**(9), 2075–2084 (2011)

20. Nakamura, K., Shimizu, S., Umemoto, S., Thierry, A., Lotz, B., Okui, N.: Temperature dependence of crystal growth rate for α and β forms of isotactic polypropylene. Polym. J. **40**, 915 (2008)
21. Janeschitz-Kriegl, H.: Crystallization Modalities in Polymer Melt Processing. Springer, Wien (2010)
22. Ziabicki, A.: Generalized theory of nucleation kinetics. IV. Nucleation as diffusion in the space of cluster dimensions, positions, orientations, and internal structure. J. Chem. Phys. **85**(5), 3042–3057 (1986)
23. Hoffman, J.D., Davis, G.T., Lauritzen, J.I.: The rate of crystallization of linear polymers with chain folding. In: Hannay, N.B. (ed.) Treatise on Solid State Chemistry. Springer, New York (1976)

Investigations on the Influence of High Pressures on the Curing Behaviour and Material Properties of Composite Structures for the Development of a Material Model

Nadine Gushurst[1(✉)], Tim Frerich[1], and Axel S. Herrmann[1,2]

[1] Faserinstitut Bremen e. V, Bremen, Germany
gushurst@faserinstitut.de
[2] Universität Bremen, Bremen, Germany

Abstract. Former investigations have shown, that high pressure influences the cure behaviour and mechanical properties of epoxy resins. Simulation approaches for manufacturing processes of composite parts developed so far do not consider pressure dependency, since process pressure is generally low. In recent developments, curing of epoxy resin under high pressure occurs, for example during the high-pressure resin transfer moulding (HP-RTM) process. In this study, a model approach is presented, which considers the effect of pressure by augmenting an existing simulation approach using free-volume theory. It could be demonstrated, that a description of the material properties based on the fractional free volume is possible and thus the pressure dependency could be brought into the process simulation via free volume, in theory. In order to verify the practical relevance, experimental investigations were conducted to confirm an effect of pressure on dielectric properties, conversion and glass transition temperature of epoxy resins. An increase of process pressure leads to a higher final degree of cure, glass transition temperature and ion viscosity. The influence of high pressure on mechanical properties, such as elastic modulus or fracture stress could not be proven conducting tensile tests.

Keywords: Process simulation · Material modelling · Pressure dependency · High-pressure RTM

1 Introduction

Process simulation offers valuable insights for composite applications with high requirements concerning quality and structural performance as well as tight shape tolerances, such as aerospace components. Process simulation does not only reduce time and costs during the process design, but also leads to a profound process understanding which benefits process stability and robustness and consequently the part quality. In most of the manufacturing processes of composite parts, low pressure (<10 bar) is

C. Hopmann and R. Dahlmann (Eds.): *Advances in Polymer Processing 2020*,
pp. 313–323, 2020. https://doi.org/10.1007/978-3-662-60809-8_26

used and its influence is considered negligible throughout most of the literature. In recent developments, curing of epoxy resin under high pressure, up to 120 bar, occurs, for example during the high-pressure resin transfer moulding (HP-RTM) process, where high pressure inside the mould is a secondary effect from fast injection. Using the mixing technology from polyurethane industry and highly reactive resins, this method results in shorter injection times as well as in the reduction of curing times and process costs in comparison to the traditional RTM process, which makes the process attractive for aerospace applications. Former investigations have shown that high process pressures have a significant influence on the curing reaction and the material properties of epoxy resins. Consequently, the influence of pressure should no longer be neglected within simulation of the HP-RTM process.

For the development and optimization of the RTM-process physical based process simulation tools have been developed and published, amongst others, in [1–3], which are capable to predict temperature and degree of cure distribution as well as process induced stresses and distortions. This simulation approach is a knowledge-based method and models the occurring chemical and physical processes during cure. This is achieved by a simulation consisting of two coupled modules:

1. A thermo-chemical simulation, which calculates the progression of the resins cure.
2. A thermo-mechanical simulation, which simulates the build up of stresses and distortions during cure.

This simulation procedure for the RTM-process is schematically displayed on the left side of Fig. 1.

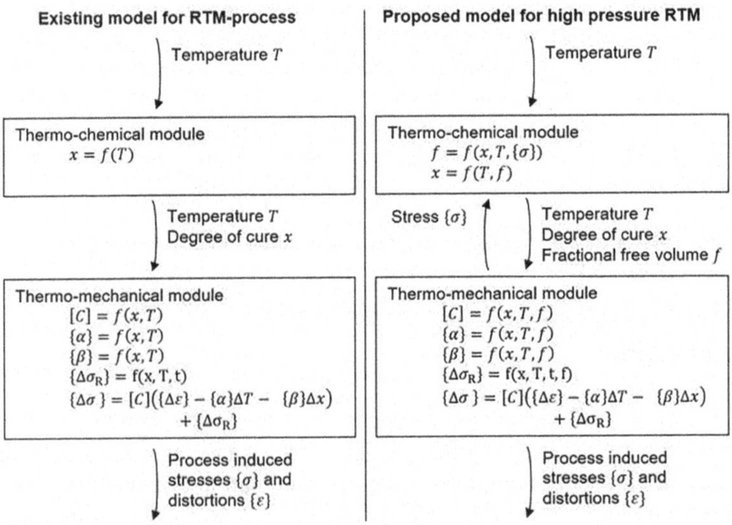

Fig. 1. Schematics of the existing simulation of the RTM-process (left) and the proposed model for high-pressure RTM (right).

The thermo-chemical module calculates the degree of cure x based on the applied temperature T. The degree of cure is used as a status variable to determine all other material properties. The thermo-mechanical module calculates the stiffness matrix $[C]$, the thermal expansion coefficients α, the chemical shrinkage β and the stress relaxation $\Delta\sigma_R$, based on the degree of cure and the temperature. On the basis of the material parameters of the matrix, the material parameters for the composite material are determined by the use of mixture rules. Based on this material parameters an equation system, which gives the dependency of the stresses on the strains, is created. This equation system is used as a material model for the FEM-calculation.

For the RTM-process, where low pressures of approximately 8 bar are used, the influence of the pressure is considered negligible throughout most of the literature, therefore only the dependencies from the degree of cure and temperature are considered in the model.

For high pressure RTM, the pressure has a far more significant influence on the material properties. For example, Cruz and Osswald [4] showed that higher pressures lead to an increase of the final degree of cure. Also reducing effect of increased pressure on the chemical shrinkage has been noted [5]. Both authors attribute the effect to the decreased distance between reacting molecules, as the increased pressure compresses the material.

The aim of this study is to find a process simulation approach for the HP-RTM, which takes into account the effects of high pressure on cure kinetics and the mechanical properties. In addition, the practical relevance of the mathematical derivation shall be verified by selected experimental investigations on pressure dependency of an epoxy resin. At the same time, these experiments serve as basis for the creation of the material models presented.

2 Model Approach

In this study, the effect of high pressure on curing behaviour and properties of an epoxy-resin in process simulation shall be considered by augmenting the existing simulation approach using free-volume theory. The proposed model approach is shown on the right side of Fig. 1. For this purpose a temperature, pressure and cure dependent formulation of free volume as well as free volume dependent models for all material parameters are necessary to develop. The free-volume theory considers polymers to consist of two phases: the free volume υ_f, in which no molecules are present and the occupied volume υ_o, where polymer molecules are present. The free volume is integrated into the model approach via f, the fractional free volume, which is defined as free volume divided by the specific volume.

In the following, existing free volume dependent formulations for the material properties of polymers shall be discussed and it will be demonstrated, that a description of the material properties based on free-volume-theory is possible.

Within the thermo-chemical module, a free volume dependent cure-kinetics model for an epoxy-resin is required. This has been suggested in [6] for instance. The basic equation to describe the reaction rate is presented in Eq. 1, where x is the degree of cure, and K_1 and K_2 are Arrhenius-constants to describe the temperature dependence of the reaction

$$\frac{dx}{dt} = [K_1 + K_2(1-x)]x^2. \quad (1)$$

In [6] the following formulation is used to describe the Arrhenius-constants, where E_a is the activation energy, R is the universal gas constant and A_i as well as δ, are constant factors. D is the molecular diffusivity of the material

$$K_i = \frac{A_i \exp(-E_a/RT)}{1 + (\delta/D)\exp(-E_a/RT)}. \quad (2)$$

By substituting the molecular diffusivity D according to the following equation, where D_0 and b_d are constant factors, [6] incorporates the fractional free volume f into the cure-kinetics model

$$D = D_0 \exp(-E_a/RT) exp(-b_d/f). \quad (3)$$

Therefore the reaction rate can be formulated as a differential equation dependent on the degree of cure x, the temperature T and the fractional free volume f.

Within the thermo-mechanical module, free-volume dependent models for stiffness, thermal expansion and chemical shrinkage are needed. The mechanical behaviour of polymers cured to a specific degree of cure is visco-elastic. The properties can be described by a master-curve, which provides the elastic modulus in dependence of time for constant loads. For a modulated load, the master curve provides the storage and loss modulus in dependence of the frequency. Both master curves contain the same information and may be transferred into each other. Master curves are obtained, by measuring the modulus over a certain time or frequency domain at different temperatures. The curves for all but one temperature are then shifted, in a logarithmic time or frequency domain, until all different curves form one continuous master-curve. The shift in the logarithmic time or frequency domain is described by a shift-factor $\log(a_T)$. This master-curve describes the behaviour of the material at the temperature of the un-shifted curve. Different approaches are available to describe the master-curve mathematically. One commonly used approach in the literature, for instance [7], is the following equation, called Prony-series

$$\mathrm{E}(t, T, x) = E_\infty(x) + \sum_{i=1}^{n} E_i \exp\left(\frac{-t}{\tau_i(T, x)}\right) \quad (4)$$

In [7] the increase of the equilibrium rubbery module $E_\infty$is described in dependence of the degree of cure x only, the value is independent of temperature in the investigated temperature range. Therefore, the rubbery equilibrium module will be considered as only influenced by the increase of interconnection in the forming network, but not by the change in free volume. The relaxation time τ_i can be described by the relaxation time at a reference temperature and reference degree of cure and the previously mentioned shift factor a_T, which describes the shift of the curve on the logarithmic time scale. A free-volume based description of the shift factor has been presented in [8], which is based on the Doolittle-equation for the dependence of viscosity on free volume. In this equation f_g is the fractional free volume at glass transition temperature

$$\log(a_T) = (1/2.303)\left(1/f - 1/f_g\right). \tag{5}$$

A model for the thermal expansion based on free volume theory can be deduced from the definition of free-volume used by several authors, for example [8, 9]. They define the temperature dependent free-volume as the difference in the thermal expansion coefficient above and below the glass transition temperature. Below the glass transition temperature, thermal expansion is considered to be caused by an increase in occupied volume only, while above the glass transition temperature free volume increases as well. This can be described by Eq. 6, where α_{aTg} is the thermal expansion coefficient above the glass transition temperature and α_{bTg} is the thermal expansion coefficient below the glass transition temperature

$$f = f_g + \left(\alpha_{aTg} - \alpha_{bTg}\right)\left(T - T_g\right). \tag{6}$$

The chemical volume shrinkage is usually described in the literature by a simple, linear dependency from the degree of cure as in [2, 5]. However a free-volume dependent description should be possible. The occupied volume will not be influenced by the reaction of the monomers with each other, only the free volume will diminish. Therefore, the chemical shrinkage should be equivalent to the reduction of free volume during the reaction.

A temperature, pressure and cure dependent formulation of free volume, which can be inserted into the other material models, in order to implement pressure dependency into the material model is possible to derivate from dielectric measurements. The measurement of most material properties at high pressure is not possible using standard equipment. While high pressure DSCs do exist, in order to characterize the curing behaviour at high pressure, for example the viscoelastic behaviour after gelation of polymers under high pressure cannot be measured. However, measurement of the dielectric properties are possible using the dielectric analysis (DEA), whose sensors can be implemented into the mould for high pressure RTM with relative ease. The Dielectric Analysis (DEA) is usually used for dielectric cure monitoring to track the cure state of a resin in real time. It works by measuring the dielectric properties conductivity σ and permittivity ε, which arise from ionic current and dipole rotation in a stimulated sample. With growing cross-linked network during cure, the mobility of the ions and thus the conductivity decreases while resistivity increases. Commonly the ion viscosity ρ, which is the frequency independent resistivity (or DC resistivity), is observed when tracking the advancing cure [10]. This measurand is related to both mechanical viscosity and cure state of epoxy resin. In [6] a free-volume dependent model for the conductivity has been suggested, by which the relationship between dielectric properties and mechanical properties can been explained. The model considers the effect of the fractional free volume f as well as the effect of the molecular mass M, which can be connected to the degree of cure. All other parameters are constant factors.

$$\sigma = A_\sigma M^{-m_\sigma} \exp\left(-E_a/RT\right) exp(-b_d/f). \tag{7}$$

The equation suggested in [6] can be used to evaluate the fractional free volume by dielectric measurements.

The presented models are deduced from different materials than the one, which shall be investigated within this study. Therefore, the models will have to be modified to describe the examined material adequately. In principle, however, it could be demonstrated that a free volume-dependent formulation of the material models and thus the implementation of the suggested simulation approach is possible in theory.

3 Experimental Investigations on Pressure Dependency

The experimental investigations on pressure dependency are divided into two sections. On the one hand, the effect of internal mould pressure on the resin during the manufacturing process is investigated online by dielectric measurements. On the other hand, the dependency of the final material properties on the process pressure is investigated. Results of final parameters of cure and mechanical properties are presented.

For the experimental study on the pressure dependency samples, including pure resin and carbon fibre reinforced plastic (CFRP) samples, were produced using the HP-RTM manufacturing process. Process conditions like curing temperature, curing time and in-mould pressure were varied. The samples were cured using isothermal temperatures of 140 °C (40 min curing time) or 180 °C (25 min curing time) and pressures of 8, 20, 40, 60 and 80 bar, respectively. Pressure was adjusted by pressure sensors, which regulate the amount of injected resin.

These studies concentrate on an aircraft certificated 2-component epoxy resin, developed for HP-RTM. For the composite samples, preforms made of UD carbon fibre fabric G1157 from Hexcel were inserted. The fibre volume content of the samples was about $55 \pm 3\%$.

3.1 Pressure Effects During Curing

In this study, an instrument type DEA 230/10 from Netzsch Gerätebau GmbH in combination with a tool mount sensor, integrated into the mould, for recording the dielectric properties during manufacturing, was used. The evaluation shows a negligible frequency dependence of the ion viscosity for low frequencies up to 10 Hz, which was chosen for the analysis of the results.

The development of ion viscosity during manufacturing of the pure resin samples at 180 °C is illustrated in Fig. 2. The different stages during cure of the resin, derived from the curves, are labelled using the 80 bar curve as an example. The measured values of the repeated measurements at 8 bar, show a large scattering range, despite identical process parameters (expect of curing time). The curves with an early drop of the slope cannot be explained from a chemical-physical point of view. This may result from sensitivity of the measuring method to environmental influences. Small fluctuations in mould temperature or electric fields from devices such as the machine press can have an influence on the measurement. The measurement results with higher pressures are very close and follow the same trend. A pressure dependent trend can be observed towards the end of the curing time, where the ion viscosity of 60 and 80 bar is slightly higher than in the 20 and 40 bar curves. However, there is only one measurement curve per pressure stage, which is why the composite samples are included in the consideration.

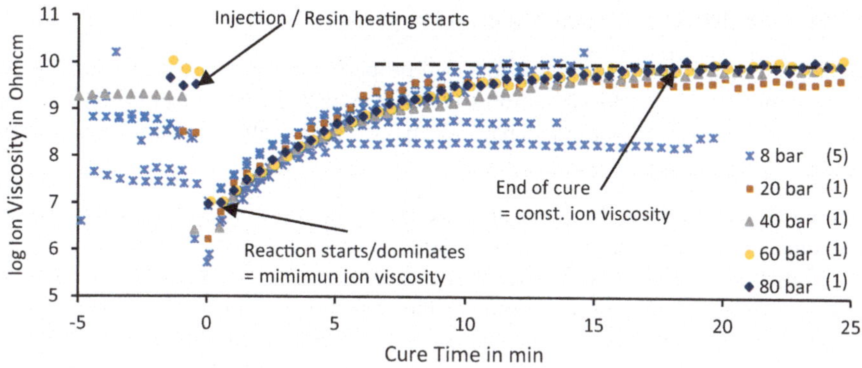

Fig. 2. Logarithmic ion viscosity of pure resin samples.

Figure 3 shows the ion viscosity recorded during curing of the manufactured composite samples at 180 °C (left axis) and 140 °C (right axis), respectively. Two axes were chosen so that the curve progression can be better represented. For both process temperatures the ion viscosity at 80 bar is higher than at 40 bar, where at 180 °C the effect is more distinct.

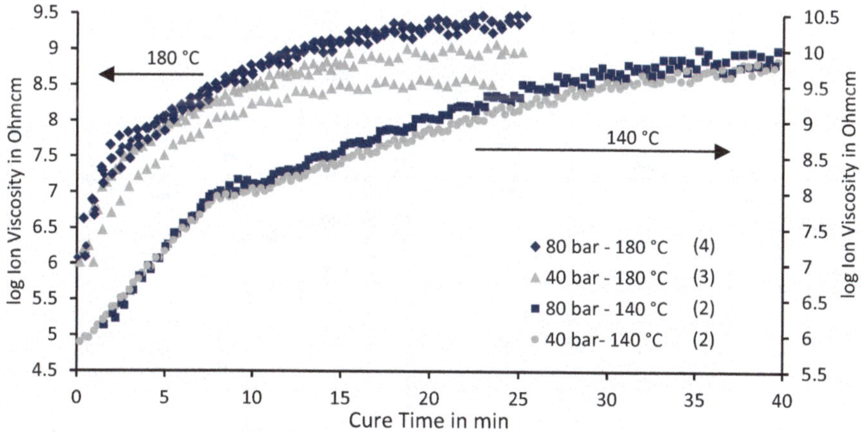

Fig. 3. Logarithmic ion viscosity of composite samples.

This result arises out of the restriction of the flow of ions due to its limited mobility at higher pressure. There are two possible explanations for this phenomenon, that apply either separately or in combination: (1) The free ion mobility is limited due to the smaller available free volume left at high pressure. (2) The ion flow is restricted due to the better developed cross-linked network at high pressure. This second possibility includes the first, since the free volume decreases with increasing degree of cure (see Eq. 6). In order to quantify the influence of pressure on the degree of curing, the final degree of cure was determined by means of a DSC analysis.

3.2 Pressure Effects on Final Material Properties

Since in-situ measurements of most material properties is not possible, the final material properties of the produced samples were investigated. In the following, the final cure parameters as well as the mechanical characteristic values in dependence of the production pressure are presented.

Final cure parameters. In order to determine the final degree of cure and the glass-transition temperature of the samples manufactured under different process conditions, dynamic Modulated Differential Scanning Calorimetry (MDSC) measurements with heating rates of 10 K/min in the temperature range of 25–260 °C were conducted. Temperature was modulated with a temperature amplitude of 1.06 °C and a period of 40 s to obtain reversible and non-reversible signals. The measurements took place in a DSC 250 Discovery with RCS 90 cooler from TA Instruments.

Figure 4 displays the final conversion (left) and the glass-transition temperature (right) of the resin samples after 25 min curing at 180 °C depending on manufacturing pressure. The conversion results show that pressure indeed influences the degree of cure with otherwise same curing conditions. By increasing the in-mould pressure from 8 to 80 bar, the average final degree of cure increases from 97.4 to 98.5 %. A similar trend to the conversion can be observed for the results of the glass-transition temperature, which correlates with the degree of cure. Moreover, the scatter of the results decreases at high pressure (see Fig. 4)

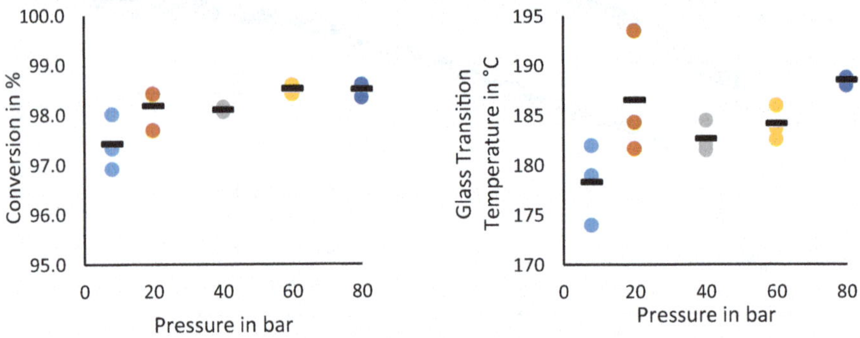

Fig. 4. Conversion (left) and glass-transition temperature (right) of pure resin samples.

The tendency of increasing degree of cure with increasing pressure corresponds to the results of the DEA. The measured raise of conversion by pressure increase is slightly below the perceived effect of [4] (approximately 2% with 69 bar pressure increase), but it should be noted that the absolute value of the conversion is also about 10% higher in this case. [4] has shown that with extend periods of curing time the degree of cure will finally converge to the same level. It can therefore be assumed that the influence of the pressure on the degree of cure will be greater at an earlier point in cure time.

Mechanical properties. The final mechanical properties such as Young's modulus and fracture stress of the resin samples dependent on different manufacturing pressures are examined by tensile tests. The tests were performed using a Zwick 1474 with an applied cross-head speed of 2 mm/min and strain gauges, measuring the strain during the experiment. The sample geometry of 2 mm × 10 mm × 140 mm deviates from testing standards, since tooling size limits test specimen size.

The stress–strain curves in Fig. 5 represent two of five tensile test results per pressure level, which enclose the results area. The curves do not differ greatly and no pressure-dependent trend can be detected either.

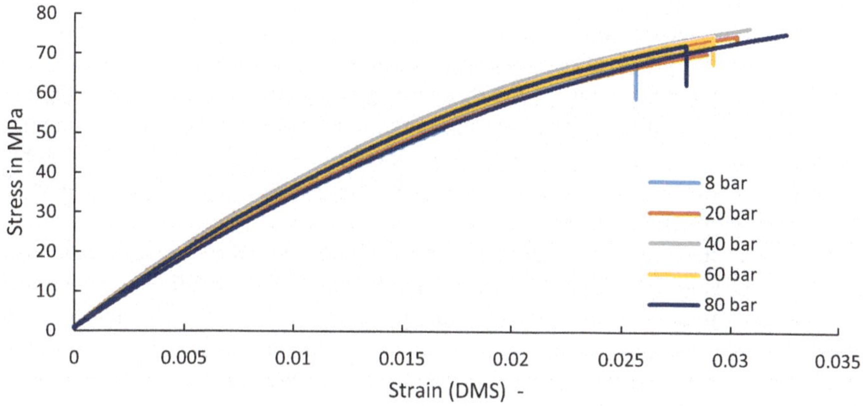

Fig. 5. Stress–strain curves of pure resin.

A closer look at the mechanical properties elastic modulus and fracture stress in Fig. 6 reveals that no significant difference can be observed, except the results of the 8 bar sample. One reason for the lower values, especially for the Young's modulus, may be the poor surface quality of the test specimens at 8 bar.

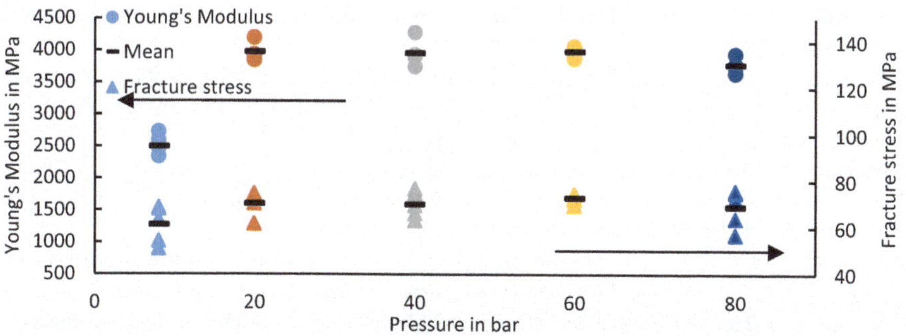

Fig. 6. Young's modulus and fracture stress of pure resin.

4 Conclusion and Outlook

The high pressure occurring during HP-RTM demands the consideration of pressure in process simulation and the underlying material models accordingly. A simulation approach, based on the free volume theory was introduced to consider the pressure within simulation. It was shown, that a model for free volume dependent from pressure, temperature and degree of cure can be deduced from dielectric measurements in theory. In addition, all necessary material parameters could be expressed in dependence of the free volume by adequate material models. Thus, the pressure dependency could be brought into the process simulation via free volume theoretically.

First experimental investigations on pressure dependency have shown an effect of pressure on dielectric properties, conversion and glass transition temperature. This not only shows the practical relevance of consideration of pressure in process simulation, but also constitutes part of the basis for the application of the model approach described above. The influence of high pressure on mechanical properties could not be proven. Further investigations with dynamic mechanical analysis (DMA) are scheduled. Moreover, additional measurements with high-pressure DSC are planned in order to validate the DEA results.

In the next step, further experiments must be carried out to adapt or extend the presented models to determine the required model parameters and thus to fit the models to the investigated resin. This essentially includes: thermomechanical analysis, dynamic mechanical analysis, measurements of rheological properties and shrinkage and additional DSC analysis. Finally, the aim is to demonstrate the practical implementation of the presented approach.

Acknowledgements. The authors would like to thank Deutsche Forschungsgemeinschaft DFG, which funded the work in the course of the project Pressure dependent modelling of thermoset epoxy resins DFG HE 2574/40-1.

References

1. Brauner, C., Block, T.B., Herrmann, A.S.: Meso level manufacturing simulation of sandwich structures to analyse viscoelastic dependent residual stresses. J. Compos. Mater. **46**(7), 783–799 (2012)
2. Brauner, C., Block, T.B., Purol, H., Herrmann, A.S.: Microlevel manufacturing process simulation of carbon fiber/epoxy composites to analyze the effect of chemical and thermal induced residual stresses. J. Compos. Mater. **46**(17), 2123–2143 (2012)
3. Brauner, C.: Analysis of process induced distortions and residual stresses of composite structures. Dissertation at the University of Bremen, Logos Verlag Berlin, Germany. ISBN 978-3-8325-3528-5 (2013)
4. Cruz, J.C., Osswald, T.A.: Monitoring epoxy and unsaturated polyester reactions under pressure – reaction rates and mechanical properties. Polym. Eng. Sci. **49**, 2099–2108 (2009)
5. Nawab, Y.: Effect of pressure and reinforcement type on the volume chemical shrinkage in thermoset resin and composite. J. Compos. Mater. (2013) (published online 13. September 2013)

6. Sanford, W.M., McCullough, R.L.: A free-volume-based approach to modeling thermoset cure behavior. J. Polym. Sci. Part II Polym. Phys. **28**, 973–1000 (1990)
7. Adolf, D., Martin, J.E., Chambers, R.S., Burchett, S.N., Guess, T.R.: Stresses during thermoset cure. J. Mater. Res. **13**, 530–550 (1998)
8. Williams, M.L., Landel, R.F., Ferry, J.D.: The temperature dependency of relaxation mechanisms in amorphous polymers and other glass-forming liquids. J. Am. Chem. Soc. **77**, 3701–3707 (1955)
9. Turnbull, D., Cohen, M.H.: Free-volume model of the amorphous phase: glass transition. J. Chem. Phys. **34**, 120–125 (1961)
10. Lee, H.L.: The Handbook of Dielectric Analysis and Cure Monitoring. Lambient Technologies, LLC, Boston (2014)

Digital Twin of the Polyurethane Rotational Moulding Process

Michael Stanko(✉) and Markus Stommel

Chair of Plastics Technology, TU Dortmund University, Leonhard-Euler-Str. 5, 44227 Dortmund, Germany
{michael.stanko,markus.stommel}@tu-dortmund.de

Abstract. The development of a digital twin for the engineering process of rotational moulded polyurethane components is presented in this contribution. The digital twin includes the coupling of FE analyses with a novel simulation environment of the rotational moulding process. For the setup of the numerical process simulation, the C++ based open-source software package OpenFOAM is used, which is widely established in research and development for the numerical solution of fluid mechanics problems (Computational Fluid Dynamics, CFD). Based on the Volume-of-Fluid method, a simulation environment called interRoMoFoam representing the rotational moulding process according to the Rock'n'Roll method has been developed. Within the simulation, the resin distribution calculation is carried out taking the chemo-rheological processes during the curing of the polyurethane into account. Among other determined process-related properties the wall thickness distributions of the hollow bodies can be exported and evaluated in appropriate FE analyses under structural-mechanical aspects, for example. In this way, the results of the process and structure simulations can be used to perform precise design adjustments and an optimisation of the part, the mould for manufacturing the part or the process parameters.

Keywords: Digital twin · Rotational moulding · Polyurethane

1 Introduction

Digital twins of products, processes and services are important items of current industry 4.0 strategies. The digital representation of a production process contributes to the reduction of time and effort in development and production. Digital twins are of particular importance in the design of time- and cost-intensive processes such as the polyurethane rotational moulding. The polyurethane rotational moulding is a specific process used in plastics processing to produce large-volume, thin-walled hollow bodies. A general overview regarding the rotational moulding process, the plastics used and the processing techniques can be found in [1]. Detailed information on reactive rotational moulding, which is discussed in this contribution, are given in [2, 3]. The main advantage of using a digital twin for rotational moulding is the reduction of experimental effort required to design and optimise the manufacturing process.

C. Hopmann and R. Dahlmann (Eds.): *Advances in Polymer Processing 2020*, pp. 324–335, 2020. https://doi.org/10.1007/978-3-662-60809-8_27

The wall thickness distribution, which exhibits the most significant influence on the component stiffness, is evaluated virtually and iteratively using the digital twin. In practice, roto-moulded polyurethane components are several meters long, so that experimental investigations require a large amount of additional material, which can be saved by using the developed digital twin. Digital twins modelling a manufacturing process are always based on a numerical process simulation. In contrast to processes such as injection moulding, there are no commercially available simulation environments, which can already be relied upon. First approaches for modelling the rotational moulding process can be found in the investigations of Ianakiev and Lim [4, 5] or Attaran et al. [6]. Riviere et al. [7] present the numerical simulation of the polymer flow for thermoplastic materials processed in rotational moulding using the so-called Smoothed Particle Hydrodyanamics method (SPH). The SPH method is also applied in [8, 9] and extended by the chemo rheological properties of reactive moulding resins. The SPH method is often used to minimise the computational effort compared to CFD methods. But also, within the SPH method a correspondingly high number of parts is required to generate results with high quality. Furthermore, due to the stable algorithm the SPH method is able to calculate results which can be physically wrong despite wrong assumptions and models. For this reason, this contribution pursues the development of a simulation environment using the Volume of Fluid method (VOF) for a two-phase flow. In this context, reference is made to the solution proposed by Dissanayake et al. [10] for the simulation of particle flow within a rotating cylinder. In contrast to this work the OpenFOAM solver twoPhaseEulerFoam is used, not taking into account the rock movement of the moulding machine and the chemo-rheological properties of a reactive resin. Therefore, in the following a new approach for a process simulation and fundamental aspects of Computational Fluid Dynamics (CFD) will be discussed, which are used to develop the process simulation environment of the later described digital twin of the entire engineering process.

2 Basics of Computational Fluid Dynamics

Flow processes can be described mathematically by the conservation laws of mass, momentum and energy, which can be summarised with the following transport model equation:

$$\frac{\partial}{\partial t}(\rho\,\phi) + \nabla \cdot (\rho\,\underline{u}\,\phi) = \nabla \cdot (D\,\nabla\,\phi) + Q_\phi, \tag{1}$$

Where ρ describes the density and ϕ defines a flow quantity e.g. the flow velocity $\underline{u}$ or the temperature T. The term $\nabla \cdot (\rho\,\underline{u}\,\phi)$ describes the convective flux of the flow quantity ϕ, which is caused by the transport of ϕ due to a current flow. The expression $\nabla \cdot (D\,\nabla\,\phi)$ with D as diffusion coefficient, depicts the diffusive flux of ϕ, which is the result of concentration differences in the spatial distribution of ϕ. All other sources and sinks of the flow process are described by Q_ϕ. By specifying the flow quantity ϕ

and the terms used to describe the convective and diffusive flow, the equations of the general mathematical model of a flow can be derived. This results in the continuity equation:

$$\frac{\partial \rho}{\partial t} + \nabla \cdot (\rho \, \underline{u}) = 0 \tag{2}$$

and the momentum conversation principle:

$$\frac{\partial}{\partial t}(\rho \, \underline{u}) + \nabla \cdot (\rho \, \underline{u} \, \underline{u}) = \nabla \cdot \underline{\underline{\tau}} - \nabla p + \rho \underline{g} \tag{3}$$

where $\underline{\underline{\tau}}$ describes the shear stress tensor, p the pressure and $\underline{g}$ the acceleration due to gravity [11].

The law of conservation of energy is defined in general form as follows:

$$\frac{\partial}{\partial t}(\rho \, h) + \nabla \cdot (\rho \, \underline{u} \, h) = -\nabla \cdot \underline{q}'' + \frac{\partial p}{\partial \mathrm{t}} + \nabla \cdot \left(\underline{\underline{\tau}} \cdot \underline{u}\right) \tag{4}$$

where h is enthalpy and $\underline{q}''$ is the heat flux density [11].

In addition to the conservation laws, thermodynamic equations of state as well as material-specific material laws are used to describe a fluid mechanical problem, which define further relationships between the flow quantities ρ, $\underline{u}$, $\underline{\underline{\tau}}$, p, h and $\underline{q}''$. Known examples are the thermal equation of state of an ideal gas:

$$p = \rho \, R \, T \tag{5}$$

with R as gas constant and the caloric equation of state:

$$dh = c_p \, dT \tag{6}$$

where c_p describes the specific heat capacity for a constant pressure [11].

Examples of material-specific laws are, on the one hand, the Fourier law of heat conduction:

$$\underline{q}'' = -k \, \nabla \, T \tag{7}$$

with the thermal conductivity k and on the other hand Newton's flow law:

$$\underline{\underline{\tau}} = \eta \left[2\underline{\underline{S}} - \frac{2}{3}(\nabla \cdot \underline{u})\underline{\underline{\delta}} \right] \tag{8}$$

where η defines the material-specific dynamic viscosity of the fluid, which describes the relationship between the shear stress tensor $\underline{\underline{\tau}}$ and the velocity gradient $\underline{\underline{S}}$. The variable $\underline{\underline{\delta}}$ describes the so-called Kronecker delta [11].

In current CFD simulation environments, the Finite Volume Method (FVM) is the most commonly used solution method for the numerical calculation of the abovementioned equations. The basic principle of FVM is based on the local balancing of the fluid mechanical model equations within discrete sub domains (control volumes, CV), which are defined by a grid. Figure 1 shows the basic approach of the FVM schematically for a two-dimensional case.

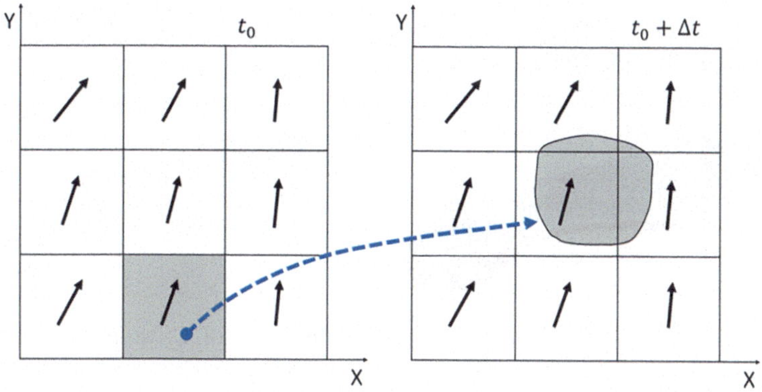

Fig. 1. Schematic two-dimensional illustration of a spatially discretised flow area and the time-dependent displacement of a fluid element

During the displacement due to an acting flow, the fluid element transports the quantities mass, momentum and energy according to the conservation laws from the initial position at time t_0 to the new position at time $t + \Delta t$. Balancing of the convective and diffusive fluxes of the given conservation quantities is performed on the surfaces of the sub domains by integrating Eq. 1 over the control volumes CV:

$$\int_{CV} \frac{\partial}{\partial t}(\rho\,\phi)dV = \int_{CV} \left[\nabla \cdot (D\,\nabla\,\phi) - \nabla \cdot (\underline{u}\,\rho\,\phi) + Q_\phi\right] dV \tag{9}$$

By applying the Gaussian theorem, the volume integrals of the convective and diffusive terms are substituted by surface integrals, so that the calculation of the convective and diffusive flows can be performed through the surface of the control volume:

$$\int_{CV} \frac{\partial}{\partial t}(\rho\,\phi)dV = \oint_A (D\,\nabla\,\phi)\cdot\hat{n}\,dA - \oint_A (\underline{u}\,\rho\,\phi)\cdot\hat{n}\,dA + \int_{CV} Q_\phi\,dV \tag{10}$$

The integration over the time interval Δt results in the following:

$$\int_{CV}\left(\int_{t_0}^{t+\Delta t}\frac{\partial(\rho\phi)}{\partial t}dt\right)dV = \int_{t_0}^{t+\Delta t}\left(\int_A \hat{n}\cdot(D\,\nabla\,\phi)dA\right)dt - \int_{t_0}^{t+\Delta t}\left(\int_A \hat{n}\cdot(\underline{u}\,\rho\,\phi)dA\right)dt + \int_{t_0}^{t+\Delta t}\left(\int_{CV} Q_\phi\,dV\right)dt \tag{11}$$

In OpenFOAM the resulting equations are discretised using numerical methods such as the explicit and implicit Euler method and the Crank–Nicolson procedure and solved iteratively under specified boundary conditions.

3 Rotational Moulding Simulation with OpenFOAM

The solver interRoMoFoam described here is based on the solver interFoam provided by OpenFOAM. With this solver a transient flow of two incompressible immiscible phases is calculated using the Volume of Fluid method. Both phases are separated from each other by an interface or a free surface as it is formed in rotational moulding between the air and the resin (Fig. 2).

Fig. 2. Phase fraction distribution for the initial time step of a two-phase simulation in OpenFOAM including the phase interface: The red region represents the polyurethane whereas the blue zone depicts the air within the rotation mould

Within the interFoam solver a modified convection equation is used to determine the relative volume fractions of the phases in the discrete cell volumes of the calculation grid for the considered time steps of the simulation.

$$\frac{\partial \alpha}{\partial t} + \nabla \cdot (\alpha \underline{u}) + \lambda_m \nabla \cdot \left(a \cdot (1 - a)\underline{u}_c\right) = 0 \tag{12}$$

The first two terms are similar to Eq. 2 and represent the components of a simple convection equation to describe the transport of α. A transport equation consisting only of these terms would result in a strong numerical diffusion at the interface due to its numerical discontinuity. This has a negative effect on the quality of the simulation results. For this reason, the simple convection equation in OpenFOAM is extended by a third term which represents the so-called interface compression and counteracts the numerical diffusion at the phase boundary. Equation 12 is solved using the Multidimensional Universal Limited Explicit Solver (MULES), which is designed for solving purely convective transport equations in multiphase simulations and guarantees the limitation of the phase fraction to values between 0 and 1. InterFoam is an isothermal equation solver computing with a uniform global temperature. In the initial configuration of the solver the implementation of the curing kinetics and the related effects on the viscosity and temperature of the resin are also missing. Accordingly, the interFoam solver is extended by program modules to take the chemo-rheological properties of the polyurethane and corresponding subroutines into account to depict a non-isothermal moulding process with locally different temperatures and progresses of cross-linking reaction (conversion c_α). The latter is particularly required in order to control the wall thickness of roto-moulded components by applying locally different temperatures within the mould. Furthermore, the characteristic Rock'n'Roll kinematics of the rotational moulding machine has to be considered within the two-phase simulation. More detailed explanations regarding multiphase simulations or the structure of the basic solver interFoam can be found in [12, 13].

In order to consider a locally different reaction progress in the simulation, the conversion c_α can no longer be determined globally for the entire computation domain and has be integrated as a scalar transport routine using a convection equation. The interRoMoFoam solver contains an energy equation as well as a transport equation for the conversion, which are used for the calculation of a temperature distribution and locally different reaction progresses. For single-phase flows the transport of one scalar μ can be described as follows:

$$\frac{\partial \mu}{\partial t} + \nabla \cdot (\mu \underline{u}) - \Delta\big(D_\mu \cdot \mu\big) = Q_\mu \qquad (13)$$

with D_μ as diffusion coefficient and Q_μ as source term. For two-phase simulations the calculation of the scalar transport is numerically even more demanding. In the interRoMoFoam solver, the transport of conversion is performed by integrating the phaseScalarTransport function of OpenFOAM. The calculation for the transport of the scalar quantity μ for one phase α_i of the two-phase flow is defined according to Eq. 13 within this subroutine as follows:

$$\frac{\partial \mu}{\partial t} + \nabla \cdot \big(\alpha_i \underline{\phi}\, \mu\big) - \nabla \cdot \big(D_\mu \, \nabla(\alpha_i C)\big) = Q(\alpha_i, \mu) - R_1 D(\alpha, \mu) - R_2(\alpha, \mu) \qquad (14)$$

with C as the fraction of the scalar within the phase α_i and two stabilisation terms R_1 and R_2 [14, 15].

Within the interRoMoFoam solver, the scalar μ is replaced by the conversion c_α. The source term $Q(\alpha_i, \mu)$ describes the change of the reaction progress according to the material-specific conversion rate $\delta c_\alpha / \delta t$ and the resulting dissipated reaction heat, which again is also included in the equation implemented in interRoMoFoam for the calculation of the temperature distribution according to [16]:

$$\frac{\partial T}{\partial t} + \nabla \cdot (T\underline{u}) - T \cdot \nabla \cdot \underline{u} - \Delta(\alpha \cdot T) = \frac{h_{total}.\alpha_i}{c_{p,i}} \frac{\partial c_{\alpha,i}}{\partial t} \qquad (15)$$

with h_{total} total enthalpy released by the cross-linking reaction and c_p the specific heat of the polyurethane (both determined by means of appropriate DSC tests) [15].

Detailed information regarding the determination of these quantities for fast curing polyurethane resins can be found in [17]. The calculation of conversion c_α at a time step Δt and given temperature T is done in interRoMoFoam according to the model of Chiacchiarelli et al. [18], while the viscosity of the polyurethane resin corresponding to this conversion is calculated according to the model of Kim et al. [19]. However, in principle any model can be implemented in the interRoMoFoam solver by adapting small parts of program code, which represents one of the main advantages of an open source platform such as OpenFOAM.

The machine-motion of the roto-moulding machine is implemented by a time- and speed-controlled manipulation of the entire discretised calculation area, which means that the calculation grid (or mesh) of the CFD simulation is moved over the entire simulation according to user-predefined motion profiles. For this purpose, a so-called dynamic motion solver of the OpenFOAM development environment is integrated into the simulation setup. Taking into account the Rock‘n’Roll kinematics of the manufacturing process, the mesh is rotated about the lengthwise axis and rocked about the transverse axis of the roto-moulded component.

4 Digital Twin for the Engineering Process of Roto-Moulded Components

The interRoMoFoam solver has been developed to calculate the wall thickness distribution of thin-walled hollow polyurethane bodies. Figures 3 and 4 show typical geometries applied in the rotational moulding process on the example of a roto-moulded white water kayak and a rotor blade used for the construction of ventilation fans.

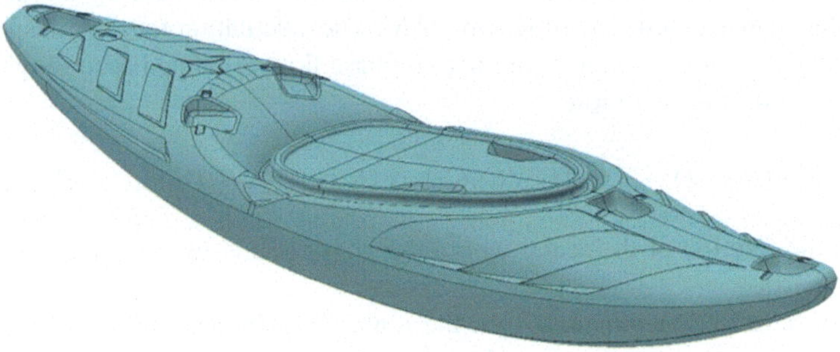

Fig. 3. 3D-Model of a roto-moulded white water kayak made by Lettmann GmbH

Fig. 4. 3D-Model of a roto-moulded rotor blade used for the construction of ventilation fans

As in the case of the kayak, the geometries show several important constructive and design details. Their exact reproduction during rotation moulding is of particular importance. Geometries such as those of the rotor blade in Fig. 4 are technically demanding due to the tapered cross-section and flat, non-rotationally symmetrical shape.

Without the use of the process simulation the determination of the process parameters is associated with a large experimental and material-related effort. This effort has to be minimised, especially for small series and the production of low quantities. The digital image of rotational moulding constitutes a tool for minimising effort and costs in production planning.

The resin distribution determined by simulation (Figs. 5 and 6) is used for a systematic optimisation of the production process and rotation mould geometries. The rotational moulding process according to the Rock'n'Roll method includes like mentioned above a rolling movement around the longitudinal axis and a rocking movement around the transverse axis of the hollow body.

Fig. 5. Illustration of the resin flow during a time step of the rotational moulding simulation on the example of a white water kayak

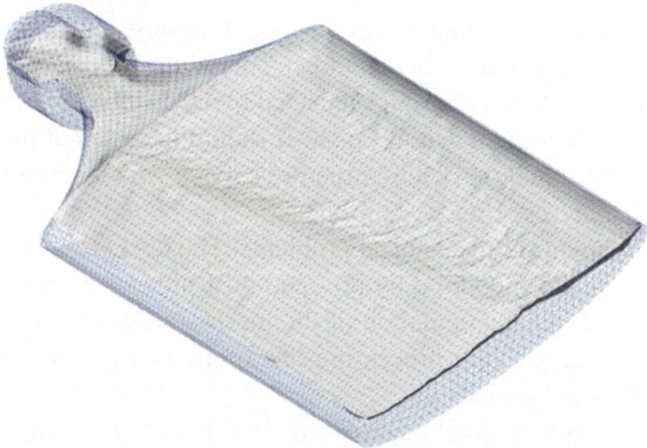

Fig. 6. Illustration of the resin flow during a time step of the rotational moulding simulation on the example of a rotor blade used for the construction of ventilation fans

Due to the overlapping of these movements, an incorrect choice of rotation speeds and angles can lead to quality defects caused by the fact that specific areas are not flooded with the resin during the entire manufacturing process shown in Fig. 7. By using the process simulation in the early phases of the development process, motion profiles can be determined that lead to such local defects and excluded for practical applications. In addition, by generating a CAD model from the results of the process simulation, taking into account the wall thickness distribution, the usual process-specific asymmetries and local thinning can be evaluated in subsequent FE analysis regarding the influence on the mechanical behaviour.

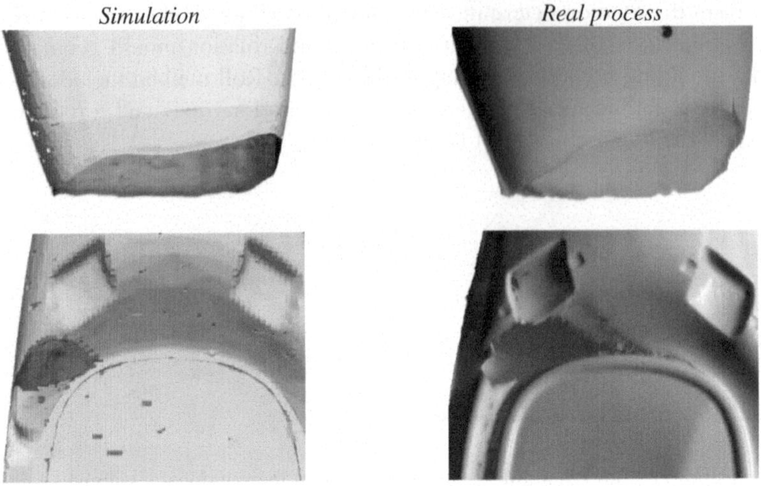

Fig. 7. Typical defects of roto-moulded components due to incorrect machine movements [20]

By coupling process simulation and structural analyses a digital image of the entire development process for roto-moulded polyurethane components is defined. The process simulation shown above is enhanced by CAD environments and structure solvers such as ABAQUS as shown in Fig. 8. It should be noted at this point that also the OpenFOAM environment provides structure solvers which can be used for subsequent mechanical analyses. However, in this case, the Finite Volume Method (FVM) is used to consider structural-mechanical problems. The CAD conception, process and structural simulations are supported by pre- and post-processors such as HyperMesh. These enable an extensive manipulation of the meshes according to the requirements of a CFD simulation or the post-processing of exported geometries, which has to be prepared for subsequent FE analyses.

In addition to computer-aided engineering, the digital twin can be used to determine parameters for monitoring the real manufacturing process. This means that at chosen areas of the component, tolerance ranges can be specified for the wall thickness, which should not be exceeded, as otherwise a negative influence on the mechanical properties of the product can be expected. The quality criteria can be integrated into an online process monitoring of a rotation mould. Therefore, the digital image shown here also represents a quality control tool.

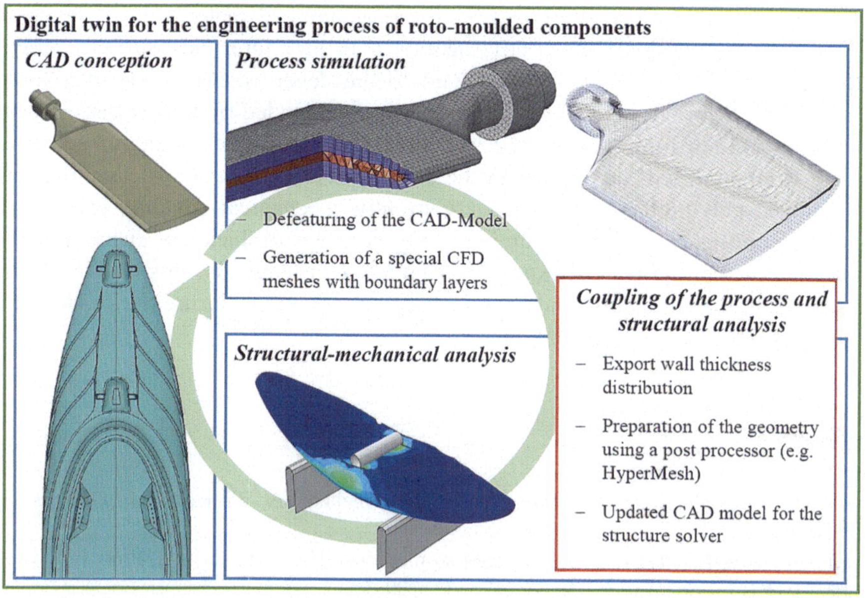

Fig. 8. Graphical overview of the digital twin representing the computer-aided engineering procedure for the rotation moulding process

5 Conclusion

In the scope of this work, the development of rotational moulding process simulation and its integration into a digital twin representing the computer-aided engineering procedure for the rotation moulding process is discussed. The novel interRoMoFoam simulation environment enables the calculation of the resin distribution taking into account the chemo-rheological processes during the curing of the polyurethane and an approach to consider locally different reaction progresses due to a temperature distribution and the resin flow inside the rotation mould. The process simulation developed using OpenFOAM can be easily integrated into various CAD and FEM systems due to the system openness of the open source environment. It also enables the flexible transfer of simulation results via several interfaces and data formats. By integrating the interRoMoFoam simulation into a design procedure in connection with CAD conception and subsequent FE analyses, a digital image of the entire rotational moulding process is obtained. The resulting digital twin can be used for a computer-aided iterative design of rotational moulded polyurethane components as well as for the definition of quality criteria within the framework of quality prediction and online process monitoring. The considerations of this contribution are currently limited to the numerical simulation of the rotational moulding process for reactive moulding materials. An extension to thermoplastic materials is possible taking into account appropriate viscosity models, crystallisation models (semi-crystalline materials) and the pvT behaviour (for correspondingly high process pressures). In addition, the application of the OpenFOAM solver interFoam for the simulation of manufacturing processes of thermoplastic materials is already mentioned in literature and can be combined with the results of this work. With regard to the development of the interRoMoFoam solver, the program code will be successively extended by further viscosity and reaction models in future work. The wall thickness distributions determined by the process simulation will be applied in the development of a design methodology for rotomoulded polyurethane components. Currently, mechanical structure calculations are carried out based on the assumption of an average wall thickness due to a lack of knowledge of the real wall thickness distribution. Consequently, this leads to deviations between the simulation and experimental investigations during the validation of the structural mechanical calculations.

References

1. Crawford, R.J., Throne, J.L.: Rotational Molding Technology. William Andrew Publishing, Norwich (2002)
2. Harkin-Jones, E., Crawford, R.J.: Rotational moulding of liquid polymers. Proc. Inst. Mech. Eng. Part B J. Eng. Manuf. **210**(5), 437–447 (1996)
3. Viale, J., Nony, F., Mazabraud, P., Gérard, J.-F., Tcharkhtchi, A., Doulin, G.: Rotational moulding of thermosets: understanding of a reactive forming process. Int. J. Mater. Form. **1**(1), 803–806 (2008)
4. Ianakiev, A., Lim, K.K.: Modelling of rotational moulding process: analysis of process parameters and warpage on cycle times. Plast. Rubber Compos. **36**(10), 455–462 (2007)
5. Lim, K.K., Ianakiev, A.: Modeling of rotational molding process: multi-layer slip-flow model, phase-change, and warpage. Polym. Eng. Sci. **46**(7), 960–969 (2006)
6. Attraran, M.T., Wright, E.J., Crawford, R.J.: Computer modelling of the rotational moulding process. J. Reinf. Plast. Compos. **17**(14), 1307–1318 (1998)
7. Riviere, S., Khelladi, S., Farzaneh, S., Bakir, F., Tcharkhtchi, A.: Simulation of polymer flow using smoothed particle hydrodynamics method. Polym. Eng. Sci. **53**(12), 2509–2518 (2013)
8. Hamidi, A., Khelladi, S., Illoul, L., Tcharkhtchi, A.: Simulation of the reactive rotational molding using smoothed particle hydrodynamics. Int. J. Mech. Mechatron. Eng. **9**(4), 679–683 (2014)
9. Nguyen, H.T., Cosson, B., Lacrampe, M.-T., Do, T.A.: Numerical simulation of reactive polymer flow during rotational molding using smoothed particle hydrodynamics method and experimental verification. Int. J. Mater. Form. **11**, 583–592 (2018)
10. Dissanayake, S., Karunarathne, S., Lundberg, J., Tokheim, L.-A.: Conference Proceedings of the 58th Conference on Simulation and Modelling (SIMS 58) Reykjavik, Iceland (2017)
11. Schwarze, R.: CFD-Modellierung – Grundlagen und Anwendungen bei Strömungsprozessen. Springer Vieweg, Berlin (2013)
12. OpenFOAM User Guide version 7. http://foam.sourceforge.net/docs/Guides-a4/OpenFOAMUserGuide-A4.pdf. Accessed 28 Aug 2019
13. Deshpande, S.S., Anumolu, L., Trujillo, M.F.: Evaluating the performance of the two-phase flow solver interFoam. Comput. Sci. Discov. **5**, 014016 (2012)
14. GitHub, Inc. https://github.com/OpenFOAM/OpenFOAM-dev/commit/7b1840c7d-3619da7e9efaa4de561205476f8b42. Accessed 20 Aug 2019

15. Bünker, M.: Weiterentwicklung einer Prozesssimulation des Polyurethan-Rotationsgießens in OpenFOAM. TU Dortmund University, Chair of Plastics Technology (2019)
16. Adding the Energy Equation to interFoam (OF 2.4.0). https://www.cfd-online.com/Forums/openfoam-programming-development/162577-adding-energy-equation-interfoam-2-4-0-a-2.html. Accessed 20 Aug 2019
17. Stanko, M., Stommel, M.: Kinetic prediction of fast curing polyurethane resins by model-free isoconversional methods. Polymers **10**(7), 698 (2018)
18. Chiacchiarelli, L.M., Kenny, J.M., Torre, L.: Kinetic and chemorheological modeling of vitrification effect of highly reactive poly(urethane-isocyanurate) thermosets. Thermochim. Acta **574**, 88–97 (2018)
19. Kim, D., Macosko, C.: Reaction kinetics and chemorheology of a highly reactive PU system. Korea Polym. J. **4**(1), 54–60 (1996)
20. Fischer, F.: Entwicklung einer CFD-Simulationsumgebung zur Abbildung des Polyurethan-Rotationsgießverfahrens in OpenFOAM. TU Dortmund University, Chair of Plastics Technology (2018)

Author Index

C. Hopmann and R. Dahlmann (Eds.): *Advances in Polymer Processing 2020*,
pp. 337–338, 2020. https://doi.org/10.1007/978-3-662-60809-8

The manufacturer's authorised representative in the EU is Springer Nature Customer Service Centre GmbH, Europaplatz 3, 69115 Heidelberg, Germany. If you have any concerns regarding our products, please contact ProductSafety@springernature.com

Printed and bound by CPI Group (UK) Ltd, Croydon, CR0 4YY
25/09/2026
02222894-0011